ACOUSTICAL HOLOGRAPHY
Volume 5

ACOUSTICAL
HOLOGRAPHY

A Continuation Order Plan is available for this series. A continuation order will bring delivery of each new volume immediately upon publication. Volumes are billed only upon actual shipment. For further information please contact the publisher.

ACOUSTICAL HOLOGRAPHY

Volume 5

Edited by

Philip S. Green

Program Manager, Ultrasonics
Information Science and Engineering Division
Stanford Research Institute
Menlo Park, California

PLENUM PRESS • NEW YORK AND LONDON

The Library of Congress cataloged the first volume of this title as follows:

International Symposium on Acoustical Holography.
    Acoustical holography; proceedings. v. 1–
New York, Plenum Press, 1967–

        v. illus. (part col.), ports. 24 cm.

    Editors: 1967–   A. F. Metherell and L. Larmore (1967 with
H. M. A. el-Sum)
    Symposiums for 1967–   held at the Douglas Advanced Research
Laboratories, Huntington Beach, Calif.

    1. Acoustic holography—Congresses—Collected works.   I. Meth-
erell. Alexander A., ed.   II. Larmore, Lewis, ed.   III. el-Sum, Hussein
Mohammed Amin, ed.   IV. Douglas Advanced Research Laboratories.
v. Title.

QC244.5.I 5   69–12533

Library of Congress   [r70n3]   rev

Library of Congress Catalog Card Number 69-12533
ISBN 0-306-37725-X

Proceedings of the Fifth International Symposium on Acoustical Holography
and Imaging held in Palo Alto, California, July 18-20, 1973

© 1974 Plenum Press, New York
A Division of Plenum Publishing Corporation
227 West 17th Street, New York, N.Y. 10011

United Kingdom edition published by Plenum Press, London
A Division of Plenum Publishing Company, Ltd.
4a Lower John Street, London W1R 3PD, England

Printed in the United States of America

2-MHz acoustic shadow image of 45-mm high expanded polystyrene letters. The image was formed with a real time imaging
system employing a 256 X 256 matrix of electrostatic transducers. (See P. Alais, Chapter 40).

PREFACE

This volume contains the Proceedings of the Fifth
International Symposium on Acoustical Holograhy and Imaging,
held in Palo Alto, California, on July 18-20, 1973. The
title of this Symposium differed from that of the previous
four by the addition of the word "Imaging," reflecting an
increase in emphasis on nonholographic methods of acoustical
visualization. For convenience, no change has been made in
the title of this published series.

The 42 Symposium papers cover a wide range of theoretical and applied topics, and effectively define the state-of-
the-art in this rapidly developing field. Many of them
relate to applications of acoustic visualization in such
diverse fields as nondestructive testing, medical diagnosis,
microscopy, underwater viewing, and seismic mapping.

The papers presented at the Symposium were selected
with considerable assistance from the Program Committee. The
Editor wishes to thank the following persons for serving as
members of this committee: P. Alais, University of Paris,
France; B. A. Auld, Stanford University; D. R. Holbrooke,
Children's Hospital of San Francisco; A. Korpel, Zenith
Radio Corporation; J. L. Kreuzer, Perkin Elmer Corporation;
A. F. Metherell, Actron Industries, Inc.; R. K. Mueller,
Bendix Research Laboratories; B. Saltzer, U. S. Naval Under-
sea Research and Development Center; F. L. Thurstone, Duke
University; and G. Wade, University of California, Santa
Barbara.

That this volume has reached the press at all is due to
the gentle but unrelenting efforts of Mrs. Nancy Driesbock
in persuading the Editor to apply himself to the task. For

this, and for her tireless help in setting up and running
the Symposium, he is most grateful.

The Symposium was sponsored by the IEEE, the Acoustical
Society of America, and Stanford Research Institute.

CONTENTS

TRW ACOUSTO-OPTICAL NONIMMERSION FLAW-IMAGING SYSTEM

R. A. Smith, N. H. Doshi,
R. L. Johnson, and P. G. Bhuta
TRW Systems Group
One Space Park
Redondo Beach, California 90278

BACKGROUND

Several investigators[1,2,3] have described techniques
and experiments where objects in media opaque to light have
been made visible by Bragg diffraction of light due to ultrasound scattered from the object viewed. These authors
immersed the object viewed in water to aid the conduction
of scattered sound into a light/sound interaction region.
More recent published work also concerned immersed objects.[4,5] The work reported here extends previous results
to formation of images of nonimmersed solid objects which
require access from only one side. This is accomplished by
using pulsed sound reflected from discontinuities within a
test object composed of solid material. Nonimmersion is
accomplished by transmitting and receiving sound through a
membrane separating water in the light/sound interaction
region from the test object.

INTRODUCTION

Figure 1 shows the essential elements of the Bragg
imaging processes which are used. Collimated light from a
laser is formed into a cylindrically convergent wedge of
light with a line of focus just outside the Bragg cell.
The Bragg cell is defined as the region where sound waves
scattered by an object meet cylindrically convergent light.
The Bragg cell is located between the two cylindrical lenses

1

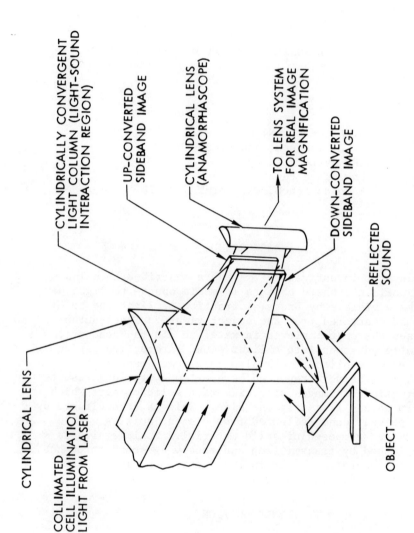

Figure 1   Bragg imaging elements.

shown in Figure 1. A basic characteristic of the Bragg imaging method is that sound from the object diffracts light into two sidebands. The position of these sidebands is determined by the Bragg formula for diffraction. The selective property of Bragg diffraction makes the sidebands possess a light distribution which is a replica of the object viewed with one important difference. The sideband image is greatly demagnified in one direction and full scale in the other direction. In order to form an image which is orthoscopic, a cylindrical lens (anamorphascope) is positioned on the right side of Figure 1 to magnify the sideband image in one direction. Careful selection of the focal length of the anamorphoscope and distance between the object to be viewed and the Bragg cell makes it possible to form a high quality image at some distance to the right in Figure 1.[5]

New work reported here concerns the formation of an image from pulsed sound reflected from an object which is not immersed in water that fills a standard Bragg cell. Flaws existing in solid material which must be kept dry can be discovered by projecting sound in the material and subsequently intercepting this sound by a Bragg cell positioned to receive reflected sound. Reflected sound diffracts light forming an image of flaws within the test object. The best images are obtained by positioning a TV pick-up tube in the image plane. The use of repetitively pulsed sound results in the image forming sound waves being present within the light field for only a small fraction of time. Cell illumination light is also pulsed to reduce unnecessary noise and scatter of light which would degrade the desired image.

A block diagram of the system is shown in Figure 2. Electric energy which generates the sound is also sent to a delay line that feeds an electronic shutter controlling the emission of laser light. In this way, light and sound are caused to be simultaneously present at only the time when the desired image can be formed. The laser light is off at times when it would simply pass through the Bragg cell (without diffraction) except for scattering from imperfections such as dust particles. By pulsing the sound at some given time, subsequent to the generation of sound, reflected sound will be returned from a specific distance within the object viewed. Objects in a plane at this distance will be imaged. For example, at a later time, the sound present in the Bragg cell would be reflected from a greater distance if

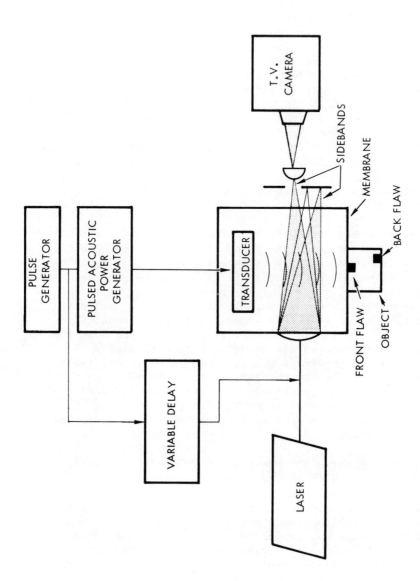

Figure 2   Block diagram of the Pulsed Bragg Imaging System.

there is a flaw or other reflecting surface at a greater
distance. Consequently, the image will be a replica of
objects located at ever increasing distances as time pro-
ceeds. The specific image distance is determined by the
time selected for viewing as defined by the variable delay
line which operates laser light. The duration of the laser
light pulse is made sufficiently short to have a negligible
effect on the size of the range interval viewed. In this
case, all flaws will be superimposed within a range interval given by one half the duration of the sound pulse multiplied by the velocity of sound in the material inspected.
Flaws outside this range interval will be rejected from
view. The inspection interval can sometimes be positioned
where there is only one flaw near the limit of the interval.
In these cases, flaws may be separated which are as close
together as pulse rise or fall times will allow. A complete
inspection of an object would require repetitive pulsing of
a sound transducer with repetitive illumination from the
laser using sound pulses and laser pulses which are repeated
with increasing delay after a given image plane has been
inspected.

To be more specific, consider Figure 3 which is representative of the typical situation. Figure 3 was drawn to
represent the case where sound would be reflected from
(1) the front surface of the object to be inspected, (2) a
flaw just below the front surface, (3) a flaw near the back
surface, and (4) reflection from the back surface. Reflection from these sources would appear successively in time
as illustrated in Figure 3. In order to view the flaw near
the front surface, the laser pulse would be timed to occur
coincident with the return of sound from the front surface
flaw. The whole process would be repeated at a constant
repetition rate to obtain a continuous image.

Experimental Results

Figure 4 shows two objects which were used to demonstrate the performance of this system. Object A is a solid
piece of aluminum with cut-outs about 1/4" deep in the top
surface. These cut-outs are in the form of a triangle,
square, rectangle, and trapezoid. Object B has two strips
of silicon rubber deposited on the top surface in the form
of an "x." These objects were positioned up against a
polyethylene membrane which is used to contain the water

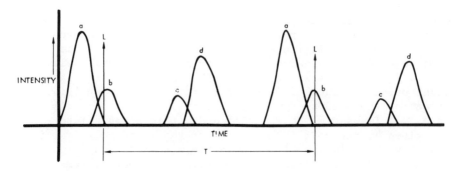

a   FRONT SURFACE REFLECTED SOUND

b   FRONT FLAW REFLECTED SOUND

c   BACK FLAW REFLECTED SOUND

d   BACK SURFACE REFLECTED SOUND

L   LASER PULSE

T   LASER SOUND PULSE REPETITION PERIOD

Figure 3.   Sound and light pulses as a function of time.

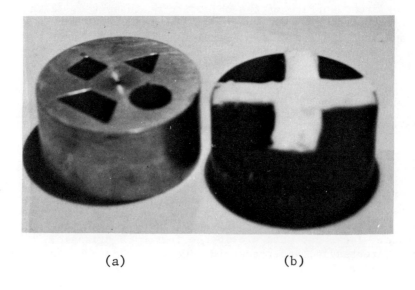

(a)   (b)

Figure 4.   Test objects.

(which is essential to operation) within the Bragg cell.
Sound from a transducer from within the water volume of the
Bragg cell is made to propagate through this membrane and
to the object to be tested. In order to facilitate the
coupling of the sound into the test object, a drop of water
was placed on the inner face between test specimen and membrane. The object then appeared as shown in Figure 5 when
this object was illuminated by pulsed sound at a frequency
of 18.7 megahertz. The image shown in Figure 5a was made
with the object positioned with flaws next to the membrane.
An image of the test specimen shown in Figure 4b appears in
Figure 5b. The image in Figure 5b was made with the test
specimen positioned so that the "x" was at the maximum
distance from the membrane. The smooth surface of the
aluminum block supporting the "x' was next to the membrane.
A replica of the "x" shaped pattern of extruded rubber is
visible in Figure 5b.

A number of experiments were run with an object similar
to that in Figure 4a. These experiments were performed with
four similar test specimens of different dimensions in the
axial direction. This placed the flaws at various distances
from the membrane. Images formed with these test specimens
are shown in Figure 6. The test specimens used to make the
images in Figures 6a,b,c and d were 1", 3", 1/4" and 1/2"
thick, respectively. Figure 6b shows that flaws can be
recognized through a 3" thickness of aluminum. The formation of images through a 3" section of aluminum required a
transducer input power of 200 w.

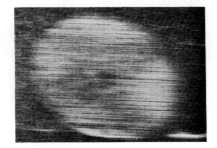

IMAGE OF OBJECT (a) AT 18.7 MHz   IMAGE OF OBJECT (b) AT 18.7 MHz

Figure 5   Images from sound reflected within test objects
             shown in Figure 4.

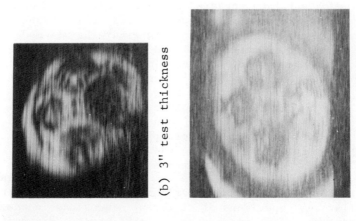

(a) 1" test thickness   (b) 3" test thickness

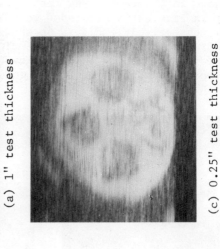

(c) 0.25" test thickness   (d) 0.5" test thickness

Figure 6   Images of the test specimens of thickness ranges from 0.25" to 3" at acoustic frequency of 18.7 MHz in a nonimmersion pulse-echo mode of operation.

The system used to make images in Figures 5 and 6 is shown in Figure 7. The Bragg cell is located in the center of this figure. It appears as a cube which is about 7" on a side. Laser light is passed through the central part of the cell from left to right. The resulting sideband image is magnified by the cylindrical lens shown just at the right of the cell. The wedge forming the cylindrical lens is visible on the left side of the Bragg cell. Test specimens are shown in position against the membrane which is facing the observer. Images were formed on an image plane under black cloth at the left which keeps stray light out.

DISCUSSION

Possible image quality is determined by the number of resolution cells which are within a field of view. For example, high quality TV images contain as many as 500 cells

Figure 7   Pulse-echo imaging system.

within the width of the viewing screen. Images of equal
quality should contain just as many resolution cells. The
number of resolution cells possible with the Bragg imaging
system is no more than the width of the collimating light
expressed in sound wavelengths. To simplify the presentation of why this is the case, the line of focus formed by
the wedge is considered vertical. This occurs when the
wedge-forming-cylindrical lens is oriented with its power
direction horizontal.

Previously published analysis and experiments show that
vertically divergent sound rays diffract illumination light
rays into rays that converge in the same way that focused
rays converge. The highest quality two-dimensional images
have been shown possible only by causing horizontal and
vertical components to be in focus in the same plane. This
is easily accomplished while magnifying the down converted
sideband image in the horizontal direction (without magnification in the vertical direction). However, the object
in the sound field which is imaged cannot be moved in distance from the light/sound interaction region without
causing the vertically convergent ray components to come to
a focus in a different plane than the plane where horizontal
components of the same rays are in focus. Both directional
components can be brought back into focus by readjustment
of the relative position of a cylindrical lens oriented with
the power direction vertical. Focusing characteristics of
vertical components become more evident as the object imaged
is moved to greater distances.[5]

When the Bragg system is in focus, the size of a resolution in either direction is given by one expression,
namely[5]

$$\Delta = \frac{\Lambda}{2 \sin\beta} \tag{1}$$

where $\sin\beta$ is the numerical aperture for sound waves
          which contribute to the image

When the object to be imaged is quite close to the
light column, the numerical aperture in the horizontal
direction is determined by letting β equal half the horizontal angular extent of convergent light 2α. When the
object to be imaged is at a distance R greater than
$D/(2 \sin\alpha)$, the numerical aperture is limited by the length

of the interaction in the horizontal direction D. In that case the numerical aperture is D/(2R) so that the resolution in the horizontal direction is given by[5]

$$\Delta_h \cong R \frac{\Lambda}{D} \qquad \text{for} \quad R > \frac{D}{2 \sin\alpha} \qquad (2)$$

$$\Delta_h = \frac{\Lambda}{2 \sin\alpha} \qquad \text{for} \quad R < \frac{D}{2 \sin\alpha}$$

The numerical aperture in the vertical direction is given approximately by the sine of half the angle which subtends the height of the light column in the vertical direction. This relation applies at any range if the system is kept in focus. The resolution in the vertical direction is given approximately by[5]

$$\Delta_v \cong R \frac{\Lambda}{H} \qquad \text{for} \quad R > H$$
$$\Delta_v \cong \Lambda \qquad \text{for} \quad R \ll H \qquad (3)$$

where H is the height of the light column in the vertical direction

The maximum number of resolution cells occurs when the object is adjacent to the light column. In that case, the number of resolution cells in the vertical direction is [from Eq.(3)] just the height of the column of cell illumination light expressed in sound wavelengths.

The width of the light field in the horizontal direction in a full scale image is, of course, just the width of the interaction length D. The number of resolution cells in the horizontal direction is then

$$N_H = \frac{D}{\Delta_H} \leq \frac{2D \sin\alpha}{\Lambda} < \frac{2D}{\Lambda} \frac{W_h}{2D} \qquad (4)$$

or simply

$$N_H < \frac{W_h}{\Lambda}$$

where W_h is the maximum width of cell illumination light in the horizontal direction.

Equations (3) and (4) show that the maximum number of resolution cells is just the maximum dimension of cell illumination light measured in the respective direction when expressed in sound wavelengths.

Experiments reported here were performed at sound wavelengths slightly longer than .075 millimeters. Cell illumination light was about 25 millimeters in the horizontal direction (100 millimeters in the vertical direction). There are, therefore, less than 333 resolution cells in the horizontal direction. Images formed cannot be equal to commercial TV quality without increasing the maximum horizontal dimension of cell illumination light with a fixed focal length lens forming the cylindrically convergent wedge of cell illumination light.

The number of resolution cells in the horizontal direction decreases with increasing distance to the object imaged after the object is beyond $D/(2 \sin\alpha)$. All experiments reported here were within this distance to the object imaged so that reasonably good images could have been anticipated. An experimental check on system resolution showed that resolution in the horizontal direction was .08 cm at 20 MHz at the time images in the included figures were made.

A probable major cause for the poor quality images shown in Figure 6 is due to the requirement that sound must pass through solid material supporting shear waves. When it is recognized that longitudinal waves may be converted to shear waves, and vice versa by irregularities within a solid and at a liquid/solid interface, it is surprising that images were obtained as good as those in Figure 6. Sound waves described by all but the lowest spatial frequencies can be lost at the solid/liquid interface.

CONCLUSION

The Bragg imaging system has been operated with pulsed sound and shown to be a convenient method for forming images of flaws at a predetermined depth by echo ranging. Recognizable images were formed from flaws within solid aluminum at a depth of 3". It was shown that images of good quality are possible without immersion of the test specimen, which is accessed from one side only, but limitations on quality are imposed by characteristics of the sample tested.

Advantages of the nonimmersion pulse-echo acousto-optical imaging system can be summarized as follows:

o Does not require mechanical stability and can be utilized in a production type environment.

o Ability to produce visual images of the interior of optically opaque bodies.

o Operates where access is limited to one side only.

o Operates in real time.

o Eliminates the need for highly trained operators (because optical images are obtained and data interpretation is not required).

o Operates in both transmission and pulse-echo sound modes.

o Operates without immersion of the component.

ACKNOWLEDGMENTS

This work was sponsored by the Defense Advanced Research Projects Agency under Contract No. DAAG46-70-C-0103 for the Army Materials and Mechanics Research Center, Watertown, Massachusetts 02172.

REFERENCES

1. A. Korpel, "Visualization of the Cross-Section of a Sound Beam by Bragg Diffraction of Light," Appl. Phys. Letters, 9:425 (1966).

2. G. Wade, C. J. Landry and A. A. de Souze, Acoustical Transparencies for Optical Imaging and Ultrasonic Diffraction, Acoustical Holography, Vol. I, Metherell, El-Sum and Larmore, Editors, Plenum Press (1969), pp. 159-172.

3. J. Landry, J. Powers and G. Wade, "Ultrasonic Imaging of Internal Structure by Bragg Diffraction," Appl. Phys. Letters, 15, (15 September 1969), pp. 186-188.

4. R. Aprahamian and P. G. Bhuta, "NDT by Acousto-Optical
    Imaging," Materials Evaluation ASNT, Vol. XXIX, No. 5,
    (May 1971), pp. 112-116.

5. R. A. Smith, G. Wade, J. Powers and J. Landry, "Studies
    of Resolution in a Bragg Imaging System," J. Acoust.
    Soc. Am., Vol. 49, No. 3 (part 3) (1971), pp. 1062-1068.

RECENT DEVELOPMENTS WITH THE SCANNING LASER ACOUSTIC MICROSCOPE

L. W. Kessler, P. R. Palermo and A. Korpel

Zenith Radio Corporation
6001 W. Dickens Avenue
Chicago, Illinois 60639

A scanning laser acoustic microscope that simultaneously produces acoustic and optical images of a specimen has already been described.[1,2,3] This paper reports recent developments regarding improving the resolution and regarding operating the microscope in the "interference" mode. The principle of operation of the acoustic microscope is based upon optically measuring the localized dynamic displacements of a mirrored surface which are caused by incident sound waves that are scattered from an object.[4] For further details concerning the basic instrument the reader is referred to the literature cited above.

The resolution capability of this acoustic imaging system is dependent upon the characteristics of the sample-mirror interface. For aqueous-like specimens and typical plastic materials used for the mirror, the smallest discernible detail is about 1.3 wavelengths of sound. Since the wavelength varies inversely with frequency, proportional resolution improvements are achieved with frequency increases. Illustrating this, Fig. 1 shows the acoustic images of similar wire honeycomb screens made at three frequencies, 96, 220 and 465 MHz. The resolutions, deduced from known dimensions of the grid, are, respectively, 20μ, 9μ and 4μ. The grid dimensions are as

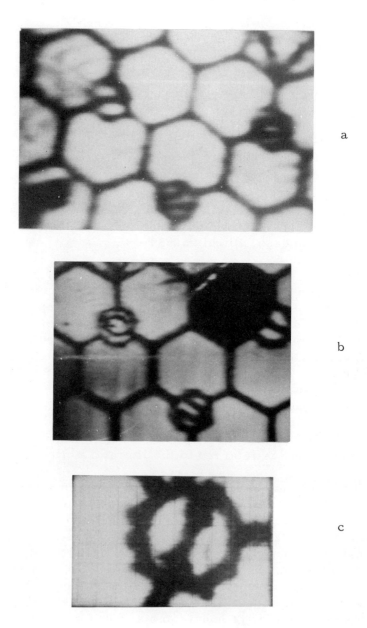

a

b

c

Fig. 1. Ultrasonic images of a wire mesh at various
frequencies a) 96 MHz, b) 220 MHz, c) 465 MHz.

follows: The center-to-center spacing between parallel
faces of the hexagons is 225 μm and the circle diameters
are 100 μm. Note that the field of view decreases with increasing frequency and this is analogous to the situation
for ordinary optical microscopy. In the case of point-by-
point scanning, the field of view is determined by compromising the number of resolvable image elements and the
frame time to be consistent with an adequate signal-to-
noise ratio. The frame times employed in these studies
are 1/30 sec at 96 MHz and 1/5 sec at 220 and 465 MHz.
The sensitivity is of the order of 10^{-3} W/cm^2 and is dependent upon a number of factors such as laser power, acoustic
frequency, etc.[5]

Optical interference microscopy is a well-established
and important technique for quantitative determinations of
index of refraction variations within a specimen. Analo-
gously, acoustic interference microscopy may be used to
deduce the spatial variations in the velocity of sound. This
may be an important consideration for specimen characterization. An in-focus acoustic hologram, formed with an
off-axis plane wave reference signal, can be treated as a
simple interferogram. In the case of the acoustic microscope, however, because of geometrical considerations,
it does not appear to be practical to generate the interferogram with an acoustic reference beam per se; rather an
electronic signal, with a particularly chosen relationship
to the acoustic signal is more suitable. In the usual mode
of operation of the scanning laser acoustic microscope,
insonification of the object occurs at an angle of 10° with
respect to the normal. An electronic signal coherent with
the insonification produces an acoustic hologram with a
simulated normal incidence reference beam. An in-focus
acoustic hologram produced in this manner of a section of
a larva of the fruit fly, Drosophila melanogaster, is shown
in Fig. 2a. Figures 2b and 2c are the corresponding
acoustic and optical images of the same specimen. The
horizontal field of view is of the order of 1.5 mm, and the
insonification is at 96 MHz.

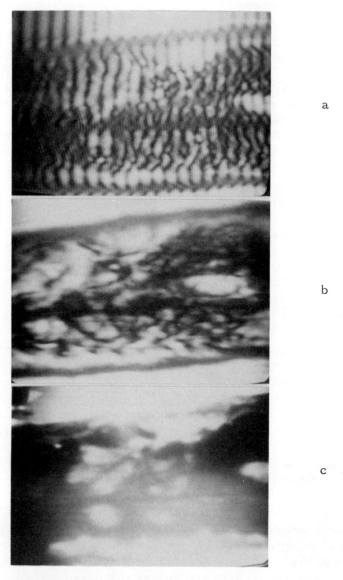

Fig. 2. Images of a section of a fruit fly larva a) in-focus
         acoustic hologram (interferogram) with an electron-
         ically simulated normal incidence reference beam,
         b) acoustic image of same, c) corresponding optical
         image.

For a homogeneous sample, the spacing between the fringes is dependent upon the angle between the object and the simulated reference beam. Localized variations in the velocity of sound cause lateral shifts of the fringes which may be measured to quantify the velocity change. In order to improve the spatial localization of velocity of sound determinations it is necessary to space the fringes closer together. This may be accomplished by increasing the angle between the insonification and the simulated reference beam. If the insonification angle is increased, leaving the simulated reference beam at normal incidence, image degradation results from side shadowing of object detail along the lateral component of the direction of sound. Therefore, it is preferable instead to increase the angle of the simulated reference beam well beyond that of the insonification.

The generation of a simulated off-axis reference signal, electronically, is not a new concept; see for example references 6, 7, 8. The method employed here is shown in Fig. 3; details of the sample cell, laser scanning method

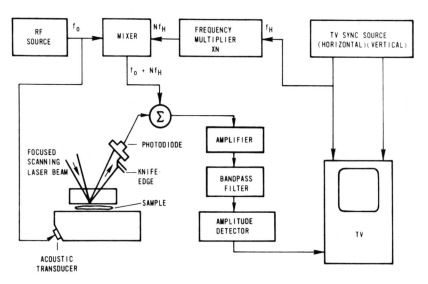

Fig. 3. Experimental setup for electronically simulating an off-axis acoustic reference beam.

and optical detection methods were described in ref. 3.
In our method, the reference signal is derived by frequency upshifting the sound source signal by an amount which
is coherent with the TV horizontal rate. The amount of
frequency shift can be varied to simulate reference beams
at various angles. The reference signal is added to the
photodiode output and the composite signal is processed
in the conventional way. In general, the fringe spacing,
M, on an interferogram, is given by the relationship

$$M \ = \ \frac{\Lambda}{|\sin \varphi - \sin \beta|} \qquad (1)$$

where Λ denotes the wavelength of the sound, φ if the angle
of the reference beam and β is the angle of the insonification, both angles measured from the normal. For the interferogram shown in Fig. 2a $\varphi = 0°$, $\beta = 10°$; therefore
M is of the order of 6 wavelengths; that is to say, the spatial discrimination of acoustic phase is limited to 6 Λ whereas the detail on the acoustic image is of the order of 1.3 Λ.
For an electronically simulated reference beam φ is determined by the frequency multiplier N depicted in Fig. 3
in the following way:

$$\sin \varphi \ = \ \frac{N \Lambda}{1.2 \ell} \qquad (2)$$

where ℓ is the scan length observed on the TV. Note the
factor 1.2 which comes from the fact that 20% of the horizontal line time is used for blanking purposes in the conventional TV system. The reference beam angle may be
changed in one of two ways; by choosing an appropriate N
or by changing the electronic magnification of the image,
therefore, ℓ. Furthermore, it is entirely possible to
choose $\sin \varphi > 1$; however, detail cannot be resolved beyond
the resolution limit. Relating to this method the interferogram shown in Fig. 2 corresponds to N = 0. Figure 4 is
an acoustic interferogram of the same specimen shown in
Fig. 2; however, the "reference beam" is at 28° instead
of 0°, here, N = 72. Note that in addition to the closer
fringe spacing, compared with Fig. 2, there is also a

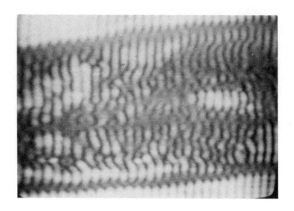

Fig. 4.   Acoustic interferogram of same subject in Fig. 2
with an electronically simulated reference beam
at 28° from the normal.

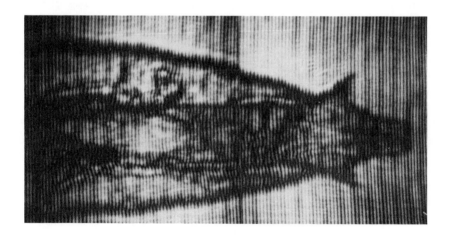

Fig. 5.   Acoustic interferogram of a fruit fly larva at
94 MHz with ß = 10° and sin φ = 1.05 (see text).

difference in the directions of the fringe shifts. That this occurs is a consequence of the fact that the direction of the simulated reference beam has changed sides with respect to the 10° insonification. One further illustration of the simulated reference beam is shown in Fig. 5. This is an interferogram of a fruit fly larva in which $N = 144$, $\ell = 1.83$ mm, $\Lambda = 16$ μ and $\beta = 10°$. Hence, $\sin \varphi = 1.05$.

Heretofore, acoustic holograms have only been useful to the extent that optical reconstruction has been possible. Here we emphasize that the interferometric aspects of acoustic holography may also be useful since quantitative information is extractable without optical reconstruction.

REFERENCES

1) A. Korpel, L. W. Kessler and P. R. Palermo "Acoustic Microscope Operating at 100 MHz", Nature 232, 110 (1971).

2) L. W. Kessler, A. Korpel and P. R. Palermo "Simultaneous Acoustic and Optical Microscopy of Biological Specimens", Nature 239, 111 (1972).

3) L. W. Kessler, P. R. Palermo and A. Korpel "Practical High Resolution Acoustic Microscopy", pp. 51-71, Vol. 4 "Acoustic Holography" ed. by G. Wade, Plenum Press, New York (1972).

4) A. Korpel and P. Desmares "Rapid Sampling of Acoustic Holograms by Laser Scanning Techniques", J. Opt. Soc. Amer. 45, 881 (1969).

5) A. Korpel and L. W. Kessler "Comparison of Methods of Acoustic Microscopy", pp. 23-43, Vol. 3 Acoustical Holography, ed. by A. F. Metherell, Plenum Press, 1971.

6) A. F. Metherell and H. M. A. El-Sum "Simulated Reference in a Coarsely Sampled Acoustical Hologram", Appl. Phys. Lett. 11, 20-22, 1967.

7) A. F. Metherell, S. Spinak and E. J. Pisa "Temporal Reference Acoustical Holography", Appl. Phys. Lett. 13, 340 (1968).

8) H. M. A. El-Sum "Progress in Acoustical Holography", pp. 7-22, Vol. 2 Acoustical Holography, ed. by A. F. Metherell and L. Larmore, Plenum Press, New York (1970).

BRAGG-DIFFRACTION IMAGING: A POTENTIAL TECHNIQUE FOR MEDICAL DIAGNOSIS AND MATERIAL INSPECTION, PART II

Hormozdyar Keyani, John Landry, and
Glen Wade

University of California, Santa Barbara
Department of Electrical Engineering
    and Computer Science
Santa Barbara, California 93106

ABSTRACT

In terms of image degradation, one of the
most serious problems encountered in low-frequency
Bragg-diffraction imaging is caused by multiple
acoustic reflections within the Bragg cell. The
effect of these reflections can be reduced by frequency sweeping or by using properly-located acoustic absorbers. This paper discusses experiments
in this regard and describes several different
types of absorbers for a 3.58 MHz system.

Another source of substantial difficulty at
the low frequencies is spherical aberration in the
cylindrical converging lens used to form the laser
beam. The design of such a lens and its effect on
the system resolution are discussed.

INTRODUCTION

Last year, at this symposium, we described
our newly constructed 3.58 MHz Bragg-diffraction
imaging system[1]. We pointed out a number of problems which were encountered with the low frequency
system. In this paper, some of these problems and
their solutions will be discussed.

A schematic diagram of the basic components
of the system is shown in Figure 1. The light
from the He-Ne laser is first expanded, then collimated, and finally converged into a vertical
line focus by means of a cylindrical converging
lens. In so doing the converging wedge of light
passes through the water-filled acoustic cell.

The zero-order and the two first-order diffracted light beams are projected by means of
another cylindrical lens to plane P_2, and the down-
shifted first-order and zero-order beams are blocked
off (at plane P_2) using a masking stop. A cylindrical lens, with its axis in the horizontal direction, is employed to correct the aspect-ratio
of the up-shifted image. The image is then focused onto the face of a vidicon camera tube and
is viewed on the TV monitor. It is of interest to
note that the image can be viewed directly on a

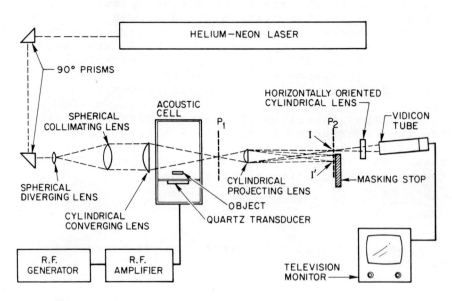

Figure 1 - Schematic Diagram of the System.

white screen without the use of the vidicon and
the TV monitor. The basic simplicity of a Bragg-
diffraction imaging system is important from a
cost and operational point of view.

The acoustic cell is an aluminum box with
glass windows. The cell is separated into two
compartments (interaction region and specimen
tank) by means of a .254 mm thick polyethylene membrane. Both compartments are filled with distilled
water. The compartment where the light wedge passes
through the water, i.e., the interaction region, is
completely sealed off from the laboratory environment using a lucite cover plate.

Since the Bragg angle is less than 2 milli-
radians in a low-frequency system such as this,
the optical image is bathed by a large quantity
of forward scattered light from the incident wedge
of laser light. It is therefore essential to reduce the scattering particles in the interaction
region by isolating this region from the sources
of impurities.

In addition to isolating the interaction region from impurities, the polyethylene membrane
partially eliminates the turbulance and streaming
in the water. If not eliminated, the turbulance
and streaming can seriously degrade the image.

The low-frequency system, as it was originally
designed, used a 10.16 cm x 15.24 cm active area
transducer. The large area transducer was made up
of six 5.08 cm x 5.08 cm x-cut quartz transducers
connected in parallel and matched to 50Ω for proper operation with our RF generator and amplifier.
No special care was taken to perfectly align the
six transducers.

The first images obtained, with this transducer arrangement, contained undesired inhomogeneities, i.e., intensity variations across the
optical image. Our initial conclusion was that
the inhomogeneities were a result of misalignment
of the six transducers. We had also noticed that
slight variation in the driving RF frequency changed

the position of the inhomogeneous regions. We
therefore frequency modulated the RF input to the
transducer (10 KHz deviation, modulation frequency
50 KHz).

Further experiments with the same transducer
mosaic and a 7.62-cm-diameter circular x-cut quartz
transducer (Figure 2) of fundamental frequency 3.58
MHz indicated that the observed inhomogeneities
were caused by multiple acoustic reflections in
the Bragg cell and not caused by misalignment of
transducers. Mr. Soo-Chang Pei of our department
has examined theoretically the question of the inhomogeneities and his calculations quantitatively
confirm our experimental results[2].

Since multiple reflections are thus shown to
be a major source of image degradation, we have
therefore spent considerable effort in searching
for a means to eliminate these reflections.

Figure 2 - 7.62-cm x-cut quartz transducer
and its matching network.

ELIMINATION OF ACOUSTIC REFLECTIONS

The test object used in our experiments (at
3.58 MHz with incident acoustic power intensity
of approximately 10 mw/cm^2) is shown in Figure 3.
It is an aluminum grid of approximately 8 mm x 8 mm
spacing.

The Bragg-diffraction image of this test grid,
without any means to reduce the multiple reflections, is shown in Figure 4a. Figure 4b is the
image of the same test object with a flat plate
of lucite at the far end of the water tank. The
lucite plate is oriented at a small angle such that
it is not parallel with the transducer. As a result, part of the reflected acoustic waves are at
angles that do not meet the diffraction requirements and therefore do not degrade the image.

Figure 3 - Test Grid of 8mm x 8mm Spacing.

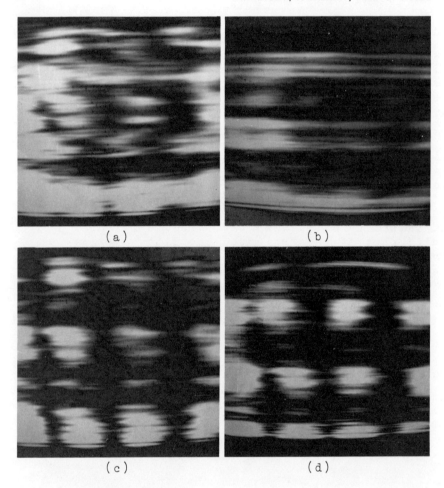

Figure 4 - Image of Test Grid at 3.58 MHz
          with Incident Acoustic Power In-
          tensity of 10 mw/cm^2.

          (a)   No absorber
          (b)   Lucite plate
          (c)   Lucite structure
          (d)   Castor oil absorber

Figure 4c is the image obtained with a lucite
structure as the absorber. The lucite structure is
shown in Figure 5. It consists of .48 cm x 3.8 cm
x 22.8 cm pieces of lucite which are tapered down
to an angle of 20°. Then these pieces are stacked
together to construct a 3.8 cm x 19 cm x 22.8 cm
slab.

This structure serves a dual purpose when used
in the acoustic cell. Some of the acoustic waves
continue to bounce back and forth between the tapered edges and will not reflect back into the interaction region. The portion which is reflected
back is dispersed in different directions and has
a reduced degrading effect on the image.

Figure 4d is the image obtained with a castor
oil absorber in the Bragg Cell. The oil absorber
is a box of dimensions 5 cm x 14.6 cm x 19 cm made
of lucite with a .254 mm thick polyethylene membrane

Figure 5 - Lucite Structure.

on one side (Figure 6). The container is filled
with castor oil which has the same acoustic impedance as water but its absorption coefficient is
approximately two orders of magnitude larger than
that of water. Since the acoustic impedance of
the polyethylene membrane is larger than that of
water the absorber is rotated in such a way that
it is not exactly parallel with the transducer.

A comparison of the images shown in Figure 4
indicates that the best low-frequency image is obtained when the oil absorber is in place.

Figure 7 shows the results of the same experiments performed at 10.74 MHz with incident
acoustic power intensity of approximately 10 mw/cm^2.
It should be noted that the effect of different absorbers at 10.74 MHz is not as pronounced as it is
at 3.58 MHz. This is an indication of the greater
necessity for the use of absorbers with low-frequency systems.

Figure 6 - Castor Oil Absorber.

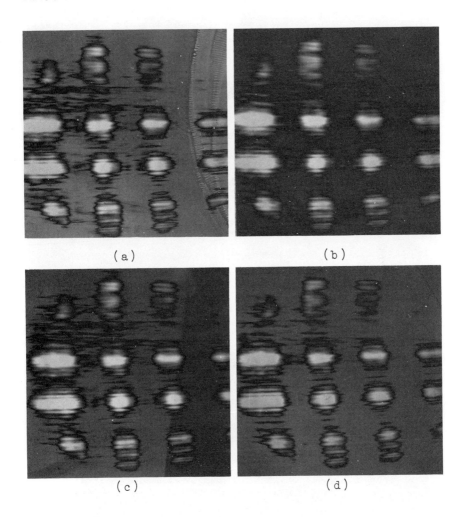

(a)   (b)

(c)   (d)

Figure 7 - Image of Test Object at 10.74 MHz
           with Incident Acoustic Power In-
           tensity of 10 mw/cm^2.

           (a)   No absorber
           (b)   Lucite plate
           (c)   Lucite structure
           (d)   Castor oil absorber

Figure 8 shows an aluminum test grid of 2.3 mm x 2.3 mm spacing. Its image at 10.7 MHz is shown in Figure 9.

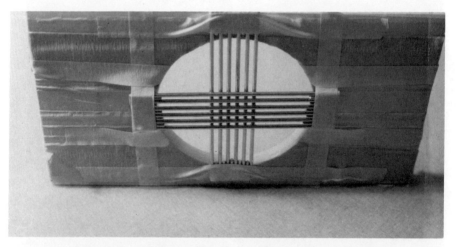

Figure 8 - Test Grid of 2.3 mm x 2.3 mm Spacing.

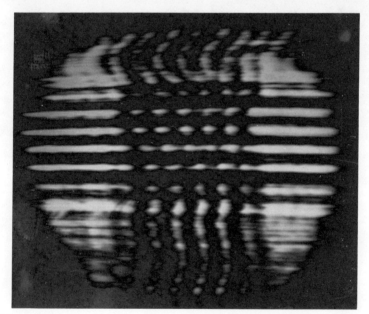

Figure 9 - 10.74 MHz Image of the Test Grid of Figure 8.

The theoretically calculated resolution of the system at 3.58 MHz is 3.36 mm or 8 wavelengths of sound. At 10.74 MHz it is .64 mm or 4.5 wavelengths of sound. These theoretical results agree well with the experimental determinations.

IMPROVED CYLINDRICAL OPTICS

In the design of a cylindrical converging lens for a Bragg-diffraction imaging system it is necessary to correct only for spherical aberration. Coma, astigmatism, distortion, and chromatic aberration do not occur since the coherent beam is parallel, on-axis, and monochromatic. However, it must be borne in mind that at large numerical apertures the water in the cell itself makes a significant contribution to the spherical aberration, though opposite in sign to that contributed by the lens. For this reason, a lens designed for minimal spherical aberration in air will be found to be overcorrected when used with the Bragg-diffraction system.

In order to keep down the cost of fabrication, a basically simple design of a cemented doublet lens was made. In such a design, there is less flexibility than with other designs since the two inside surfaces must have identical radii of curvature but a lens system of this type has the advantage of being less critical to mount and also less sensitive to mechanical shock than other systems.

In order to find an optimum design for a cemented doublet, a ray-trace program was implemented for the configuration shown in Figure 10. The water thickness, t_5, was taken to be 16 inches and t_3 was set at 1.0 inch. The window thicknesses t_4 and t_6 were assumed to be 0.125 inch. Actually, it turns out that t_3 is not critical provided that t_7 does not become negative. The Bragg cell can be moved along the optical axis without affecting the aberrations. A negative value of t_7 would simply indicate that the focus falls within the water, in which case the spherical aberration is greatly increased. This is the cause of the familiar distortion seen by observing fish in an aquarium when the

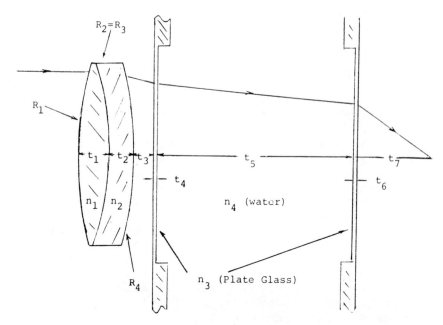

Figure 10 - Parameters for Ray-trace Design
of Beam Converging Lens.

line of sight is at a grazing angle to the glass.

As a further simplification, the back surface
of the lens was assumed plane (i.e. $R_4 = \infty$). In a
large-aperture lens, such as would be required for
a medical imaging system, each additional curved
surface represents considerable expense, so it is
advantageous, if possible, to minimize the number
of surfaces which must be ground and polished.

Two readily available glasses were chosen
from the Schott optical glass catalog. These are
a Barium crown, BK-7, with $n_d = 1.516800$ and a
flint glass, F-2, with $n_d = 1.620040$. Using the
interpolated refractive indices for these glasses
at 5145 Angstroms (Argon laser light), the computer

ray-trace program was employed to find optimum designs for lenses of relative aperture f/2.5, f/3.0,
f/3.5, and f/4.0, the f/# in this case referring
to the actual relative aperture of the converging
wedge in the water. Within the restrictions already made, it is not possible to design a lens
for the largest aperture, say f/2.5, which will
also be optimal when stopped down to smaller apertures. No effort was made to design these lenses
with chromatic correction since they were intended
for use at 5145 Å only. The f/4.0 design provided
the smallest spherical aberration blur at full aperture and thus was the best choice. On the basis of
numerical aperture, the resolution of the 3.6 MHz
system would then be improved by a factor of 3.0,
a sufficient amount to make the fabrication of the
doublet lens well worthwhile. This design is shown
in Table 1.

$$n_1 = 1.516800 \qquad R_4 = \infty$$

$$n_2 = 1.620040 \qquad t_1 = 1.10"$$

$$R_1 = 6.709" \qquad t_2 = 0.50"$$

$$R_2 = -7.129" \qquad FL = 18.632"$$

$$R_3 = R_2$$

Table 1 - f/4.0 Doublet

The lens described was designed for a nominal focal length of 18 inches when used with a thickness of water equal to 16 inches. The lens radii, and element thicknesses for these designs, may be scaled down provided that the same scaling is applied to the Bragg cell thickness. For any resonable value of scaling likely to be used, the window thickness can be held at 0.125 inches without introducing significant additional spherical aberration.

If better resolution were required, there would not, in general, be a great deal to be gained by giving a curvature to the rear surface of the doublet. The blur angle could probably be reduced to about one third of its value for those lenses with a plane rear surface, but this would result in an improvement in resolution by a factor of $\sqrt[3]{3}$, or 1.4, which would not generally be worth the additional expense.

Of course this is not the last word in lens design. It is possible, by using a three-element series of air-spaced meniscus lenses to make the blur angle due to spherical aberration almost arbitrarily small. While it would be possible to design such a composite lens to a relative aperture of perhaps f/1.0, this would only be warranted in situations where the active area of the ultrasonic beam is quite small. An f/1.0 lens system of 18-inch focal length would be extremely expensive.

As an alternative to multielement lenses, it has been proposed that holographic lenses might be used, either as complete cylindrically converging units or as corrector plates to be combined with a conventional inexpensive single-element lens. This remains a possibility, but there is some doubt as to the practicality of the approach.

CONCLUSION

Significant improvements of Bragg-diffraction imaging quality at bio-medically usable frequencies has been realized by greatly reducing the multiple reflections of acoustic energy that occur within the water-filled cell. This was not a problem at frequencies of 15 MHz or more because of the relatively high absorption in the water itself. Acoustic pulsing and optical range gating should provide additional benefit for this problem.

The limited resolution resulting from the small numerical aperture of readily available cylindrical lenses is a weakness of the present low-frequency systems. A simple cemented doublet lens as described here will improve the resolution by a factor of about three yielding an overall image resolution of about three acoustic wavelengths. This could be further improved with more sophisticated lens designs.

ACKNOWLEDGEMENTS

The authors wish to thank Professor Jorge Fontana for his direction and assistance during the past year. The work was supported by funds from the United States Public Health Service and from the Quantum Institute of the University of California at Santa Barbara. Our thanks to Ruth Smith for her effort in typing the manuscript.

REFERENCES

1. John Landry, Hormozdyar Keyani, and Glen Wade, "Bragg-Diffraction Imaging: A Potential Technique for Medical Diagnosis and Material Inspection," Acoustical Holography, Vol. IV, ED. Glen Wade, Plenum Press, 1972.

2. Soo-Chang Pei, "Elimination of Bragg Image Speckle by Frequency Modulation," August 1972, unpublished.

LINEARIZED SUBFRINGE INTERFEROMETRIC HOLOGRAPHY

A. F. Metherell

Actron Industries, Inc.
A Subsidiary of McDonnell Douglas Corporation
700 Royal Oaks Drive
Monrovia, California 91016

ABSTRACT

This paper describes a number of methods for recording suboptical wavelength displacements or vibrations of a surface as a linear function of the irradiance of the image of the surface. The methods are an extension of conventional interferometric holography, including time averaged holography of sinusoidal vibration, double pulsed holography of sinusoidal vibrations or general displacements, time averaged holography of linear motion, and real time interferometric holography. The basic concept described here was first presented at the Second International Symposium on Acoustical Holography and published in the proceedings. The theory was later expanded and presented at the Annual Meeting of the Optical Society of America in October 1969. This paper presents the theoretical analysis of linearized subfringe interferometric holography and some experimental results. The theory is expanded to show how this can be used to record acoustical holograms.

1. INTRODUCTION

Holographic interferometry[1-6] has proven to be a useful and sensitive tool for measuring the motion of objects and changes in the refractive index of fluids. Motions and changes of an order of magnitude of a few optical wavelengths are usually observed by counting the number of interference fringes in the image of the subject starting from a point of known zero motion or change. Holographic interferometry in its basic form when used

41

on vibrating objects is insensitive to the phase of the vibrations but is sensitive to the absolute amplitude of vibrations.

The purpose of this paper is to describe a modified form of holographic interferometry using a phase modulated reference or object beam which is sensitive to much smaller (subfringe) changes and motions, is sensitive to both phase and amplitude of motion, and where image intensity is linearly related to phase and amplitude. The basic idea was first presented at the Second International Symposium on Acoustical Holography[7]. The entire theory presented here was later developed and presented at the annual meeting of the Optical Society of America[8-10] in October 1969 but was never published. Since then the use of phase modulation of one of the beams in holography[11, 12] has been discussed and developed further, notably by Aleksoff[12], to cover more general forms of phase modulation.

This paper restricts the discussion of phase modulation to the "subfringe" region. The sensitivity, characteristics, and application to the detection of acoustical holograms are discussed.

2. CONVENTIONAL TIME AVERAGED HOLOGRAPHY OF A SINUSOIDALLY VIBRATING OBJECT

Powell and Stetson[1] were the first to propose and demonstrate a holographic technique for performing vibration analysis. Consider a point at coordinates (x_0, y_0) on a planar object which is vibrating sinusoidally with angular frequency Ω. We wish to make a conventional time average optical hologram of the vibrating planar object. If the peak amplitude of the object vibration is $m(x_0, y_0)$ and its fixed phase is $\mu(x_0, y_0)$, then the light incident at film coordinates (x, y) from the object point (x_0, y_0) has the time varying phase modulation:

$$\phi(x, y, t) = \frac{2\pi}{\lambda} (\cos\theta_1 + \cos\theta_2) m(x_0, y_0) \cos\left[\Omega t + \mu(x_0, y_0)\right] \qquad (1)$$

where

λ = optical wavelength of illuminating source

θ_1 = angle between the vector displacement and the direction of propagation of the light incident at (x_0, y_0)

θ_2 = angle between the vector displacement and the line joining (x_0, y_0) with (x, y)

If the exposure time in making the hologram is much longer than the vibration period (i.e., $T \gg 2\pi/\Omega$) and if the exposure is linearly recorded as an amplitude transmittance, then, as shown by Powell and Stetson, the relative intensity of the reconstructed image of the point (x_0, y_0) will be

$$I(x_0, y_0) = \frac{I_{image}}{I_0} = \left| J_0 \left[\frac{2\pi}{\lambda} (\cos\theta_1 + \cos\theta_2) \, m(x_0, y_0) \right] \right|^2 \qquad (2)$$

where

I_{image} = the intensity in the reconstructed image of the object point (x_0, y_0)

I_0 = the intensity in the reconstructed image when the object vibration is zero; and

J_0 = zeroth order Bessel function of the first kind

Equation (2) is shown plotted in Figure 1 for $\theta_1 = 0 = \theta_2$. Figure 1 shows that extinction of the image occurs at peak vibration amplitudes, m, of 0.2λ, 0.45λ, etc. In conventional applications of time average holography the vibration amplitude is typically a few optical wavelengths. By counting the number of fringes (places where the image brightness goes to zero) from a known zero amplitude point to any point in question, it is possible to determine the vibration amplitude at that point.

In view of Equation (2) and Figure 1, conventional time average holography cannot be used to record the amplitude and phase of the object surface vibrations. Equation (2) indicates that time average holography is sensitive only to vibration amplitude, m, and not to the phase of the vibration, μ. Secondly, except for surface displacement amplitudes located roughly between 0.04λ and 0.14λ, image intensity is a highly nonlinear function of

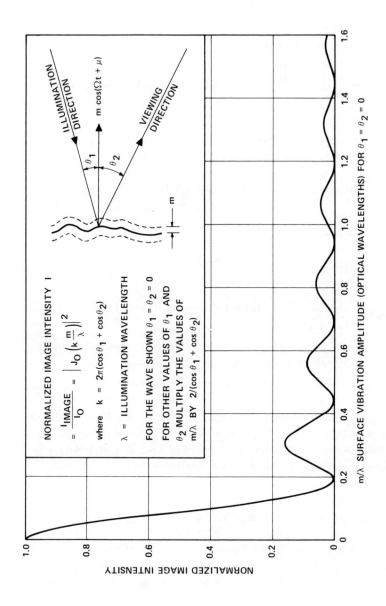

Figure 1. Time averaged (Powell-Stetson) holographic interferometry.

surface displacement amplitude. Finally, from a practical standpoint the sensitivity (dI/dm) of conventional time average holography is small for small values of displacement amplitude, m, (i.e., dI/dm $\rightarrow$ 0, as m $\rightarrow$ 0). Unfortunately, the amplitude of displacements generated at the surface boundary of a sound propagating medium by typical scattered acoustical fields is usually less than 0.01λ at acoustical frequencies in the MHz range.

3. TIME AVERAGED LINEARIZED
SUBFRINGE HOLOGRAPHIC INTERFEROMETRY

With one simple modification to the recording arrangement used in conventional time averaged holography, the limitations just described can be eliminated, and the amplitude and phase of the surface displacements can be essentially linearly recorded. If the optical illuminating beam is properly phase modulated at the vibration frequency, the zero amplitude surface vibration datum (m/λ = 0) in Figure 1 can be shifted from the I = 1.0 point (peak of the Bessel function) to the I = 0.5 point. This is the center of the linear region of the I vs. m/λ curve. The expanded portion of this curve as shown in Figure 2 has been replotted with the zero datum shifted to the I = 0.5 position for light in the green wavelength of 5145Å.

By operating about the I = 0.5 point, we obtain a linear relation between hologram image intensity and surface vibration amplitude for vibration amplitudes between 0 and roughly 0.04λ. The sensitivity, dI/dm, is a maximum at this point. To show that phase modulation of the illuminating beam also allows surface vibration phase information to be recorded, recall the description above of conventional time averaged holography. For simplicity assume $\theta_1 = 0 = \theta_2$. Then the light incident at film coordinate (x, y) from the object point (x_o, y_o) (located on the surface and vibrating at frequency Ω) has the time varying phase modulation:

$$\theta(x, y, t) \sim m(x_o, y_o) \cos\left[\Omega t + \mu(x_o, y_o)\right] + a \cos\Omega t \qquad (3)$$

because the object beam is coherently phase modulated at the acoustical frequency Ω. Or, rewriting θ we have

$$\theta(x, y, t) \sim M\cos(\Omega t + \beta), \qquad (4)$$

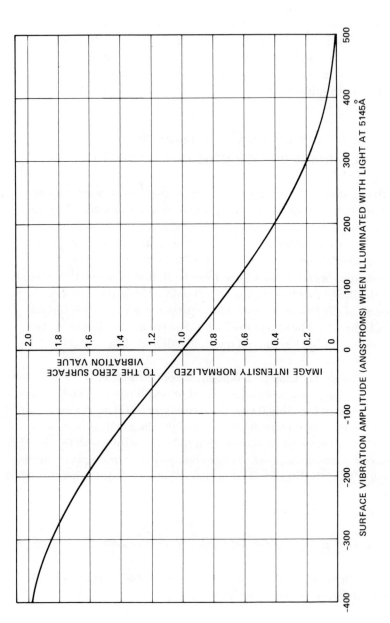

Figure 2. Linearized subfringe holographic interferometry (modified Powell-Stetson).

where

$$M = (m^2 + a^2 + 2m\,a\cos\mu)^{\frac{1}{2}} \tag{5}$$

and

$$\beta = \tan^{-1}\left[\frac{m\sin\mu}{a + m\cos\mu}\right] \tag{6}$$

Following the previous discussion, time average holography only records M and not β. Now if $m \ll a$ then $M \approx a(1 + m\,a\cos\mu)$, so both amplitude and phase are essentially linearly recorded. Under all of the foregoing assumptions the relative intensity of the reconstructed image of the point (x_0, y_0) will be

$$I = 1 - 16.94\,\frac{m}{\lambda}\cos\mu \tag{7}$$

which is the linearized version of Equation (2) with a phase-modulated illuminating beam. This modified version of time average holography is termed subfringe holographic interferometry.

4. DOUBLE EXPOSURE LINEARIZED SUBFRINGE HOLOGRAPHIC INTERFEROMETRY

Conventional double exposure interferometry[3-5] results in an image intensity which is given by

$$\begin{aligned}
I = \frac{I_{image}}{I_0} &= \left[\cos\left[\frac{\pi}{\lambda}(\cos\theta_1 + \cos\theta_2)\,\ell(x, y)\right]\right]^2 \\
&= \frac{1}{2}\left[1 + \cos\left[\frac{2\pi}{\lambda}(\cos\theta_1 + \cos\theta_2)\,\ell(x, y)\right]\right]
\end{aligned} \tag{8}$$

where $\ell(x, y)$ is the displacement of the surface between the two exposures. This equation is plotted in Figure 3 for $\theta_1 = \theta_2 = 0$.

Equation 8 can be linearized for small values of $\ell(x, y)$ by introducing an optical path length change $\lambda/4$ in either the reference or object beams. This is shown plotted in Figure 4. The zero datum has thus been shifted from

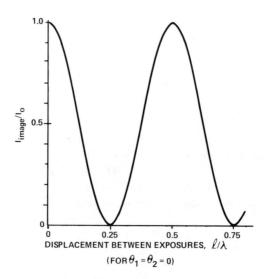

Figure 3. Double exposure interferometric holography.

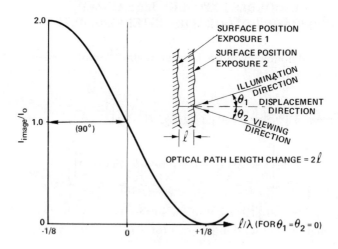

Figure 4. Datum shifted to halfway between crest and trough by adding a $\lambda/2$ change in path length between exposures.

the crest of the interference curve in Figure 3 to halfway between the crest and trough. The relationship between image brightness and displacement between exposures can thus be considered substantially linear over the region $\ell = 0$ to $\ell = \pm \lambda/16$.

Supposing now the object surface is vibrating with frequency Ω radians/ sec and amplitude m and the two hologram exposures are made as two substantially instantaneous exposures spaced half a vibration cycle apart. Let the first exposure be made at time $t = 0$ and the second exposure at $t = n/\Omega$ where n is an odd integer (preferably 1). Then the vibration of the surface can be expressed as $m\cos(\Omega t + \mu)$, where μ is the phase of the vibration measured with respect to the timing of the exposure $t = 0$. Then

$$\ell = 2m\cos\mu \tag{9}$$

With an optical phase shift of $\lambda/4$ in either the object or reference beam the image intensity relation of Equation (8) becomes

$$I = \frac{I_{image}}{I_o} = \left\{ 1 + \cos\left[\frac{\pi}{2} + 4\pi(\cos\theta_1 + \cos\theta_2)\frac{m}{\lambda}\cos\mu \right] \right\}$$
$$= \left\{ 1 - \sin\left[4\pi(\cos\theta_1 + \cos\theta_2)\frac{m}{\lambda}\cos\mu \right] \right\} \tag{10}$$

The ½ factor in front of the bracket in Equation (8) disappears because I_o (the intensity for $m = 0$) in Equation (10) is one half that in Equation (8). For $\theta_1 = \theta_2 = 0$ and $m \ll 0.1\lambda$ then

$$I = 1 - 25.13\frac{m}{\lambda}\cos\mu. \tag{11}$$

5. REAL TIME, TIME AVERAGED INTERFEROMETRIC HOLOGRAPHY

If an optical hologram of the surface of interest is recorded while the surface is stationary and the hologram is developed and replaced exactly in the recording position, the reconstructed virtual image of the stationary object can be superimposed on the actual vibrating object. Instantaneously during the vibration cycle interference fringes will be superimposed on the object as it is viewed through the hologram. These fringes will move during the vibration cycle, and if an image is recorded by time averaging over many cycles (say, with a camera or by viewing with the naked eye) then the apparent image brightness relationship is given by

$$I = \frac{I_{image}}{I_o} = \frac{1}{2}\left[1 + J_o\left[\frac{2\pi}{\lambda}(\cos\theta_1 + \cos\theta_2)\,m\right]\right] \tag{12}$$

This is shown plotted in Figure 5.

If the reference beam reconstructing the hologram (or the object illumination beam) is path length modulated at the vibration frequency and with a zero to peak amplitude equal to 0.294λ (107°), then $m = 0$ datum will be shifted from the crest to halfway between the crest and first trough at $I = 0.65$. This is shown plotted in Figure 6 where the relationship becomes

$$I = \frac{I_{image}}{I_o} = \frac{0.5}{0.65}\left[1 + J_o\left[\frac{2\pi}{\lambda}(\cos\theta_1 + \cos\theta_2)(M)\right]\right] \tag{13}$$

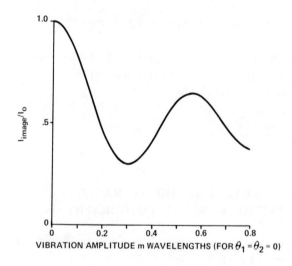

Figure 5. Real time interferometric holography of sinusoidally vibrating object.

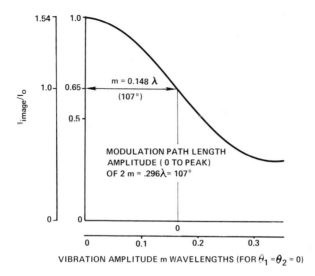

Figure 6. By phase modulating the object or reference beams at the vibration frequency the datum can be shifted to the I = 0.65 position.

where

$$M = \left[m^2 + (0.148\lambda)^2 + 2m(0.148\lambda) \cos\mu \right]^{1/2}$$

If the vibration amplitudes $m < 0.1\lambda$ then Equation (13) is substantially linear with respect to I, being

$$I \approx 1 \mp 5.63 \frac{m}{\lambda} \cos\mu. \tag{14}$$

6. TIME AVERAGED LINEAR MOTION

If an object surface is moving with linear velocity $v(x_0, y_0)$ while an optical hologram of the object is being recorded during an exposure time T, then the image intensity is given by

$$I = \frac{I_{image}}{I_0} = \text{sinc}^2 \left[\frac{2\pi \, vT}{\lambda} \right]. \tag{15}$$

Equation (15) is shown plotted in Figure 7.

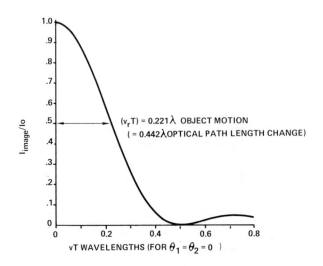

Figure 7. Time averaged hologram—constant velocity of motion v
with exposure time T.

If a constant rate of change in optical path length, v_r, is added to either
the optical reference or object illumination beam, the zero object velocity
datum can be shifted as before from the crest in Figure 7 to halfway between
the crest and trough if $v_r T = 0.442\lambda$.

7. USING TIME AVERAGED SUBFRINGE HOLOGRAPHIC
INTERFEROMETRY TO RECORD AN ACOUSTICAL HOLOGRAM

Since m as given by Equation (7) is recorded linearly, the subfringe
holographic interferometry technique can be used to record an acoustical
hologram.

The propagation of a sound wave through a fluid is the propagation of
a molecular strain (compression - rarefaction) wave. If the surface vibration
is the acoustical strain then the surface vibration is the sound wave in the
plane of the surface. If the sound wave is described by $A\cos(\Omega t + \mu)$ the
surface vibration will be proportional to $A\cos(\Omega t + \mu)$. However, only the
component of the surface vibration in the viewing/illuminating direction is
recorded. If the acoustically generated surface vibration makes an angle α

with the viewing/illumination direction (perpendicular to the surface) then the recorded vibration is $[\,m\cos(\Omega t + \mu)]$ where $m = A\cos\alpha$. The effect of this is that the method has a cosine angular sensitivity which will have negligible effect on the imaging of distant acoustical objects (high f number) and will slightly decrease the resolution of near acoustical objects (very low f number imaging).

The optical modulator in the illuminating beam will vary the phase of the illuminating beam uniformly over the entire area of the vibrating surface. This will be exactly equivalent to having an acoustical reference wave vibrating the surface where the reference wavefront conforms exactly to the shape of the surface. Thus if the surface is flat, the optical modulator will simulate a plane collimated acoustical reference impinging perpendicular to the plane. If the vibrating surface is spherical, the simulated acoustical reference is a point source located at the center of radius of the spherical surface. The term $(a\cos\Omega t)$ in Equation (3) represents the acoustical reference wave.

When illuminating the surface with an argon laser at 5145Å, Figure 2 shows that an acoustical hologram of a sound wave, impinging at a slight angle to the perpendicular which will produce a 10% depth of modulation (peak to peak) in the acoustical hologram fringe pattern, requires a surface vibration amplitude of 15Å. If the surface is the free surface to the acoustically propagating medium, the particle vibration in the medium is 7.5Å amplitude. In water at 1 MHz this represents an acoustic intensity of $0.00165\ \text{w/cm}^2$.

8. RECONSTRUCTION OF THE ACOUSTICAL HOLOGRAM

When the optical hologram is reconstructed the image of the vibrating surface will be visible with the intensity of the surface being linearly related to the surface vibration (assuming that the illuminating beam intensity is constant across the surface). From Equation (7) we can see that for small surface vibration amplitudes the intensity I is given by

$$I = 1 - 16.94\,\frac{m}{\lambda}\,\cos\mu$$

Let us assume that the vibrating acousto-optic surface is a plane optically specular surface and the optical illuminating wave U_O is a plane collimated beam falling perpendicular to the surface

$$U_O = A_O \exp i\omega t. \tag{16}$$

where ω is the light frequency and A_O is the amplitude which is uniform across the surface.

Then the reconstructed image gives the reflected wave from the surface as U_O with the intensity modulation given by Equation (7). Thus the wave reflected from the surface as reconstructed by the optical hologram is

$$
\begin{aligned}
U_s &= U_O\left(1 - 16.94\frac{m}{\lambda}\cos\mu\right) \\
&= U_O\left[1 - 16.94\frac{m}{\lambda}\tfrac{1}{2}\left(\exp(i\mu) + \exp - (i\mu)\right)\right] \\
&= U_O - 8.47\,A_O\,\frac{m}{\lambda}\left[\exp i(\omega t + \mu) + \exp\left(- i(\omega t + \mu)\right)\right]
\end{aligned} \tag{17}
$$

The first term in Equation (17) is the zero order wave from the acoustical hologram; the second term is the optical wave field analog of the acoustical wave field (which yields the reconstructed acoustical image); and the third term is the conjugate acoustical image.

Thus, when the acousto-optic surface is specular, the acoustical image may be reconstructed directly from the optical hologram without making an intermediate recording of the acoustical hologram plane.

If the acoustio-optic surface is not optically specular and capable of focusing the optical illuminating beam to a fine point then the reconstructed acoustic image is either distorted or destroyed in the direct reconstruction from the optical hologram because the spatial coherence of U_O in Equation (17) is destroyed. However, all is not lost because the acoustical hologram information can still be retrieved by reconstructing the image of the acousto-optic surface from the optical hologram and recording this on a second film. This will record the intensity pattern defined by Equation (7), which is the acoustical hologram. An example of an acoustical hologram recorded on an optically diffuse acousto-optic surface is shown in Figure 8. The circle on the right is the 12-inch diameter acousto-optic surface and the

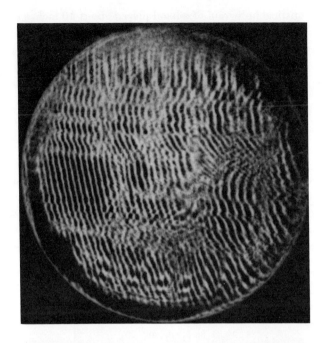

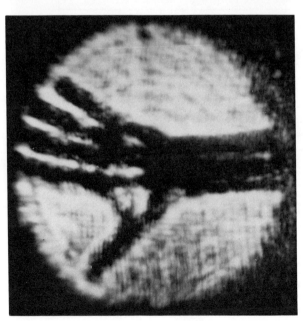

Figure 8. Acoustical hologram (right) with a 12-inch aperture recorded at 1 MHz using the time averaged linearized holographic interferometry method. Reconstructed image (left) focused on forearm.

fringe pattern is an acoustical hologram of a hand placed about 10 inches in front of the surface and acoustically illuminated with a 12-inch square 1 MHz sound source. The reconstructed image obtained from this hologram is shown on the left. The reconstruction was focused on the forearm to show the vascular structure (near the bottom of the picture). The hand itself is out of the plane of focus.

9. CONCLUSIONS

Sections 2 through 6 have described how to "linearize" the image intensity relationship for a number of forms of holographic interferometry. The extension to other forms of holographic interferometry, such as stroboscopic holography [13-15] is obvious. Tables 1 and 2 summarize the relationships for conventional and linearized subfringe holographic interferometry, respectively.

Table 1. Conventional interferometric holography.

TYPE	IMAGE INTENSITY RELATION (for $\theta_1 = \theta_2 = 0$)	
Time Av. Sinusoidal motion [object motion = $m \cos(\Omega t + \mu)$]		$= \left| J_0[4\pi m/\lambda] \right|^2$
Double exposure (object linear displacement ℓ)		$= \frac{1}{2}\left|1 + \cos(4\pi \ell/\lambda)\right|$
Time Av. real time [object motion = $m \cos(\Omega t + \mu)$]		$\frac{1}{2}\left\{1 + J_0\left|4\pi m/\lambda\right|\right\}$
Time Av. Linear motion [object velocity = v; exposure time T]		$= \text{sinc}^2\left|2\pi vT/\lambda\right|$

Table 2. Linearized subfringe interferometric holography.

TYPE	REFERENCE (OR ILLUMINATING) BEAM PATH LENGTH MODULATION WITH SHIFT OF DATUM TO LINEAR REGION	$\dfrac{dI}{d\left(\frac{m}{\lambda}\right)}$ (for $\theta_1 = \theta_2 = 0$)	LINEARIZED DISPLACEMENT –INTENSITY RELATION	OBJECT AMPLITUDE TO PRODUCE 1% CHANGE IN INTENSITY
Time Av. Sinusoidal motion [object motion $= m \cos(\Omega t + \mu)$]	$a \cos \Omega t$ $a = \pm 0.179\lambda$ ($\pm 64.5^0$)	∓ 16.94	$[1 \mp 16.94 (m/\lambda) \cos \mu]$	0.00059λ ($= 3.73\overset{\circ}{A}$ at $\lambda = 6328\overset{\circ}{A}$)
Double exposure [object motion $= m \cos(\Omega t + \mu)$]	b $a = \pm 0.296\lambda$ ($\pm 90^0$)	∓ 25.13	$[1 \mp 25.13 (m/\lambda) \cos \mu]$	0.000398λ ($= 2.52\overset{\circ}{A}$ at $\lambda = 6328\overset{\circ}{A}$)
Time At real time [object motion $= m \cos(\Omega t + \mu)$]	$a \cos \Omega t$ 1.538 $a = \pm 0.296\lambda$ ($\pm 107^0$)	∓ 5.63	$[1 \mp 5.63 (m/\lambda)\cos \mu]$	0.00175λ ($= 11.1\overset{\circ}{A}$ at $\lambda = 6328\overset{\circ}{A}$)
Time Av. Linear motion [object velocity $= v$; exposure time T]	$v_r T$ $v_r T = \pm 0.446\lambda$	∓ 12.97	$[1 \mp 12.97 \, (v_0 T)]$	0.00077λ ($= 4.88\overset{\circ}{A}$ at $\lambda = 6328\overset{\circ}{A}$)

Sections 7 and 8 describe how the time averaged linearized subfringe holographic interferometry method of Section 3 has been applied to the recording of acoustical holograms. It is straight forward to see how acoustical holograms can be recorded using the double exposure method of Section 4 with Equation (11), and the real time method of Section 5 with Equation (14). The method of time average linear motion of Section 6 can even be used to record an acoustical hologram if the surface vibration amplitude is very high which it will be if the acoustic intensity is high, or the acoustic frequency is low. By making the recording exposure time T in Section 6 much less than the acoustic cycle (less than 1/10th cycle) then this will effectively measure and record the instantaneous velocity of displacement of the surface. If the surface vibration is $m\cos(\Omega t + \mu)$ then the velocity is $\Omega m \sin(\Omega t + \mu)$, the recording of which will given an acoustical hologram. The difference between recording $\sin\mu$ and $\cos\mu$ is merely an overall 90° phase shift.

REFERENCES

1.   R. L. Powell and K. A. Stetson, *J. Opt. Soc. Am.*, 55, 1593 (1965).

2.   K. A. Stetson, and R. L. Powell, *J. Opt. Soc. Am.*, 56 (9), 1161-1166 (1966).

3.   L. O. Heflinger, R. F. Wuerker, and R. G. Brooks, *J. Appl. Phys.*, 37, 642 (1966).

4.   R. E. Brooks, L. O. Heflinger, and R. F. Wuerker, *IEEE J. Quantum Electronics*, QE-2, 275 (1966).

5.   R. E. Brooks, L. O. Heflinger, and R. F. Wuerker, *Appl. Phys. Letters*, 7, 248 (1965).

6.   J. M. Burch, A. E. Ennos, and R. J. Wilton, *Nature*, 209, 1015 (1966).

7.   A. F. Metherell, S. Spinak, and E. J. Pisa, "Temporal Reference Acoustical Holography," paper presented at Second International Symposium on Acoustical Holography, Huntington Beach, California, March 1969, *Applied Optics*, 8 (8), 1543-1550 (1969).

8.   A. F. Metherell, S. Spinak, and E. J. Pisa, *J. Opt. Soc. Am.*, 59 (11), 1534A (1970).

9.   E. J. Pisa, A. F. Metherell, and S. Spinak, *J. Opt. Soc. Am.*, 59 (11), 1534A (1970).

10. S. Spinak, E. J. Pisa, and A. F. Metherell, *J. Opt. Soc. Am.*, 59 (11), 1535A (1970).

11. D. B. Neumann, "The Effect of Scene Motion on Holography," Ph.D. Thesis, Ohio State University (1967).

12. C. C. Aleksoff, *Applied Optics*, 10, 1329 (1971).

13. E. Archbold and A. E. Ennos, *Nature*, 217, 942 (1968).

14. B. M. Watrasiewicz and P. Spicer, *Nature*, 217, 1142 (1968).

15. P. Shajenko and C. D. Johnson, *Appl. Phys. Lett.*, 13, 44 (1968).

ACOUSTICAL HOLOGRAPHY USING
TEMPORALLY MODULATED OPTICAL HOLOGRAPHY

K. R. Erikson

Actron Industries, Inc.
A McDonnell Douglas Subsidiary
700 Royal Oaks Drive
Monrovia, California 91016

This paper discusses a new technique for obtaining an acoustical hologram using optical holography. Temporal modulation in the form of phase modulation and Doppler upshifting is used in the reference beam of the optical system. This allows recording of the amplitude and phase of sound waves impinging on a mirror interface. The advantages and disadvantages of this system technique are discussed.

INTRODUCTION

Many acoustical imaging systems have been developed which use optical techniques for recording the minute displacements of a surface caused by an incident acoustical beam.[1-9]

These optical techniques may be characterized by their use of serial or parallel detection of the surface displacement. Serial detection requires some sort of scanning to cover a surface with the information transmitted over one channel or a few channels at most. Parallel detection records the displacements without any scanning.

Examples of serial detection are the scanned focused laser beam recording of a mirrored surface by Adler, Korpel, and Desmares[1] and Massey.[2] The former authors used spatial angle modulation and the latter used temporal heterodyne modulation, together with a photodiode and television-type display. These systems have very good sensitivity; however, the requirement for scanning has obvious limitations and does not exploit the tremendous spatial bandwidth inherent in optical systems.

Mueller and Sheridan[3] and Brenden[4] used an optical schlieren system as a parallel or area detector of a liquid surface responding to acoustic radiation pressure. These systems neglect the large temporal bandwidth properties of the light beam. Metherell, Spinak, and Pisa[5-8] have described several optical methods using area detectors for recording acoustical holograms. One of their techniques called "Linearized Subfringe Holographic Interferometry" (LSHI)[8] uses a phase modulated optical reference beam and is the basis of the present work. Another closely related technique for recording acoustical images called "Upshifted Reference Holography" has been described by Whitman.[9]

In this paper we will first review time average holographic interferometry which is the basis for LSHI. Then LSHI is discussed in detail. A new method using both phase modulation and Doppler upshifting of the optical reference beam is then developed as an extension to LSHI. Finally, optimization of both LSHI and the new technique is considered.

The first appendix contains the mathematical derivation of temporally modulated reference beam holography. The second appendix discusses "Upshifted Reference Holography"[9] in enough detail to show that the conditions for recording on acoustical hologram are not met. The third appendix discusses the effect of finite amplitude acoustic wave on these recording processes.

TIME AVERAGE HOLOGRAPHIC INTERFEROMETRY

Time average holography was demonstrated by Powell and Stetson[10] to be useful for analyzing the vibration of a surface. However, this technique is not sensitive to the minute (100 Å or less) surface vibration amplitudes caused by acoustic waves in the low MHz frequency range at reasonable power levels. Also, phase information, a requirement for holography, is lost in the nonlinear recording. In 1969 Metherell, Spinak, and Pisa[7] and Aleksoff[11] independently discovered a method for preserving the phase and recording small amplitudes.

Figure 1 shows the general relationships between the acoustical and optical systems. An acoustic ray scattered from the subject is incident on the acousto-optic interface at an angle θ_o. A laser or coherent light ray is reflected from the resulting dynamic surface perturbation to the hologram. Here the term dynamic refers to the surface vibrations at the acoustical frequency rather than any more slowly varying displacements resulting from radiation pressure or lower frequency mechanical vibrations caused by machinery, etc. The surface deflection amplitude A for an incident acoustic intensity I (watts/cm^2) is:

$$A(x_o, y_o, t) = \frac{2}{\omega_1}\left(\frac{2I}{\rho_1 c_1}\right)^{1/2} \cos\theta_o \sin\left(\omega_1 t - \frac{x_o \omega_1 \sin\theta_o}{C_1} + a\right) \qquad (1)$$

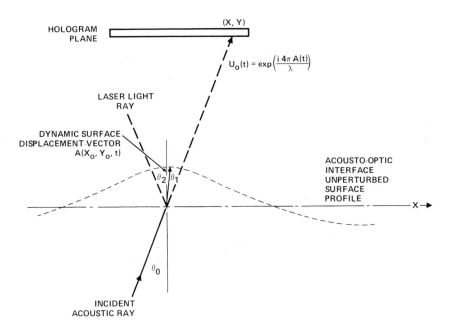

Figure 1. Acousto-optic interface recording geometry.

where

$\omega_1 = 2\pi f_1, f_1$ = is the acoustic frequency

ρ_1, c_1 = are the density and the speed of sound in the acousto-optic interface, respectively

θ_0 = is the angle of incidence of the sound wave

a = is the phase of the acoustic wave relative to some master oscillator.

The path length of the reflected light ray between the interface and the hologram plane is thus changing sinusoidally at the acoustic frequency. This phase modulated light of wavelength λ has a temporal frequency spectrum given by

$$U_0(f) = \sum_{n=-\infty}^{+\infty} J_n\left[\frac{2\pi a\,(x_0, y_0)}{\lambda}\left(\cos\theta_1 + \cos\theta_2\right)\right]$$

$$\exp\left[i_n\left(\frac{x\omega\sin\theta_0}{c_1} + a\right)\right] \delta\left(f - n\,f_1\right) \qquad (2)$$

where

J_n = nth order Bessel function of the first kind

$a(x_0, y_0) = \dfrac{2}{\omega_1}\left(\dfrac{2I}{\rho_1\,c_1}\right)^{\!1/2}\cos\theta_0$ = peak deflection amplitude

θ_1 = angle between the vector displacement and the line joining (x_0, y_0) with (x, y) as shown in Figure 1

θ_2 = angle between the vector displacement and the direction of the ray incident upon (x_0, y_0).

$U_0(f)$ is shown schematically in Figure 2 with sidebands spaced about the original laser light frequency f_0 by the acoustic frequency f_1. The relative amplitudes of the sidebands are determined by the surface displacement a.

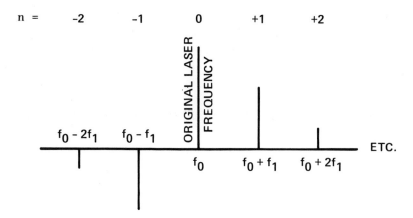

Figure 2. Temporal frequency spectrum of laser light reflected from sinusoidally moving surface f_1 is the acoustic wave frequency.

It is assumed that any other frequencies of the laser light are spaced from f_0 by many times the acoustic frequency.

If a long exposure $(T \gg 1/f_1)$ is made only the $n = 0$ term of Eq. 2 remains. Time average holography makes use of the fact that the carrier or $n = 0$ term goes to zero at points determined by the displacement. For normal incidence, the first zero occurs at

$$\frac{4\pi a}{\lambda} = 2.4 \text{ or } a = \frac{2.4\lambda}{4\pi} \simeq \frac{\lambda}{4}.$$

Thus by recording and reconstructing a long exposure holographic image of a moving subject, dark bands are seen at points whose amplitude corresponds to a zero of the zero order Bessel function. This occurs because all the light from that dark band was dynamically shifted to other frequencies where it was no longer coherent with the light in the reference beam at f_0. When combined with the angularly offset reference beam typical of off-axis holography, light at these new frequencies produces a spatial as well as temporal beat frequency with the reference beam. When averaged over a number of acoustic cycles on film, the holographic fringes wash out and only contribute to bias buildup on the film. Thus during hologram reconstruction there is nothing to diffract light into that portion of the image. If the reference beam had one or more coherent sideband frequencies, stable holographic fringes would be formed. When the hologram is reconstructed, these fringes would diffract light into the image.

LINEARIZED SUBFRINGE HOLOGRAPHIC INTERFEROMETRY (LSHI)

In LSHI the optical reference beam is put through a phase modulator driven sinusoidally at the acoustic frequency. With a long exposure, all the coherent sidebands can now form an image as shown schematically in Figure 3.

The exposure $E(x, y, t)$ on the hologram film is then given by

$$E(x, y, t) = \int_{-\infty}^{\infty} S(t') \left| U_r(t' + t) + U_0(t' + t) \right|^2 dt' \tag{3}$$

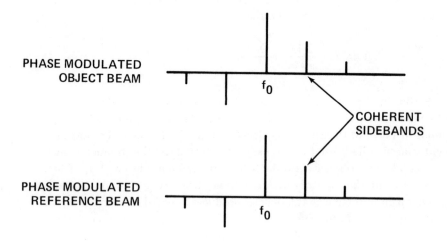

Figure 3. Linearized subfringe holographic interferometry.

where

$$S(t) = \text{generalized shutter function}[12]$$

$$U_r(t) = \exp\left[i\,\Gamma V \sin\left(\omega_1 t + \gamma\right)\right] = \text{reference beam modulation}$$

$$U_0(t) = \exp\left(i 4\pi\,A/\lambda\,\right) = \text{reflected object beam modulation at normal incidence.}$$

The optical carrier frequency, common to both beams, is implicit in Eq. 3.

The ΓV term in the reference beam results from the phase modulation being conveniently implemented with a single crystal electro-optic material such as lithium niobate. The modulation is a function of the drive voltage V, applied across the crystal and Γ, which relates the crystal dimensions, optical wavelength, indices of refraction, and the electro-optic coefficients.

Assuming that the long exposure is linearly recorded as an amplitude transmittance, reconstruction of the hologram will produce four terms. Common to all off-axis holography the relevant term is either $U_r^*(t)\,U_0(t)$ or its complex conjugate because of the angular offset of the reference beam. (The asterisk denotes the complex conjugate.) The image brightness density (IBD)[12] which is defined as the energy density per second is proportional to $|E|^2$. Then as Metherell[7] and Aleksoff[12] have shown (cf. Appendix I) for an exposure time $T > 1/f_1$:

$$\text{IBD} \propto J_0^2(c) \tag{4}$$

where

$$c^2 = \left(\frac{4\pi a}{\lambda}\right)^2 + (\Gamma V)^2 - \frac{8\pi a}{\lambda}\,\Gamma V \cos\left(a - \gamma\right).$$

If

$$\Gamma V \gg \frac{4\pi a}{\lambda}$$

then

$$c \simeq \Gamma V\left(1 - \frac{4\pi a}{\lambda \Gamma V}\,\cos(a - \gamma)\right).$$

If the zeroes and peaks of the Bessel function are avoided through proper choice of ΓV, this constraint also linearizes J_o^2. Thus a quasilinear recording of amplitude and phase required for an acoustical hologram can be made as shown in Figure 4.

Figure 5 is a schematic diagram of the experimental system used. It is a conventional holographic system with the exception of the modulator in the reference beam. See Chapter 4 of this volume for further details.

In practice an optimum modulator voltage is chosen to place the "bias" at some point c_1 (Figure 4) without any acoustic displacement. An acoustically generated displacement then swings the brightness of the "acoustical hologram" in a quasilinear manner about this point.

If this acoustical hologram is recorded on film and in turn reconstructed, an acoustical image will be obtained with the desirable properties of a holographic image such as large depth of field and high resolution. Thus the temporal and spatial bandwidth of the light beam have been simultaneously exploited to great advantage.

Materials which are acoustically suitable for the acousto-optic interface (mirrored surface) are generally difficult to fabricate into high quality optical

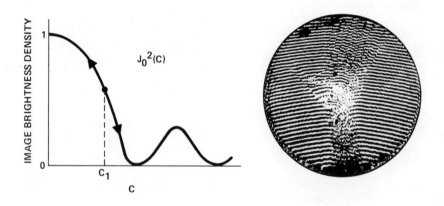

Figure 4. Linearized subfringe holographic interferometric acoustical hologram. Bias point c_1 is determined by modulator voltage.

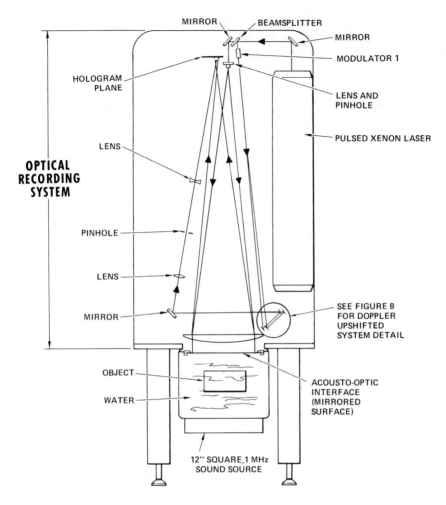

Figure 5. Experimental apparatus. For LSHI, Modulator 1 is a phase modulator. For Doppler upshifting, it is an Acousto-Optic Modulator.

surfaces. One of the advantages of the system of Figure 5 is its tolerance to a less than ideal optical surface. This results from the fact that the large lens and mirrored optical surface are only required to return the light to the film. There are also no special mechanical stability requirements because the laser pulse can be as short 5 to 10 μs for 1 MHz sound.

At frequencies near 1 MHz a large aperture system is needed for medical imaging purposes in the human abdomen to provide adequate resolution. For instance, the system of Figure 5 uses a 12-inch diameter acousto-optic interface. Fabricating a high quality Fourier transform lens of that diameter as required in a one-step acoustical image reconstruction technique is a formidable task indeed. The present method trades this fabrication job for an intermediate reconstruction system in which the hologram and required lenses are scaled down in size and cost. For some applications this two-step process may not be an acceptable trade-off because it can make image access time quite long.

UPSHIFTED, PHASE MODULATED REFERENCE ACOUSTICAL HOLOGRAPHY

The technique proposed in this work is an extension of LSHI in that the reference beam is upshifted by m times the acoustic frequency and phase modulated at the acoustic frequency. The resulting sideband relationships are shown in Figure 6. In this case the IBD of the reconstructed hologram (as shown in Appendix I) is given by

$$\text{IBD} \propto J_m^2(c) \tag{6}$$

where once again $\quad c \simeq \Gamma V\left(1 - \frac{4\pi a}{\lambda \Gamma V} \cos(a - \gamma)\right)$

for $\Gamma V \gg 4\pi a/\lambda$. Preliminary experimental results have demonstrated the validity of Eq. 6. The implications of these results are explored in detail in the next section.

Modifications of the system of Figure 5 for single sideband acoustical holography are shown schematically in Figures 7 and 8. For convenience two stable RF oscillators were used rather than deriving both frequencies from one source. Only occasionally fine tuning was required to keep the frequencies within tolerance. The double balanced mixer produces sum and difference frequencies with the carrier suppressed. Only the difference frequencies were within the bandpass of the level comparator which allowed simple digital division circuits to be used.

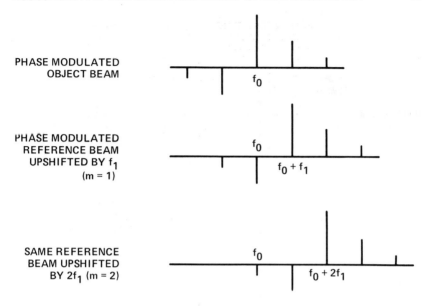

PHASE MODULATED
OBJECT BEAM

f_0

PHASE MODULATED
REFERENCE BEAM
UPSHIFTED BY f_1
(m = 1)

f_0

$f_0 + f_1$

SAME REFERENCE
BEAM UPSHIFTED
BY $2f_1$ (m = 2)

f_0

$f_0 + 2f_1$

Figure 6. Upshifted, phase modulated reference beam holography.

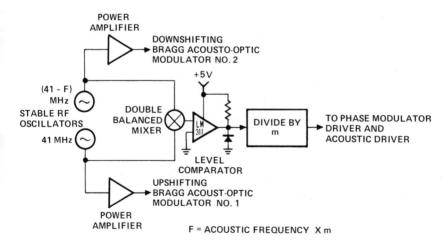

Figure 7. Single sideband electronics

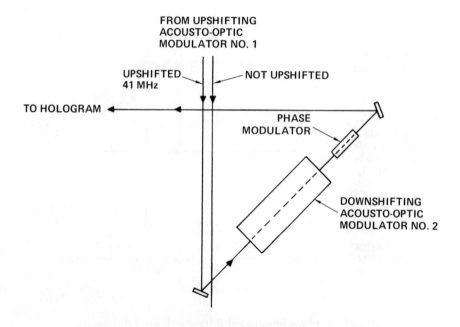

Figure 8. Details of reference beam modulation for Doppler
upshifted system (see Figure 5).

OPTIMIZATION

Image Brightness Density (IBD) curves of the $m = 0$ (LSHI), and
$m = 1$ and 2 systems are shown in Figure 9. The parameter c contains the
ΓV term as an independent parameter which allows optimization, subject to
the constraint $\Gamma V \gg 4\pi a/\lambda$. Several points are readily apparent.

1. The points in the curves where the first derivative changes sign will
   produce a phase ambiguity and are to be avoided.

2. The steepest portion of the curve will produce the greatest brightness
   change for a given acoustic displacement.

3. The steepest portion of the curve generally occurs with a large background or "bias" brightness. Setting $a = 0$ gives this brightness. This
   bright background will lead to fringes of reduced visibility.

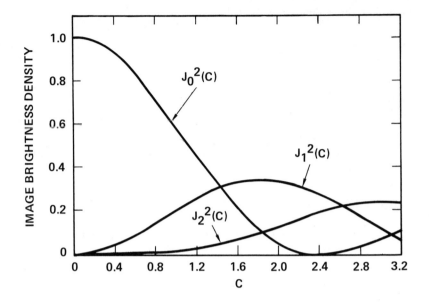

Figure 9. Image brightness versus c.

For optimization we use Michelson's definition of fringe visibility ν [13]:

$$\nu = \frac{I_{max} - I_{min}}{I_{max} + I_{min}} \cdot \qquad (7)$$

This can be rewritten as

$$\nu = \frac{\Delta I}{I_{av}}$$

where

$$\Delta I = \frac{(I_{max} - I_{min}}{2}$$

and

$$I_{av} = \frac{I_{max} + I_{min}}{2} \cdot$$

We now define an incremental visibility

$$\nu^* \equiv \frac{dI}{I} \tag{8}$$

or as applied to Eq. 6,

$$\nu^* = \frac{1}{J_m^2(c)} \frac{d J_m^2(c)}{dc} = \frac{2}{J_m(c)} \frac{d J_m(c)}{dc} . \tag{9}$$

ν^* is shown in Figure 10 with the scale factor of 2 deleted.

These curves show that the incremental visibility is optimized by biasing close to the zeros of the Bessel functions. This leads to a number of problems. The zero of the functions produce a phase ambiguity, so the effect of working closer and closer to that zero is to limit the dynamic range of the system.

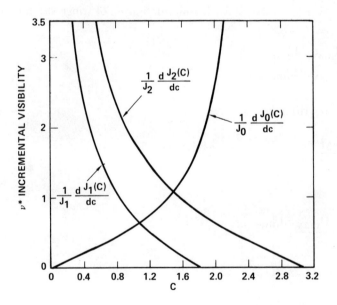

Figure 10. Incremental visibility versus c.

Secondly, for the $m = 1$ and 2 cases, the $\Gamma V \gg 4\pi a/\lambda$ constraint also limits dynamic range. Thirdly, the functions become very nonlinear near the zeros. Finally, a most important practical matter of image brightness must be considered. Any optical reconstruction system will have some finite amount of optical background noise due to lens reflections, photographic base and emulsion scattering, etc. The image should be at least ten times brighter than this noise level for an adequate signal to noise ratio.

Figure 11 shows portions of the incremental visibility vs. brightness curves with the parameter c along the curve for the three systems. The $m = 0$ (LSHI) curve always lies above the other two and is therefore usually the preferred method. Note, however, that with $m = 0$ small changes in c produce large changes in ν^* for $c > 2$ when compared to $m = 1$. This implies nonlinearity as well as limited dynamic range. In fact in the range of $\nu^* > 3$, the $m = 1$ system is very competitive with $m = 0$.

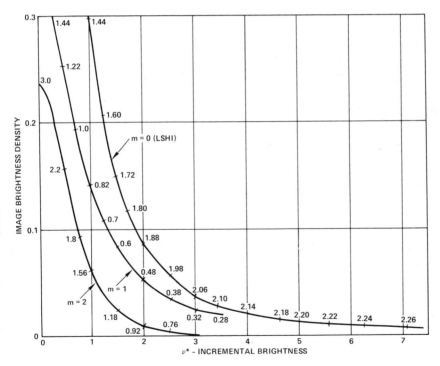

Figure 11. Image brightness versus incremental visibility with c as a parameter.

Choosing between the $m = 0$ and more complex $m = 1$ system is then a matter of system design trade-offs as well as the frequency spectrum of the incident acoustic wave.

As discussed in Appendix II there are situations in which there may be appreciable amounts of harmonic distortion present in acoustic waves. In this case the $m = 0$ system which requires a large bias term for optimum visibility will also have a large percentage of upshifted light at higher harmonics. For instance at $c = 1.8$, a typical operating point, $J_2/J_1 \sim 0.53$. Thus the $m = 0$ system is sensitive to the second harmonic of the acoustic frequency, which is a noise-like term. Also any nonlinearities in the phase modulator will be more pronounced at high voltage. The $m = 1$ system produces a similar visibility at $c = 0.4$ in which case $J_1/J_0 = 0.2$ resulting in less sensitivity to the second acoustic harmonic.

SUMMARY AND CONCLUSIONS

A general discussion of phase modulated and frequency shifted optical detection techniques suitable for acoustic hologram formation has been presented. The $m = 2$ system (upshifted by twice the acoustic frequency and phase modulated at the acoustic frequency) was shown to be a nonoptimum choice because of low image brightness. The $m = 0$ LSHI system (not up-shifted but phase modulated at the acoustic frequency) was shown to have possible sensitivity to spurious noise terms. The $m = 1$ system may offer some performance improvement at the expense of greater complexity.

ACKNOWLEDGMENTS

I would like to thank Jim Pisa for many helpful discussions and his comments on the manuscript, John Wreede for his help with the experimental work, Bob Norton for building the necessary electronics, Bob Watts for designing the optical modulator geometry, Bob Greer for designing the optical fixtures, and finally Alice Hultsman for typing the manuscript.

APPENDIX I
DERIVATION OF IBD FOR FREQUENCY UPSHIFTED
PHASE MODULATED SYSTEMS

Following Aleksoff [12] we define

$$\chi(f) = \text{F.T.}\left(U_r^* \cdot U_o\right) = \int_{-\infty}^{\infty} u_r^*(f') u_o(f'+f) df'$$ (10)

where **F.T.** stands for the Fourier Transform. Substituting Eq. (2) into Eq. (10), assuming normal incidence

$$\chi_m(f) = \int_{-\infty}^{\infty} df' \left[\sum_{p=-\infty}^{+\infty} J_p\left(\Gamma V\right) \exp(i\,p\,\gamma)\, \delta\left(f' - (p+m)f_1\right)\right]^*$$

$$\left[\sum_{n=-\infty}^{+\infty} J_n\left(\frac{4\pi a}{\lambda}\right) \exp(i\,n\,a)\, \delta\left(f' - nf_1 + f\right)\right]$$ (11)

where m is the number of times the acoustic frequency the reference beam is upshifted. $m \geqslant 0$. Then

$$\chi_m(f) = \sum_{p=-\infty}^{+\infty} J_p(\Gamma V) \exp(-i\,p\gamma) \sum_{n=-\infty}^{+\infty} J_n\left(\frac{4\pi a}{\lambda}\right)$$

$$\exp(i\,n\,a)\, \delta\left(f - (n-p-m)f_1\right).$$ (12)

Then if $q = n - p$

$$\chi_m(f) = \sum_{p,q=-\infty}^{\infty} J_p(\Gamma V) J_{p+q}\left(\frac{4\pi a}{\lambda}\right) \exp\left(ip(a-\gamma)\right)$$

$$\exp(iqa)\, \delta\left(f - (q-m)f_1\right).$$

(13)

Using Graf's addition theorem [14]

$$\chi_m(f) = \sum_{q=-\infty}^{\infty} J_q(c) \exp\left(iq(\xi+a)\right) \delta\left(f - (q-m)f_1\right) \qquad (14)$$

where

$$c^2 = \left(\frac{4\pi a}{\lambda}\right)^2 + (\Gamma V)^2 - \frac{8\pi a\, \Gamma V}{\lambda}\, \cos(a-\gamma)$$

with the geometrical relationship shown in Figure 12.

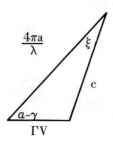

Figure 12. Graf's addition theorem.

Then taking the relevant term from Eq. 3, Ref. 12 shows that

$$E(x,y,t) = \int_{-\infty}^{+\infty} S(t')U_r^*(t'+t)\, U_0(t'+t)\, dt'$$

where $S(t)$ is the shutter function. This is simply the cross correlation of the shutter function and the holographic signal. Equivalently in the frequency domain

$$e(x, y, f) = s^*(f) \cdot \chi(f)$$

where $s^*(f)$ is the Fourier transform of $S(t)$ and is in general complex.

Thus the frequency spectrum of the shutter function picks out the same frequencies from the object and reference beam cross correlation spectrum, allowing the other frequencies to smear in the hologram.[15] In the limit of infinitely long exposure

$$e(x, y, f) = \lim_{T \to \infty} s^*(f) \cdot \chi(f) = \chi(0) \, \delta(f).$$

Thus, $e(x, y) = J_m(c) \exp\left(im(\xi + a)\right).$ \hfill (15)

APPENDIX II

The acoustic wave represented by Eq. 1 is an oversimplification since so-called "finite amplitude" effects are not included. Finite amplitude effects or nonlinear distortion have been widely discussed in the literature.[16] They occur because the velocity of sound is pressure dependent. This causes acoustic wave compressions to travel faster than rarefactions depending on the amplitude of the wave. An acoustic wave which is initially sinusoidal then becomes increasingly sawtooth with propagation. The point at which the leading edge of the sawtooth wave becomes vertical is called the discontinuity distance. This discontinuity distance ℓ is shown in Figure 13 for water at various frequencies as a function of acoustic intensity. The B/A parameter is simply the ratio of the first two terms in the power series for the pressure dependent velocity of sound. Fortunately, water appears to be the least nonlinear fluid at room temperature.

It should be noted from Figure 13 that especially at higher frequencies the discontinuity distance is relatively short for modest acoustic intensities.

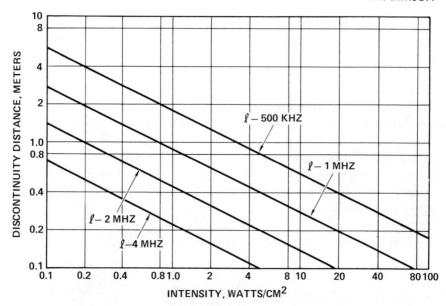

Figure 13. Discontinuity distance ℓ for water (B/A = 5).

The sawtooth waveform results in a frequency spectrum with apprec-iable amplitude at multiples of the acoustic frequency as shown in Figure 14. Another interesting point is that there is an upper limit to the acoustic power which can be transmitted over a given path at a given frequency.[17] Beyond a certain point any additional energy simply goes into harmonics.

APPENDIX III

"ACOUSTIC HOLOGRAM FORMATION WITH A FREQUENCY SHIFTED REFERENCE BEAM"[9]

In this technique the optical reference beam is upshifted by the acoustic frequency with respect to the object beam. No phase modulation is used. An acoustic wave is incident at an angle to the interface. The resulting surface ripple acts as a diffraction grating allowing a spatial filter in the back focal plane (Fourier transform plane) of a lens to select only the dynamically scattered light. This phase modulated light produces moving fringes. By

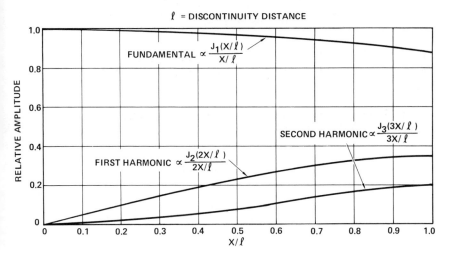

Figure 14. Finite amplitude (nonlinear) distortion of a sinusoidal acoustic wave — Fubini's solution.

introducing the upshifted reference beam, stationary fringes are obtained. The combined beams can then be used to form an image of the surface directly without the need for an intermediate film as in the systems discussed above. The resulting image is an image plane hologram.

In this method the IBD (as shown in Ref. 12) is

$$\text{IBD} \propto J_1^2\left(\frac{4\pi a}{\lambda}\right) \tag{16}$$

which can also be derived by setting $\Gamma V \equiv 0$ in Eq. 6. Equation 16 has a phase ambiguity because it is quadratic. Thus the IBD is the same for a phase angle a or $a \pm \pi$ and produces an image with fringes of twice the correct spatial frequency. This violates one of the requirements for acoustic hologram recording.

Shifted reference holography is in great detail by Aleksoff.[12]   More recently, Zambuto and Fischer[18 - 19] have used similar shifted reference methods for measuring small displacements.

REFERENCES

1. R. Adler, A. Korpel and P. Desmares, "An Instrument for Making Surface Wave Visible," *IEEE Trans. Sonics and Ultrasonics,* SU-15, 157-161 (1968).

2. G. A. Massey, "An Optical Heterodyne Ultrasonic Image Converter," *Proc. IEEE,* 56, 2157-2161 (1968).

3. R. K. Mueller and N. K. Sheridon, "Sound Holograms and Optical Reconstruction," *Appl. Phys. Letters 9,* 328-329 (1966).

4. B. B. Brenden, "A Comparison of Acoustical Holography Methods" in *Acoustical Holography Volume 1* (A. F. Metherell, H. M. A. El-Sum and L. Larmore eds.), Plenum Press, New York (1969) Chapter 4.

5. A. F. Metherell, "Temporal Reference Holography," *Appl. Phys. Letters 13,* 340-343 (1968).

6. A. F. Metherell, S. Spinak and E. J. Pisa, "Temporal Reference Acoustical Holography," *Appl. Opt. 8,* 1543-1550 (1969).

7. A. F. Metherell, S. Spinak and E. J. Pisa, "Subfringe Interferometric Holography for Linearly Recording Small Displacements" (Abstract), *J. Opt. Soc. Am. 59,* 1534(A) (1969).

8. E. J. Pisa, A. F. Metherell and S. Spinak, "Acoustical Holograms Recorded by CW Interferometric Holography" (Abstract), *J. Opt. Soc. Am. 59,* 1534-1535(A) (1969).

9. R. L. Whitman, "Acoustic Hologram Formation with a Frequency Shifted Reference Beam," *Appl. Optics 9,* 1375-1378 (1970).

10. R. L. Powell and K. A. Stetson, "Interferometric Vibration Analyses by Wavefront Reconstruction," *J. Opt. Soc. Am. 55,* 1593-1598 (1965).

11. C. C. Aleksoff, "Time Averaged Holography Extended," *Appl. Phys. Letters 14,* 23-24 (1969).

12. C. C. Aleksoff, "Temporally Modulated Holography," *Appl. Optics 10,* 1329-1341 (1971).

13. M. Born and E. Wolls, *Principles of Optics*, 2nd Edition, The MacMillan Company, New York, page 267 (1964).

14. M. Abramovitz and I. A. Stegun, *Handbook of Mathematical Functions*, Dover Publications, New York, page 363 (1965).

15. J. W. Goodman, "Temporal Filtering Properties of Holograms," *Appl. Optics*, 6, 857-859 (1967).

16. R. T. Beyer and S. V. Letcher, *Physical Ultrasonics*, Academic Press, New York (1969), Chapter 7.

17. W. W. Lester, "Experimental Study of the Fundamental-Frequency Component of a Plane, Finite-Amplitude Wave," *J. Acoust. Soc. Am. 40*, 847-851 (1966).

18. W. K. Fischer and M. Zambuto, "Optical Holographic Detection of Ultrasonic Waves," in *Acoustical Holography Volume 3* (A. F. Metherell ed.), Plenum Press, New York (1971).

19. M. H. Zambuto and W. K. Fischer, "Shifted Reference Holographic Interferometry," *Appl. Optics 12*, 1651-1655 (1973).

HIGH-RESOLUTION ACOUSTIC IMAGING BY CONTACT PRINTING*

J. A. Cunningham[†] and C. F. Quate

Stanford University

Stanford, California 94305

INTRODUCTION

In this paper we will present a technique which allows us to record high-resolution acoustic images of biological specimens. The image is detected in the extremely near field of the object and the imaging, itself, is accomplished in a thin liquid layer containing 1 μm polystyrene spheres. The response of the spheres to the "radiation pressure" of the acoustic fields traversing the liquid layer permits the spheres to condense into a pattern which produces the acoustic image. The force on an individual sphere as used in our device resulting from the radiation pressure can be calculated and by means of this calculation it is easy to optimize the magnitude and show that it is linearly proportional to the acoustic intensity -- an effect necessary for our system of imaging.

With this technique, we are now able to obtain the first acoustic images of biological specimens with resolution capabilities approaching 5 μm (~ 3λ). The present field of view is 2.5 mm in diameter and this is limited only by obtainable transducer diameter. This could

*This work was supported by the John A. Hartford Foundation, Inc.

[†]Now with TRW Systems Group, Redondo Beach, California

be improved (as detailed later) to the point where the
resolution approximates the wavelength in the object space
if we replace the mylar film which is now used to support
the liquid viewing layer.   Image formation times are on the
order of one second or less with only 10^{-3} watts/cm^2 of
acoustic power incident upon the object.   Finally, and most
importantly, we are able to demonstrate that certain
biological specimens have an acoustic image which is quite
different from the optical image.

This work is an outgrowth and continuation of earlier
work which we presented in a paper at the Acoustical
Holography Symposium, 1972.[1] There we discussed the utility
of the spheres in recording the diffraction rings that are
generated in the far field of an acoustic transducer of
moderate diameter.   We, also, discussed the use of this
method in the recording of acoustic holograms.   In this
report we will concentrate on a more simple technique for
recording images which are suitable for direct viewing and
do not require the reconstruction associated with holograms.
If we are to record quality images over a field of view
which is reasonable in size it is necessary to generate an
illuminating beam with a uniform intensity over a reasonable
area.   This is done here by using an acoustic transducer
with a large area and by locating the object in the near
field of this large transducer.   The details of how this is
done will be presented in a later section together with
some examples of high quality images that are now possible.
But, first we want to present some material on the forces
that can arise from acoustic radiation pressure.

RADIATION PRESSURE

The concept of radiation pressure as a method for
condensing the polystyrene spheres into a pattern which
reproduces the acoustic image has been introduced
previously.[1,2] Here we want to present a more refined
treatment of the force exerted on a homogenous sphere in a
liquid by the acoustic wave.

As in an optical wave, an acoustic wave can be considered to carry momentum as well as energy.   When this
wave strikes an object as shown in Fig. 1 there is a
momentum transfer from the incident wave to the object.   If
we consider only that momentum which is carried in the

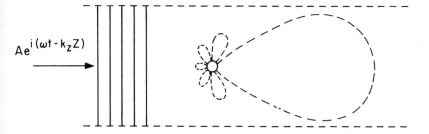

PLANE WAVE INCIDENCE

$$\langle \vec{P} \rangle_p = Y_p \pi a^2 \, I/c$$

FIG. 1--Illustration of a spherical scattering object in a plane wave. The dotted pattern shows the relative direction for the momentum transfer and $\langle \vec{P} \rangle_p$ is the resultant average force on the sphere.

direction of propagation of the wave we note that part of this forward momentum is absorbed by the object. Another part is scattered in a complex manner by the object with an additional loss of momentum from the forward directed wave. The total amount of momentum removed from the incident beam and not returned to the forward direction is imparted to the scattering object. A force arising from this radiation pressure is therefore exerted on this object in the direction of wave propagation.

For a spherical scattering object of radius   a   in water, the acoustic radiation pressure created by an incident plane wave is most easily calculated by finding the time average of the excess pressure variation over the surface of the sphere.   King[3] was the first to carry out such a computation for a rigid sphere in water.   His results were later extended to include the case of a homogeneous, elastic sphere by Hasegawa and Yosioka.[4]   More recently, one of the authors[5] has computed the radiation pressure on polystyrene spheres in water.   The computations are long and tedious, but the resultant force due to this radiation pressure for a plane progressive wave, $\langle\vec{P}\rangle_p$, may be stated as

$$\langle\vec{P}\rangle_p = Y_p(\pi a^2)\, I/c \tag{1}$$

> I = intensity of incident wave
> πa^2 = geometrical cross-section of sphere
> Y_p = efficiency factor dependent upon the size of sphere and the elastic properties of the sphere and water
> c = velocity of sound propagation in water.

A curve of Y_p vs ka (= $2\pi\, a/\lambda$) is shown in Fig. 2 and the values listed in Table 1.   As can be seen from Fig. 2, Y_p rises very rapidly for values of ka less than 2 and remains approximately constant thereafter.   This same general trend is followed for any solid elastic sphere; however, the maximum value of Y_p will depend upon the elastic parameters of the sphere.   For most solid materials this value never exceeds 1, at which point the scattering cross-section of the sphere is identical to its geometrical cross-section.

The rather large departures from a smooth curve occur at and are due to the free modes of vibration of the elastic sphere.[6]   The numbers below the arrows in the Figure indicate the order of the corresponding normal mode of the free vibration; e.g. n = 0 and n = 2 refer to the pulsation and spheroidal vibration, respectively.   The value of Y_p = 0.683 as indicated by the triangle corresponds to our 1100 MHz experiment to be discussed later.

We can now make use of Fig. 2 to select a sphere radius which will optimize the difference between the radiation pressure and the gravitational force acting on the sphere. The downward gravitational force on a sphere of radius a, counteracted by buoyancy, is defined as

$$Fg = \frac{4}{3}\,\pi a^3(\rho^* - \rho_o)g \tag{2}$$

where ρ^* and ρ_o are the densities of the solid and the liquid, respectively, and g is the downward acceleration due to gravity.   If we were to plot $Fg/\pi a^2$ on Fig. 2, then for a specific operating frequency (or wavenumber k) this

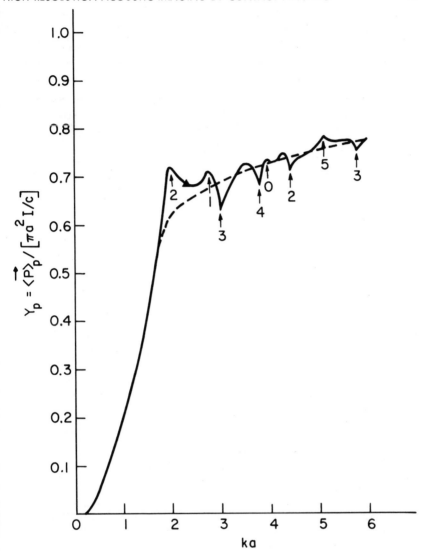

FIG. 2--$Y_p = \langle P \rangle_p / [\pi a^2 I/c]$ vs ka for a polystyrene sphere in water.

curve would be a linear function of ka, as can be seen from Eq. (2). The slope of the curve would be dependent upon the difference in densities between the sphere and the water.

TABLE 1.   The Values of Y_p for Polystyrene
Spheres in Water

ka	Y_p	ka	Y_p
0.00	0.00000		
0.25	0.00313	3.25	0.6894
0.50	0.04997	3.50	0.7236
0.75	0.1253	3.75	0.6851
1.00	0.2037	4.00	0.7533
1.25	0.2924	4.25	0.7428
1.50	0.4072	4.50	0.7252
1.75	0.5565	4.75	0.7460
2.00	0.7169	5.00	0.7649
2.25	0.6887	5.25	0.7696
2.50	0.6801	5.50	0.7227
2.75	0.7078	5.75	0.7551
3.00	0.6229	6.00	0.7756

It is evident that the difference between the radiation
pressure and the gravitational force on the sphere can be
optimized near the point ka = 2.   This statement has been
verified experimentally at 1100 MHz, by working with spheres
of various radii.   We found that a sphere of 0.5 μm radius
(ka = 2.31) produced optimum results.   With a sphere of
smaller radius, no acoustic image could be formed.   For a
sphere with a radius larger than 0.5 μm, the acoustic power
required to form an image increased almost linearly with the
radius.

CONTACT PRINTING GEOMETRY

The geometry of the contact printing interaction is
illustrated in Fig. 3.   In this situation, we have one
normally incident plane acoustic wave illuminating the object.
Considering the time factor $e^{-i\omega t}$ to be understood for all

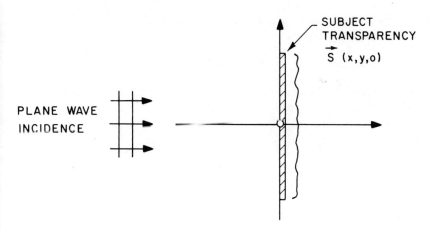

FIG. 3--Geometry of the contact printing interaction.

field quantities, we may represent the incident wave by

$$\vec{U}_i(z) \;=\; U \exp[jkz] \tag{3}$$

where $k = 2\pi/\lambda$ is the wavenumber in water. The complex transmission function of the object is assumed to be

$$\vec{s}(x,y,o) \;=\; s(x,y,o) \exp[-j\varphi(x,y,o)] \quad . \tag{4}$$

Hence, the transmitted field immediately behind the object is given as

$$U_t(x,y,o) \;=\; U \times s(x,y,o) \exp[-j\varphi(x,y,o)] \quad . \tag{5}$$

This yields a resultant intensity distribution

$$I_t(x,y,o) \;\equiv\; |\vec{U}_t(x,y,o)|^2 \;=\; \frac{1}{2} U^2 s^2(x,y,o) \quad , \tag{6}$$

where the factor of 1/2 accounts for averaging over time. Therefore, if we were to place a nonlinear detector immediately

behind the object, we could record an exact replica of that
object's intensity profile $s^2(x,y,o)$.

The case above is an idealized situation, very difficult
to achieve experimentally. In all physical situations there
is always some finite distance separating the object and the
detector. In an optical instrument the separation may be
caused by minute dust particles; in our experimental
apparatus the separation between object and detector is
simply Δz , the thickness of a mylar film. This finite
separation means that we not only have variations in the
acoustic intensity due to the differential transmission
properties of the object, but also variations in the acoustic
wavefront caused by diffraction spreading over the finite
distance Δz in the mylar. Both of these factors have
significant effects on the resultant image.

EXPERIMENTAL APPARATUS

As stated earlier, it is desirable to obtain uniform
acoustic illumination of the object. The experimental
arrangement for accomplishing this, by locating the object
in the extremely near field of a large area transducer, is
shown schematically in Fig. 4. A ZnO thin film transducer
one-quarter wavelength in thickness and operating at 1100 MHz
generates a large area, longitudinal acoustic beam into a
fused quartz platelet. This platelet is polished flat and
parallel to a thickness of 0.375 mm (70 λ) and then epoxy
bonded onto a shorted coaxial line input. To insure
efficient electroacoustic conversion and to provide a uniform
large area acoustic beam, a 3000 Å thick, 2.5 mm (460 λ)
diameter aluminum dot is evaporated onto the ZnO transducer.
Acoustic anti-reflection coatings of quarterwave gold and
silicon monoxide are deposited on the opposite side of the
platelet to pass the beam into a thin water cell which
contains the object. The object plane is therefore located
in the extremely near field of the transducer and we expect
the illumination to be uniform over most of the aperture.[7,8]

On top of this water cell is placed a 5 μm mylar film
which has been stretched taut between two brass support rings.
This mylar film serves the purpose of separating the object
space from the image space. It is also coated with 300 Å
of aluminum to avoid direct viewing of the object and at the
same time it provides a black background against which the

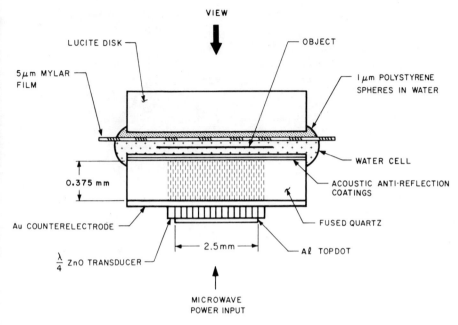

FIG. 4--Experimental arrangement for locating the object in the extremely near field of a large area transducer.

polystyrene spheres may be viewed. On top of this mylar film a lucite disk is used to form a thin emulsion of 5% by volume 1 μm polystyrene spheres in water. The lucite disk serves the additional purposes of reducing undesired acoustic reflections by 10 dB and providing a transparent medium through which we may view the image. The acoustic fields generated in the solid and perturbed by the object are passed through the mylar film with very little attenuation and into the emulsion. There, due to radiation pressure, the spheres are redistributed on a 1:1 basis with the acoustic intensity pattern present in the emulsion. Magnification is then obtained optically by viewing this distribution with a microscope. A photograph is taken for a permanent record.

PRELIMINARY RESULTS

The uniform acoustic illumination we expect can be seen clearly in Fig. 5. In this instance the detecting emulsion

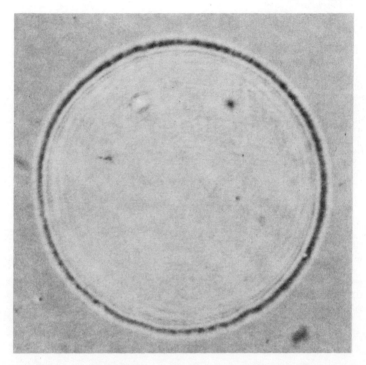

FIG. 5--Uniform acoustic illumination displayed by spheres.

is placed directly on the surface of the platelet and the
uniform illumination is displayed by the polystyrene spheres.
The dark ring at the beam perimeter is the edge effect as
detected by acoustic streaming created by the large velocity
differential at the beam edge.   Several Fresnel rings can
also be distinguished immediately inside the beam.   The
area of uniform illumination is at least 2.4 mm in diameter.
The dark spot in the upper right section of the image is an
air bubble in the detecting emulsion.

     As an estimate of the resolution capability of this
system, the first object to be imaged was a portion of the
1951 USAF Test Resolution Chart shown optically in Fig. 6(a).
This chart is made of a 0.2 mil nickel film supported by a
2 mil copper backing.   Only Groups +4 through +7 are shown
with resolution data being given for Groups +4 and +5.   As
can be seen in the acoustic image of Fig. 6(b), Group +5,

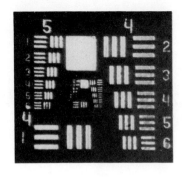

(a)   (b)

USAF RESOLUTION CHART DATA

GROUP +4		GROUP +5	
(1)	16.00 Li/mm	(1)	32.00 Li/mm
(2)	17.96 "	(2)	35.92 "
(3)	20.32 "	(3)	40.64 "
(4)	22.80 "	(4)	45.60 "
(5)	25.56 "	(5)	51.12 "
(6)	28.51 "	(6)	57.02 "

FIG. 6--1951 USAF Resolution Test Chart: Groups +4 and +5. (a) Optical image; (b) Acoustic image.

element (5) is clearly distinguishable, although the contrast is rather poor. This demonstrates a resolution capability, as determined from the data, of better than 51 lines/mm or slightly under 9 µm.

This resolution is significantly lower than anticipated for a wavelength of 1.36 µm in the object space. There are two reasons for this poor resolution: (1) the extremely minute elements of the resolution chart may not be filling with water and, consequently, respond acoustically as if they were opaque, and (2) the diffraction spreading in the mylar film imposes a lower limit on resolution capability independent of object considerations. Reference to Fig. 7

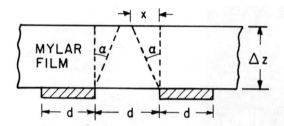

$$x = \Delta z \tan \alpha$$

TO BE RESOLVED

$$2x \lesssim d$$

OR $2\Delta z \tan \alpha \lesssim d$

WHERE $\sin \alpha = \dfrac{\lambda_{MYLAR}}{d}$

FIG. 7--Resolution limitations imposed by diffraction
spreading in the thin mylar film.

will serve to illustrate this latter point.   We consider
the mylar thickness to be Δz $(= 5 \; \mu m)$ and assume two
object details of dimension d separated by a distance d .
Each object detail will cause a beam to be diffracted into
the mylar with a half-angle α where

$$\sin \alpha \; = \; \frac{\lambda_{mylar}}{d} \quad , \qquad (7)$$

and $\lambda_{mylar} = 2.3 \; \mu m$ at 1100 MHz.   In order that these
beams do not overlap we must satisfy the inequality

$$2x \lesssim d \quad , \qquad (8)$$

where

$$x = \Delta z \tan \alpha \quad . \qquad (9)$$

Hence, the smallest resolvable object detail d is given by

$$d \gtrsim 2\Delta z \tan \alpha \qquad . \qquad (10)$$

For the values of Δz and λ_{mylar} quoted above, this analysis places a lower limit on our resolution capability of approximately 4 μm.   We are thus only a factor of two above the diffraction limited capabilities.

FILM STRUCTURE TO INCREASE RESOLUTION

In order to increase resolution capabilities we must therefore provide a thinner film to bring the detecting emulsion into closer contact with the object.   The basic requirements of this film are that it (1) be acoustically thin, (2) be acoustically transparent at the desired operating frequency, and (3) provide a dark background against which the polystyrene spheres can be viewed.

A promising film structure consists of aluminum coated, linearized polypropylene as shown in Fig. 8.   The aluminum coating is only thick enough to provide a front mirrored surface.   The linearized polypropylene has the property that all the long axes of its molecules lie in the plane of the film.   This makes it possible to stretch such a film until it is approximately 2500 Å in thickness.   We are therefore able to obtain an extremely thin film with very low acoustic losses.   Diffraction spreading in this plastic film should be very small, with a corresponding increase in resolution capability.

A large sheet of linearized polypropylene was obtained and after many attempts, a few extremely thin films were stretched over 1 cm diameter polished brass rings.   The rings had to be polished to avoid tearing the delicate films.   An aluminum coating was then evaporated onto one side of these films.   It was found that they tended to be very porous, about 1000 Å of aluminum being required to provide a mirrored surface.

Two completed structures were finally obtained and evaluated in our experimental arrangement.   Although an

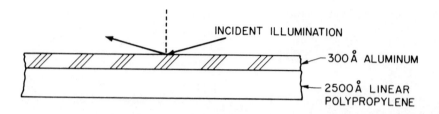

FIG. 8--Film structure to increase resolution capabilities.

increase in resolution could not be readily demonstrated, the quality of the images obtained was very good. We feel that these preliminary results are quite encouraging. If the technical difficulty of stretching the polypropylene could be overcome, we feel that the resolution capabilities of future systems would be determined by the wavelength in the water cell.

ACOUSTIC IMAGING OF BIOLOGICAL SPECIMENS

Up until now we have imaged only an object with a known structure which could therefore be used to evaluate system performance. The main goal of this research effort, however, has been the high resolution acoustic imaging of biological specimens. The first biological specimen to be imaged using this experimental technique was a 5 μm thick microtomed section of human lung tissue,[9] shown in optical transmission in Fig. 9(a). The tissue, embedded in collodian, was lifted as a unit and placed on top of the fused quartz platelet. A few drops of water were added and the detector placed on top. The acoustic image shown in Fig. 9(b) was then formed in the manner previously described. The finest structural detail which can be resolved in the image is on the order of 20 μm. The acoustic image conveys the main features of the object although the general quality is rather poor. We attribute this poor quality to the fact

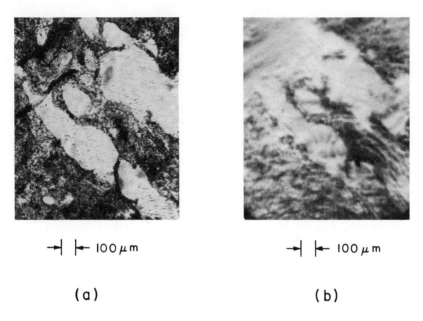

→| |← 100 μm   →| |← 100 μm

(a)   (b)

FIG. 9--Lung section at 68 × magnification.
(a) Optical image; (b) Acoustic image.

that the section was not lying flat on the fused quartz
platelet, the resultant wrinkling of the section causing
image distortion.

The biological specimen which displayed the finest
object <u>and</u> image detail was a common fly wing. A small
portion near the center of the fly wing is shown in optical
transmission in Fig. 10(a). The large rib structure of
the wing appears as white vertical stripes, the left one
having a horizontal branch. The individual cells of the
wing appear as black dots approximately 5 μm to 8 μm in
diameter. The acoustic image of this same wing section is
shown in Fig. 10(b). The rib structure is easily distinguished, but appears somewhat broader than in the optical
image. This apparent discrepancy is actually in the
optical image as determined from optical reflection photographs. More notable, however, is the fact that we can
distinguish the individual cell structure of the wing quite
clearly. A microscopic examination of these cells at
higher magnifications shows them to range from 5 μm to 8 μm
in size. This demonstrates a resolution capability

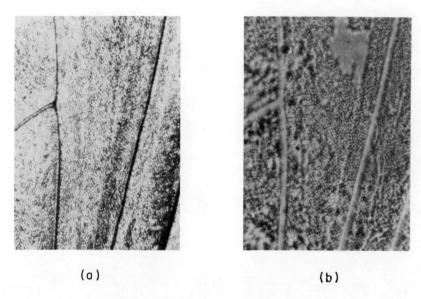

(a)   (b)

FIG. 10--Fly's wing at $68 \times$ magnification.
(a) Optical image;   (b) Acoustic image.

approaching the limit of 4μm as computed above.   It should
be made clear that it is, in fact, the cellular structure
being imaged, the smallest spot containing at least 20 polystyrene spheres.   The large white area in the top center of
the acoustic image is the result of an air bubble in the
object space.   No acoustic energy is passed through this
region and hence no image is formed there.   Another bubble,
which is in the detector space, can be seen to the left of
this region.

     Because of the wrinkling problem encountered with the
lung tissue, subsequent biological sections were prepared
and mounted in a different manner.   Each tissue to be
studied was embedded in paraffin and then microtomed into
5μm sections.   These sections were then floated on the
surface of a tank of warm water.   To insure good contact
and flatness, the section was then lifted onto the underside
of the coated mylar film.   After drying for 24 hours the
paraffin base was removed with p-xylene.   The resultant
structure was reasonably safe from damage and could easily
be inserted into the imaging system.   When desired, we
could remove the tissue section by rubbing with alcohol.

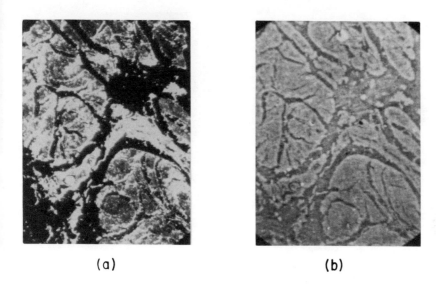

(a)   (b)

FIG. 11--Lower stomach section at 68 × magnification.
(a) Optical image; (b) Acoustic image.

One of the tissue sections imaged using this modified
technique was a section of the lower stomach[9] shown in
optical reflection in Fig. 11(a).   This particular section
was chosen because it contained both coarse and fine detail
structure.   The finest detail in this object is approximately 10 μm.   The acoustic image is displayed in
Fig. 11(b).   Notice that both the coarse and fine detail of
the object have been faithfully reproduced although the
contrast is not as good as desired.

The final, and by far the most interesting, biological
specimen to be viewed by this imaging technique is shown in
optical phase contrast reflection in Fig. 12(a).   The
specimen is a 5 μm thick section of bone from the iliac
crest (hip).[9]   The bone itself appears as the light fibrous
structure running vertically through the figure.   The lacy
structure is the surrounding marrow.   Several microtome
scratches can be seen as dark lines running diagonally
across the bone structure.   This optical photograph was
actually taken after the acoustic image had been formed.

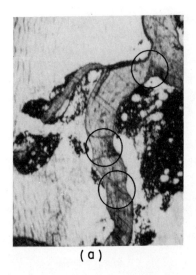 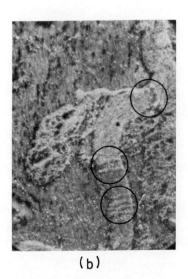

(a)   (b)

FIG. 12-- Bone section at 68 × magnification.
(a) Optical image; (b) Acoustic image.

In the process, a large section of marrow became detached
from the left branch of the bone structure.

The acoustic image of this bone section is shown in
Fig. 12(b), and is quite significant in its information
content.   First of all, the bone and surrounding marrow
are clearly revealed, although there is not much contrast
differentiation. We can see, however, that the fibrous
structure of the bone itself is revealed as well as, if not
better than in the optical counterpart. The most surprising aspect of the acoustic image is that it displays
certain object detail in a different manner than is displayed
optically. We refer specifically to three areas shown
circled in both images. The uppermost area is the joint
between two branches of the bone. In the optical image no
differentiation in structure can be detected at the joint.
In the acoustic image, however, this joint appears as a
crescent-shaped juncture between the two branches. The two
lower areas appear only as constricted regions in the optical
image. In the acoustic image these areas are now revealed
as a series of bands transverse to the long axis of the
bone.

It could be argued that the latter structure is not real, but is actually caused by acoustic interference due to the section lifting off the mylar film. We show that this argument is not justified for two reasons. First, as the section is microtomed it tends to curl up on the microtome knife blade. Therefore, when lifted off the warm water surface, any lifting of the section would occur in a direction perpendicular to the knife scratches. Since the knife scratches are diagonal across the image, we would also expect any lifting to be diagonal in the opposite direction. This is not the direction of the bar structure in the acoustic image. Second, and somewhat more convincing, is the regularity in spacing of this transverse bar pattern. If the section had lifted it would be bowed with respect to the surface, causing the spacing of the bars to vary across the region due to acoustic interference. This even spacing is a clear indication that we are truly imaging the detail of the object in a manner not obtained optically.

CONCLUSION

In this paper we have investigated the nonlinear acoustic response of polystyrene spheres in a thin water emulsion and the application of this response to the construction of a high resolution acoustic imaging system. It was demonstrated both theoretically and experimentally that this response could be optimized by the proper choice of sphere radius.

It was also shown that a uniform acoustic illumination could be obtained by locating the object in the extremely near field of a large area transducer. Using this technique, we were able to obtain the first images of biological specimens with resolution capabilities approaching 5 μm (3λ). The limit on resolution was shown to depend on diffraction spreading in a thin mylar film with a 4 μm theoretical limit at 1100 MHz. An alternate film structure has been presented for improving this resolution capability. Image formation times were on the order of 1 second or less with only 10^{-3} watts/cm^2 of acoustic power incident upon the object. More importantly, we were able to demonstrate that certain biological detail images in quite a different manner acoustically than it does optically.

Although the resolutions obtained are not yet comparable to those of an optical system, operation at higher acoustic frequencies with a thinner film separating object and detector should go far toward closing the gap. The rather slow image time, however, precludes any real-time system operation. At the present stage of development, we see this acoustic imaging system as a supplementary aid to an optical imaging system. It will provide a new, and quite different, method of observing the same object.

REFERENCES

1. Cunningham, J.A., and Quate, C.F., "Acoustic Inter-ference in Solids and Holographic Imaging", in Acoustical Holography, vol. 4, G. Wade, Editor (Plenum Press, New York, 1972) p. 667.

2. Cunningham, J.A., and Quate, C.F., "High-Resolution, High-Contrast Acoustic Imaging", J. Physique 33, Colloque C-6, Supplement, 42 (Nov-Dec. 1972).

3. King, Louis V., Proc. Roy. Soc. (London), A137, 212 (1935).

4. Hasegawa, T., and Yosioka, K., J. Acoust. Soc. Am. 46, 1139 (1969).

5. Cunningham, J.A., "High Resolution Acoustic Imaging", Ph.D. Dissertation, Stanford University (1973).

6. Faran, J.J. Jr., J. Acoust. Soc. Am. 23, 405 (1951).

7. Goodman, J.W., Introduction to Fourier Optics, (McGraw-Hill, San Francisco, 1968) Chapter 8.

8. Born, M., and Wolf, E., Principles of Optics, (Macmillan Company, New York, 1964) Appendix III.

9. Di Fiore, M.S.H., Atlas of Human Histology, 3rd edition, (Lea and Febiger, Philadelphia, 1967).

ACOUSTIC HOLOGRAPHIC INTERFEROMETRY

M. D. Fox
Electrical Engineering, Univ. of Connecticut

W. F. Ranson
Mechanical Engineering, Auburn University

J. R. Griffin and R. H. Pettey
U. S. Army, MICOM, Redstone Arsenal, Alabama

1. INTRODUCTION

Following the advent of holography, a multitude of applications were suggested arising from the ability to reconstruct wavefronts. In holographic interferometry the ability of holograms to reproduce three dimensional images was exploited to obtain for the first time interferograms which revealed displacements of rough surfaces. With recognition of its utility, the interest in holographic interferometry increased rapidly, and at present it is probably the most important application of holography.[1-5]

Acoustic holography is another application of holographic principles in which the hologram is recorded using sonic energy and reconstructed using conventional optical holography.[6] This process has the potential for three dimensional visualization of the interior of normally opaque objects, due to the propagation characteristics of vibrational energy.

Recently some experimental work has appeared on long wavelength holographic interferometry. Brenden and Hildebrand[7] showed acoustic holographic interferograms in the case of rigid body rotation of a flat plate. Kock[8] reconstructed a microwave holographic interogram, also of a rigid body rotation. The purpose of the present work is to explore the theoretical implications of holographic interferometry using a long formation wavelength, and to experimentally validate the theoretical predictions.

2. HOLOGRAPHIC INTERFEROMETRY

Gabor[9] demonstrated that the wavefront reconstruction process was not restricted to coherent optics but could be applied to any coherent wave system. Therefore, there is no fundamental difference between the formation of holograms using light and acoustical waves. The hologram formation using acoustical waves can be made into transparancies and then viewed optically.

Consider a generalized hologram formation process as shown schematically in Figure 1. Let the radiation at the detecting location due to the object wave be described by the following equation

$$\Gamma_0(x,y) = Re\{A_0(x,y)exp[i(\omega t + \phi_0(x,y))]\} \qquad (1)$$

where $A_0(x,y)$ is called the amplitude modulus and $\phi(x,y)$ is the phase and both are, in general, functions of the coordinates of the detecting surface. The circular frequency is denoted as ω. Let the reference wave be described by

$$\Gamma_R(x,y) = Re\{A_R(x,y)exp[i(\omega t + \phi_R(x,y))]\} . \qquad (2)$$

In holographic interferometry two wave fronts are recorded at the detecting surface. The first record is made with the body in a reference configuration. The second record is made with the body deformed from the reference configuration.

After the body is deformed, the radiation of the recording plane in this position is described by

$$\Gamma_0'(x,y) = A_0(x,y)exp[i(\omega t + \phi_0(x,y) + \Delta\phi_0(x,y))]$$
$$(3)$$

where $\Delta\phi_0(x,y)$ is the phase change due to the deformation of the body. It is assumed that the off axis reference wavefront is the same for both configurations of recording the hologram. In the double exposure technique $\Gamma_0'(x,y)$ will be the amplitude of the object wavefront in the deformed configuration. At the recording plane the total amplitude for the second recording is the sum of equations (2) and (3),

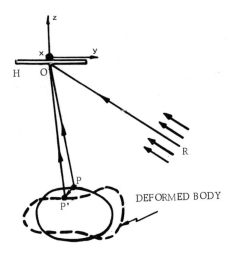

Fig. 1. An Off-Axis Hologram Configuration.

$$\Gamma_T^{\prime}(x,y) = A_o(x,y)\exp[i(\omega t + \phi_o(x,y) + \Delta\phi_o(x,y))]$$
$$+ A_R(x,y)\exp[i(\omega t + \phi_R(x,y))] .$$

$$(4)$$

The intensity for both records is then

$$I(x,y) = \Gamma_T\Gamma_T^* + \Gamma_T^{\prime}(\Gamma_T^{\prime})^* .$$

$$(5)$$

Equation (5) will reduce to

$$I(x,y) = 2A_o^2(x,y) + 2A_R^2(x,y) + A_o(x,y)A_R(x,y)\{$$
$$\exp[i(\phi_o(x,y) - \phi_R(x,y))] + \exp[-i(\phi_o(x,y) - \phi_R(x,y))]\}$$
$$+ A_o(x,y)A_R(x,y) \{\exp[i(\phi_o(x,y) + \Delta\phi_o(x,y) - \phi_R(x,y))]$$
$$+ \exp[-i(\phi_o(x,y) + \Delta\phi_o(x,y) - \phi_R(x,y))]\}.$$

$$(6)$$

For a double exposure hologram, the amplitude transmittance is

$$\Gamma_{TRANS.} = K[2A_o^2(x,y) + 2A_R^2(x,y)]A_1(x,y)\exp[i(\omega t + \phi_1(x,y))]$$

$$+KA_o(x,y)A_R(x,y)A_1(x,y)\{\exp[i(\phi_o(x,y) + \Delta\phi_o(x,y) - \phi_R(x,y) +$$

$$\omega t + \phi_1(x,y))] + \exp[-i(\phi_o(x,y) + \Delta\phi_o(x,y) - \phi_R(x,y) - \omega t -$$

$$\phi_1(x,y))]\} .$$

$$(7)$$

In a double exposed acoustical hologram, with suitable choice of reference and reconstruction waves, the change in phase $\Delta\phi_o(x,y)$ results from the change in acoustical path length. In optical holography $\Delta\phi_o(x,y)$ is related to the change in optical path length.

3. PHASE CHANGE DUE TO DEFORMATION OF A BODY

If the body is deformed or undergoes a rigid body motion between exposures of the hologram, then the phase terms of the complex wave amplitudes are different for the two exposures. This change in phase results from the deformation of the body. The phase change, denoted as $\Delta\phi_o$ in equation (3), is determined from the change in the path length of the wave. Referring to Figure 2, the length of the wave for the body in the reference configuration is denoted by L_1 and is the distance

$$L_1 = |\underline{PS}| + |\underline{PO}| , \qquad (8)$$

where S is the wave source, P is a point on the surface of the body, and O is a point in the hologram plane. The distance, $|\underline{PS}|$ between two points (S,P) is the length of the vector which connects them. Therefore the distance $|\underline{PS}|$ can be written in terms of the inner product of the point difference. The vector $\underline{PS}$ connecting the points P and S is defined as the point difference. Equation (8) can now be written in the following form

$$L_1 = \sqrt{\underline{PS} \cdot \underline{PS}} + \sqrt{\underline{PO} \cdot \underline{PO}} . \qquad (9)$$

The point P on the surface of the body is deformed into the point P´. With the body in the deformed configuration the path length of the wave, denoted as L_2, is the distance

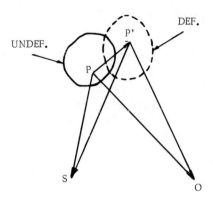

Fig. 2.   Geometry of Fringe Formation for a Deformed Body

$$L_2 = |\underline{P'S}| + |\underline{PO}| . \qquad (10)$$

Using the inner product of the point difference, this equation can be written as

$$L_2 = \sqrt{\underline{P'S} \cdot \underline{P'S}} + \sqrt{\underline{P'O} \cdot \underline{P'O}} . \qquad (11)$$

In order to determine the phase change of the wave, equations (9) and (11) are used to determine the change in path length Δ, where

$$\Delta = L_1 - L_2 . \qquad (12)$$

Substitution of (9) and (11) into (12) yields for the path length change

$$\Delta = \sqrt{\underline{PS} \cdot \underline{PS}} + \sqrt{\underline{PO} \cdot \underline{PO}} \cdot - \sqrt{\underline{P'S} \cdot \underline{P'S}} - \sqrt{\underline{P'O} \cdot \underline{P'O}} . \qquad (13)$$

The phase change $\Delta\phi_0$ is related to Δ by $2\pi/\lambda$, therefore the phase change is written as

$$\Delta\phi_0 = \frac{2\pi}{\lambda}\left[\sqrt{\underline{PS}\cdot\underline{PS}} + \sqrt{\underline{PO}\cdot\underline{PO}} - \sqrt{\underline{P'S}\cdot\underline{P'S}} - \sqrt{\underline{P'O}\cdot\underline{P'O}}\right].$$

(14)

This equation, while completely general, is not in a form convenient for numerical use. Let the point difference of the body before and after deformation be $\underline{PP'}$, then

$$\underline{PS} = \underline{P'S} + \underline{PP'}$$
$$\underline{PO} = \underline{P'O} + \underline{PP'} .$$

(15)

Therefore equation (14) can be written in the following form using equations (15) which yields

$$\Delta\phi_0 = \frac{2\pi}{\lambda}\left[\sqrt{\underline{PS}\cdot\underline{PS}} + \sqrt{\underline{PO}\cdot\underline{PO}}\right.$$

(16)

$$- \sqrt{(\underline{PS}\cdot\underline{PS}) - 2(\underline{PS}\cdot\underline{PP'}) + (\underline{PP'}\cdot\underline{PP'})}$$
$$\left.- \sqrt{(\underline{PO}\cdot\underline{PO}) - 2(\underline{PO}\cdot\underline{PP'}) + (\underline{PP'}\cdot\underline{PP'})}\right].$$

4.   IMPLICATIONS OF A LONG FORMATION WAVELENGTH

When acoustic wavelengths are used to form holographic interferograms and much shorter optical wavelengths are used for their reconstruction, the wavefronts corresponding to the two object states are duplicated to a first order approximation. As a result, interference fringes are formed corresponding to the differences between the acoustic signals from the original position and from the modified position.

The behavior of the fringes when the viewing direction is changed can then be studied to determine displacements from the reference position. Alternatively, holograms with different observation positions may be used to determine the displacement components.

The basic observation which may be made from a generalized theoretical analysis of holographic interferometry is that little change is introduced in the theory by the use of a long wavelength in the holographic interferogram formation. The explanation for this theoretical parallelism is that to a first order approximation a long wavelength hologram is equivalent to an optical hologram of the same object with the object distance increased by m/μ, where m is hologram magnification and μ is the construction-reconstruction wavelength ratio, $\lambda r/\lambda c$.

From Figure 2 it is apparent that in analyzing path length it is necessary to consider a single point 0 on the hologram. In practical geometries, it is apparent that the hologram must be viewed through some small area around the point 0. Viewing through such a limited aperture will cause a superposition of different fringe configurations, but the aberration caused by the superposition will be negligible if the area around 0 is kept small. On the other hand, if the area around 0 is made too small, the image resolution will suffer, as dictated by the Rayleigh resolution criterion: $d_{min} = .61\lambda/\sin\alpha$, where α is the half angle subtended by the hologram.

It is at this point that optical and acoustical holographic interferometry have a major difference. Since λ is so small in the optical case, a very small area of the hologram will exhibit high resolution. In the acoustic case, however, λ is often relatively large (1.5 - .15mm), and so it is often necessary to use a larger portion of the hologram to attain acceptable resolution. The use of this larger aperture may well lead to a breakdown in the assumptions used to determine the path length due to deformation, however. The permissible geometries for long wavelength holographic interferometry will be explored in the next section.

5. PERMISSIBLE GEOMETRIES FOR ACOUSTIC HOLOGRAPHIC INTERFEROMETRY

The general expression for the formation of interference fringes in double exposure holography is given in equation (16). Although equation (16) is completely general, it is not in a form which allows appreciation of the simplifications afforded by using typical geometries. Refer to Figure 3 and recognize that

$$\underline{PO} = \underline{PO'} + \underline{OO'} \quad . \tag{17}$$

Also let the hologram be positioned so that $\underline{PO}$ is normal to the hologram plane. Equation (16) can now be written in the following form:

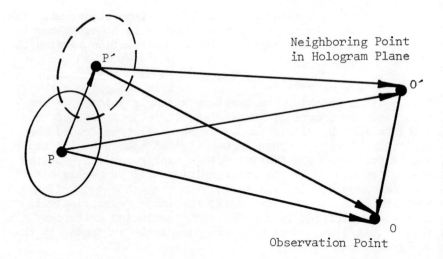

Fig. 3.   Generalized Geometry of Fringe Formation for
Higher Order Effects in Acoustical Holography

$$\Delta\phi_O = K_O\left[\sqrt{\underline{PS}\cdot\underline{PS}} + \sqrt{\underline{PO'}\cdot\underline{PO'} + \underline{OO'}\cdot\underline{OO'}}\right.$$
$$- \sqrt{\underline{PS}\cdot\underline{PS} - 2(\underline{PS}\cdot\underline{PP'}) + (\underline{PP'}\cdot\underline{PP'})}$$
$$\left. - \sqrt{(\underline{PO'}\cdot\underline{PO'}) + (\underline{OO'}\cdot\underline{OO'}) - 2(\underline{PO'} + \underline{OO'})\cdot\underline{PP'} + \underline{PP'}\cdot\underline{PP'}}\right]$$

$$(18)$$

As in optical holography, let the displacements of points between exposures be small. Therefore it will be assumed that
$|\underline{PP'}|^2 < |\underline{PP'}|$. Also assume that $|\underline{PO}| > |\underline{OO'}|max$, then the
last three items in equation (18) can be approximated by the
expansion

$$\left[\underline{PO'}\cdot\underline{PO'} + \underline{OO'}\cdot\underline{OO'}\right]^{1/2} \simeq |\underline{PO'}| + \frac{|\underline{OO'}|^2}{2|\underline{PO'}|} - \frac{|\underline{OO'}|^4\cdot}{8|\underline{PO'}|^3} + ----$$

$$\left[\underline{PS} \cdot \underline{PS} - 2\underline{PS} \cdot \underline{PP'}\right]^{1/2} \simeq |\underline{PS}| - \frac{\underline{PS} \cdot \underline{PP'}}{|\underline{PS}|} + ----$$

$$\left[\underline{PO'} \cdot \underline{PO'} - 2(\underline{PO'} \cdot \underline{PP'}) + \underline{OO'} \cdot \underline{OO'}\right]^{1/2} \simeq |\underline{PO'}| -$$

$$\frac{\underline{PO'} \cdot \underline{PP'}}{|\underline{PO'}|} + \frac{|\underline{OO'}|^2}{2|\underline{PO'}|} - \frac{(\underline{PO'} \cdot \underline{PP'})(|\underline{OO'}|^2)}{2|\underline{PO'}|^3} + \frac{|\underline{OO'}|^4}{8|\underline{PO'}|^3} + ---- .$$

$$(19)$$

Equations (19) can be used to write equation (18) in the following form

$$\Delta\phi_0 \simeq K_0\left[\frac{\underline{PS} \cdot \underline{PP'}}{|\underline{PS}|} + \frac{\underline{PO'} \cdot \underline{PP'}}{|\underline{PO'}|} - \frac{|\underline{OO'}|^2(\underline{PO'} \cdot \underline{PP'})}{2|\underline{PO'}|^3}\right] .$$

$$(20)$$

Note that all terms in the form $|\underline{OO'}|^{n+1}/k|\underline{PO'}|^n$ cancel due to the interferometric effects. Simplifying equation (20), the phase change can now be written as

$$\Delta\phi_0 = K_0\left[\underline{ps} + \underline{po'} - \frac{|\underline{OO'}|^2}{2|\underline{PO'}|^2}\underline{po'}\right] \cdot \underline{PP'}$$

$$(21)$$

where $\underline{ps}$ and $\underline{po'}$ are unit vectors. For systems in which the object points are near the hologram longitudinal axis, a hologram f number may be defined as

$$f = \frac{|\underline{PO}|}{2|\underline{OO'}|} .$$

$$(22)$$

Therefore equation (21) can be written in the following form

$$\Delta\phi_0 = K_0\left[\underline{ps} + \underline{po} - \frac{\underline{po'}}{8f^2}\right] \cdot \underline{PP'} .$$

$$(23)$$

The expression here differs from the typical fringe spacing expression in the inclusion of the term $- \underline{po'} \cdot \underline{PP'} / 8f^2$, which represents deviation from first order interferometric theory due to the spatial extent of the hologram.

To conservatively determine permissible acoustic holographic interferogram geometries, it is necessary to choose some criterion for acceptable degrations. It is convenient to choose the point at which an interferometric null will have an amplitude of 10% of the original object amplitude as the maximum allowable deviation. This means that the optical intensity, which is the observable quantity, will be 1% of the original object intensity.

From elementary geometry, a phase shift of approximately 6° between two opposing vectors of amplitude A will lead to a resultant amplitude of .1A. Thus, realizing that $6^\circ = \lambda_0/60$ choose as a criterion for permissible acoustic holographic interferometry

$$\Delta\theta \leq \frac{\lambda_0}{60} \ . \tag{24}$$

From equation (23) it is apparent that the aberration component in acoustic holographic interferometry is at worst

$$\Delta\Phi_{AHI} \neq \frac{-PP'}{8f^2} \ . \tag{25}$$

Combining (24) and (25), for aberration effects to be negligible the condition must be met that

$$|PP'| \leq .133\lambda_0 f^2 \ . \tag{26}$$

Thus the permissible displacement increases as the square of the hologram f. For example, in a typical hologram geometry, $f = 10$. Then the maximum allowable displacement would be

$$|PP'|_{max} \approx 13\lambda_0 \tag{27}$$

or thirteen acoustic wavelengths.

It should be pointed out here that the wave analysis which has been used to study acoustic holographic interferometry fails for values of f smaller than 1/2 since in this case the binomial expansion about PO will no longer converge. For larger values of f however, the theory rapidly becomes quite accurate, and should provide a convenient method of assuring that displacements will be described by the generalized interferometric analysis of fringe spacings presented in Section 3.

4. EXPERIMENTAL VALIDATION OF THEORY

The experimental validation of the acoustic holographic

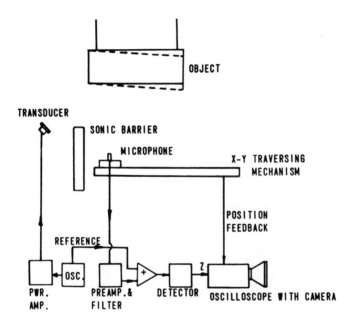

Fig. 4.   Experimental Design to Demonstrate Acoustic
          Holographic Interferometry

interferometry concept can be accomplished using readily available equipment by operating at frequencies just at the limit of human audio perception, i.e., 15,000 - 25,000 hertz. At these frequencies air can be used as the propagating medium, and common high fidelity components such as tweeters and non-directional microphones, as the sonic sources and detectors. Similar experimental schema have been employed by Metherell et al.[10] and Aoki[11].

The basic experimental procedure is illustrated in Figure 4. An audio oscillator set to the desired hologram frequency feeds a power amplifier which drives the transmitting transducers. The sonic field diffracted from the object is picked up by a non-directional microphone, which is scanned in raster fashion across the hologram plane by an X-Y translation device. The microphone output is preamplified, and high pass filtered to minimize pickup of extraneous room noises. The resulting signal is added to a signal from the audio oscillator which constitutes an electronic reference.

The output from the summing amplifier is used to modulate the Z-axis of an oscillosocope trace which is scanned in synchrony with the X-Y translator. Therefore the hologram will be imaged on the oscilloscope screen where it may be recorded by a standard oscilloscope camera.

Since holographic interferometry requires superimposed holograms of the object in reference and modified positions, it will normally be necessary to scan a second raster with the object in a deformed or displaced state to complete the recording of an acoustic holographic interferogram. In the initial experiments, the object was four tweeters mounted 5" apart on a wooden beam. Later a 30" long, 12" diameter styrofoam cylinder was utilized.

At a frequency of 19,000 HZ, sound has a wavelength of about 2cm in air. For convenience in reconstruction, the hologram must be considerably demagnified from its original meter square dimensions. This has the advantage of making the hologram a reasonable size to fit on an optical bench, and brings the reconstructed real and virtual images much closer to the hologram plane. In this study the holograms were reduced to about 2mm by 2mm for a demagnification of 1/50.

The reconstruction geometry is illustrated in Fig. 5. Due to the long distances and large wavelength ratio in the

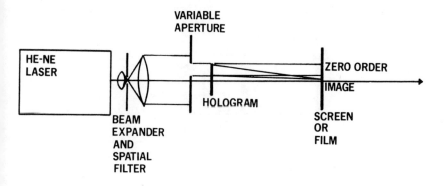

Fig. 5. Reconstruction Geometry

process, it may be necessary for the reconstructing beam to
be slightly converging to bring the image into focus in a
reasonable distance.

7. RESULTS

The first experiment performed was a rigid body rotation
of four equal point sources, each five acoustic wavelengths
apart as shown in Figure 6. Theoretically, a cancellation
of point 4 and a quadrupling of the intensity of point 2 would
be expected. The intensities of points 1 and 3 should both
be doubled.

The results are shown in Figure 7. As expected, point 4
was lowered in intensity, while points 1, 2, and 3 were increased. The fact that points 1 and 2 appear about equal in
intensity and point 3 is less intense than expected is consis-
tant with the hypothesis that the points were not rotated as
much as desired.

A second experiment, using the geometry illustrated in
Figure 8, was performed to record an acoustic holographic interferogram of a rigid body rotation about the transverse axis
of a styrofoam cylinder. Illuminating the entire hologram
aperture in the reconstruction, the image shown in Figure 9
is produced. The cylinder appears as a rectangle with three
vertical fringes to the right of the central order. Higher

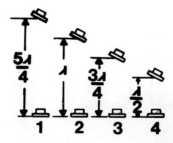

Fig. 6. Geometry for Formation of Four Point Source Acoustic
Holographic Interferogram

order images are formed due to the scanning process used in
hologram formation. The fine random structure in the reconstruction is due to speckle. The use of the entire hologram
aperture in this case results in the superposition of fringe
configurations discussed in Section 5. As predicted, the
image becomes cleaner when a small portion of the hologram is
used in reconstruction, thus producing a higher f geometry.
The results of such a reconstruction are shown in Figure 10.

8. SUMMARY AND CONCLUSIONS

A major contribution of optical holography has been in
the area of interferometry. This generalized interferometry
has extended the capabilities of experimental mechanics and
non-destructive testing, by allowing for the first time interferograms of rough objects. However, with all the applications of this technique some limitations still exist. Optical holographic interferometry can be used to determine
surface displacements of a deformed body. This technique
cannot determine deformation at points interior to the boundary of opaque solids. Interior flaw detection using optical
holographic interferometry is limited by the extent to which
the flaw communicates a deformation to the surface.

The results presented in this report indicate that interferometry theory applies to acoustic holography. In addition to indicating displacements or contour generation by
the fringe pattern, interferometry enhances the reconstruction
by superimposing a characteristic fringe pattern on the image.

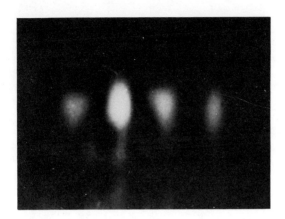

Reconstruction of Acoustic Hologram of Four Point Sources

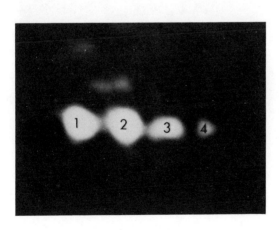

Fig. 7.   Reconstruction of Acoustic Holographic Interferogram
of Four Point Sources with Rigid Body Rotation
Between Exposures

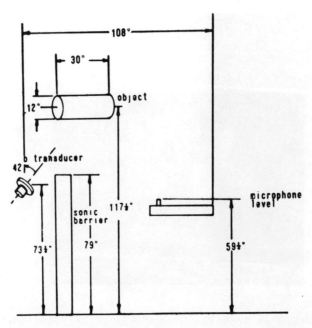

Fig. 8.   Geometry of Experimental Schema to Obtain an Acoustic
Holographic Interferogram of a Styrofoam Cylinder

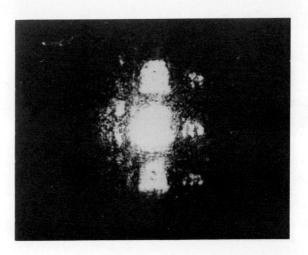

Fig. 9.   Reconstruction of an Acoustic Holographic Interferogram of a Styrofoam Cylinder Inclined 3"
Between Exposures

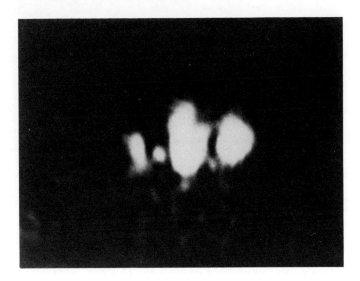

Fig. 10. Reconstructed Image of Cylinder Using a Limited
          Hologram Area in the Reconstruction, Resulting
          in a Higher f Geometry

    Application of this principle to non-destructive testing
techniques may provide a passive method of detecting interior
flaws in composite structural components. This technique may
also provide a tool for experimental mechanics with a sensitivity range between Moiré Fringe and optical holographic interferometry which is highly desirable for the study of certain non-linear problems. Finally the technique may also be
capable of leading to the determination of subsurface stress
concentrations which would be significant for fracture analysis.

    Possible biomedical applications include interferometric
vibration analysis of heart movement and in vivo measurement
of internal organ shape and volume. In order to establish
the technique as a practical non-destructive testing or medical diagnostic procedure additional experimentation with acoustic frequencies in the 1-30 mhz range must be undertaken.

                        Acknowledgments

    This work was sponsored by the Ground Equipment and
Materials Directorate of the U. S. Army Missile Command.

The authors would like to express their appreciation to
Messrs. W. A. Lewis, Virgil Irelan, and Jerry Sirote of the
Army Missile Command for their support and participation in
carrying out this research.

9. REFERENCES

(1)   Ennos, A. E., J. Sci. Instrum. (J. Phys.), Series 2,$\underline{1}$, 196

(2)   Haines, K. A. and Hildebrand, B. P., Appl. Opt., 5, (4) 19

(3)   Aleksandrov, E. B. and Bonch-Bruevich, A. M., Sov. Phys. -
      Tech. Phys., 12, (2).

(4)   Gottenberg, W. G., Exp. Mech., 8, 281-285, 1969.

(5)   Powell, R. L. and Stetson, K. A., J. Opt. Soc. Am., 55 (12
      1593-1598, 1965.

(6)   Thurstone, F. L., in Acoustical Holography Vol. I ed. by
      A. F. Metherell, et. al., Plenum Press, New York, 1969.

(7)   Brenden, B. B., and Hildebrand, B. P., An Introduction to
      Acoustical Holography, Plenum Press, 1972.

(8)   Kock, W. E., PROC. IEEE (Lett.) Vol. 61, 135-137, 1973.

(9)   Gabor, D., Proc. Roy. Soc., A197, p. 454-487, 1949.

(10) Metherell, A. F., in Acoustical Holography Vol. I, op. cit

(11) Aoki, Y., in Acoustical Holography Vol. I, op. cit.

HIGH FREQUENCY ACOUSTIC HOLOGRAPHY IN SOLIDS

H.K. WICKRAMASINGHE

Department of Electronic & Electrical
Engineering, University College London
Torrington Place, London WC1E 7JE

INTRODUCTION

Acoustic holography in solids is of direct interest for
the observation of defects within the solid. There is,
however, another strong motivation for interest in solids:
although the attenuation of sound in liquids becomes very
large at frequencies in the UHF, it is possible to work with
extremely thin liquid cells within the solid, the object
to be imaged being introduced into the cell (see Figure 1).
An imaging system using this technique at an acoustic
frequency of 50 MHz has been reported[1]. The work described
here is an extension of this technique to a frequency of
150 MHz.

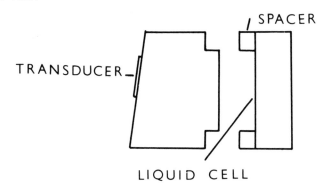

Figure 1. Sample Configuration.

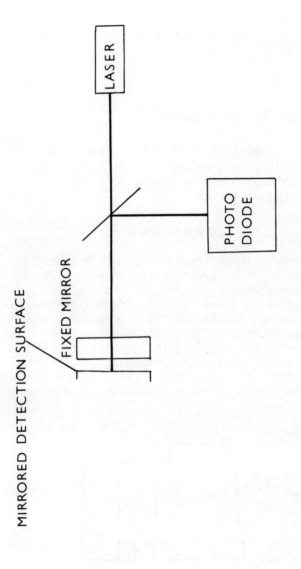

Figure 2. Basic Principle.

A detailed description of the experimental technique can be found in reference 1. The basis of the system is shown in Figure 2. The hologram is derived from the free surface of the sample by the phase modulation of a laser beam; the phase modulation being converted into amplitude modulation by making the hologram surface one mirror of a very close spaced Fabry-Perot resonator. The use of a laser beam as the detection probe offers a number of advantages, notably a virtual absence of any disturbing effect on the acoustic system, a high sensitivity, and a lateral resolution which can approach that of the illuminating wavelength. A major disadvantage of this type of laser probe is, of course, the need to scan across the entire recording aperture.

THE FABRY-PEROT SYSTEM

The basic sensitivity of the detector follows at once from the classic Fabry-Perot equations[2]. The reflected power P_r is related to the incident power P_i by

$$\frac{P_r}{P_i} = 1 - (1 + K \sin^2 \phi)^{-1} \qquad (1)$$

where $K = 4R/(1-R)^2$, R being the reflectivity of the mirrors, and $\phi = 2\pi g/\lambda$, g being the mirror spacing. In its application as a vibration detector one is concerned with a sensitivity σ defined by

$$\sigma = \frac{d}{dg} \left(\frac{P_r}{P_i}\right) \qquad (2)$$

One will normally bias ϕ to a position of maximum sensitivity ϕ_0. For highly reflecting mirrors (K>>1), $\sin \phi \approx \phi$ near this optimum bias position. Thus

$$\phi_0 = (3K)^{-\frac{1}{2}} + m\pi \qquad (3)$$

At this point the sensitivity reaches its maximum value, σ_0, where

$$\sigma_0 = \frac{9\pi}{4\lambda} \left(\frac{K}{3}\right)^{\frac{1}{2}} \qquad (4)$$

thus, provided that the optical bandwidth of the laser beam is much smaller than that of the Fabry-Perot, the maximum

sensitivity depends only on K. Equation (4) takes no
account of mirror scattering or absorption losses, but for
reflectivities up to 0.99 these assumptions are justifiable.
In practice it is experimentally convenient,for a number of
reasons, to make g as small as possible. The bandwidth of
the Fabry-Perot is therefore correspondingly large - indeed,
sufficiently so to accomodate all the axial modes of a typical multimode gas laser provided that

$$g \ll c/(\overline{\Lambda} \, \Delta\nu \, K^{\frac{1}{2}}) \tag{5}$$

where $\Delta\nu$ is the frequency spread of the multimode laser.
This result is amply satisfied by making g less than 1mm.

At the optimum bias point as given by equation (3),
$P_r/P_i = \frac{1}{4}$, resulting in a corresponding standing current in
the detector. For a sufficiently large laser power, the
dominant source of noise will therefore be the photodiode
shot current. In practice, using conventional low-noise
preamplifiers, this point is reached with laser powers around
1mW. In calculating the detection sensitivity of the apparatus we assume that the shot noise is dominant.

At the optimum bias point the diode mean current I_0 is given
by

$$I_0 = \frac{P_i e \eta}{4h\nu}$$

where η is the quantum efficiency of the diode. This gives
rise to a shot noise current I_N, where

$$I_N^2 = 2eI_0\Delta f \tag{6}$$

and Δf is the electronic system bandwidth. From equation
(4) we can obtain the signal current I_s, produced by an
excursion of the mirror having a peak amplitude Δg, as

$$Is = \frac{1}{\sqrt{2}} (3^{\frac{3}{2}} \frac{\overline{\Lambda}}{4}) (\frac{\Delta g}{\lambda}) \frac{K^{\frac{1}{2}} e \eta P_i}{h\nu} \tag{7}$$

From equations (6) and (7) we obtain the signal-to-noise
ratio.

$$\frac{S}{N} = \frac{i_S^2}{i_N^2} = \frac{27\overline{\Lambda}^2}{16} \frac{K\eta Pi(\Delta g/\lambda)^2}{\Delta f h\nu} \tag{8}$$

If a unity signal-to-noise ratio is used as the criterion, the detection sensitivity of the system is Δg_{min}, where

$$\Delta g_{min} = (\frac{4}{3^{3/2}\ \overline{\lambda}})\ \lambda\ (\frac{\Delta fh\nu}{K_n P_i})^{\frac{1}{2}} \qquad (9)$$

For a 1 mW He-Ne laser, 0.99 reflectivity mirrors, a system bandwidth of 1 Hz, and a quantum efficiency appropriate for a Si photodiode, this result suggests a detection sensitivity of just over 10^{-7}Å - that is, two orders of magnitude below the diameter of the hydrogen nucleus. Experimental results indicate that equation (9) does predict at least the order of magnitude of the detection sensitivity.

OUTLINE OF EXPERIMENTAL SYSTEM

The sample configuration shown in Figure 1 was used in these series of experiments. A quartz transducer was attached to one end of the sample. This end was cut obliquely to promote the scattering of reflected acoustic waves, while the other end formed one of the two resonator mirrors. The aperture of the resonator was 20 mm and the mirror spacing was typically 5 μm. The laser beam had a diameter of typically 10 μm at the resonator and could be scanned mechanically using one stepping motor for each of the co-ordinate directions. The electronic system included means for compensating the variation in mirror spacing as the beam was scanned across the aperture of the resonator[1].

In order to avoid spurious signals from multiple reflections, the system was used in a pulsed regime with the interval between pulses sufficiently long to ensure the decay of acoustic waves to a very low level. The main features of the electronic system are shown in Figure 3. The amplitude modulation of the laser beam is detected and compared with a reference signal from the r.f. generator in a phase sensitive detector. The output of the phase sensitive detector is given by

$$v(x,y) = A(x,y)\ cos\phi(x,y)$$

where $A(x,y)$ and $\phi(x,y)$ are the amplitude and phase distributions respectively of the surface vibration on the sample mirror, and (x,y) are the co-ordinates of the point being

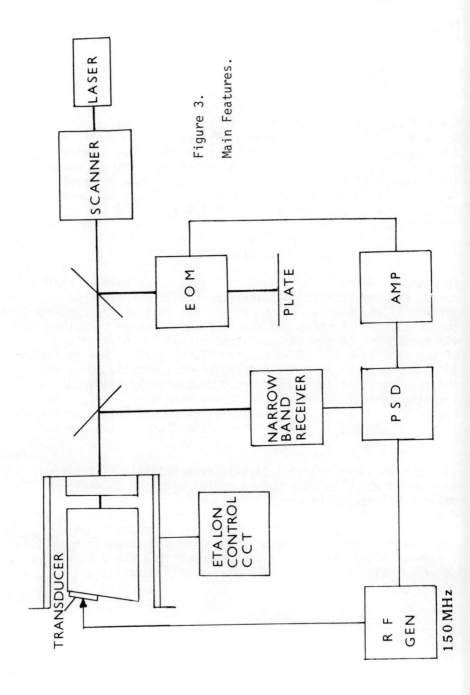

Figure 3.

Main Features.

probed by the laser. The optical record of this output on a photographic plate constitutes the acoustic hologram. However, we have restricted the recording to a binary phase hologram, (i.e., recording a black or white level depending on whether $v(x,y)$ is positive or negative respectively). In practice the restriction to phase holograms does not significantly degrade image quality[3].

The recording system described so far simulates a normally incident plane reference wave. As a result, the basic carrier spatial frequency recorded was dependent entirely on the tilt introduced on the transducer relative to the holo- gram plane. Since for practical reasons it was difficult to tilt the transducer very far from the hologram plane, this carrier frequency was rather low. Consequently, in re- constructing, the angular displacement of the first order (containing the reconstructed image) from zero order was very small, so that separation proved difficult. Although successful holograms have been recorded at 150 MHz using this technique[4], the reconstructions were not free of the un- wanted zero order and out-of-focus conjugate image.

By means of a phasing circuit synchronized with the stepping motor control, it has been possible to vary the phase of the reference wave by π between each line of scan. This is equi- valent to simulating an angled reference wave. In this way, much higher carrier spatial frequencies have been achieved, resulting in a complete separation between first and zero orders in reconstruction.

The overall gain of the system is approximately 100 dB. The S/N ratio at the input to the PSD is typically greater than 20 dB in a bandwidth of 10 KHz. The final bandwidth of the system was reduced by the PSD to 10 Hz.

ATTAINABLE RESOLUTION

The fundamental limitation upon the resolution δ that can be achieved is given by the Rayleigh criterion

$$\delta = 0.61\lambda_a/\sin(\theta/2)$$

where λ_a is the acoustic wavelength and θ is the total angle the object subtends at the hologram. Clearly, in order to achieve a resolution limit approaching the acoustic wavelength, the hologram must be recorded over a large angular aperture. This, in turn, means correspondingly reducing the laser spot size and increasing the number of scan lines. For a given scanning speed, this involves correspondingly longer recording times.

Although maximum recording times of up to 180 minutes have been used in these experiments, the scanning speed has been limited purely by the mechanical scanners used, and not by signal-to-noise ratio considerations.

DETECTION SENSITIVITY AT 150 MHz

The Fabry-Perot detector provides a method of measuring the magnitude of small displacements by direct reference to the laser wavelength[1]. Figure 4 shows a plot of PSD output against mirror spacing with a 150 MHz acoustic perturbation on the surface of one of the mirrors. Using the calibration procedure given in reference 1, it was deduced that the vibration amplitude for this case is 6.1×10^{-4}Å. We can deduce the value of Δg_{min} by relating the measured value to the recorded noise level; allowing for the fact that the mean noise voltage off-resonance ($P_r/P_i \approx 1$) is twice its value at the maximum sensitivity point ($P_r/P_i = \frac{1}{4}$). We find that $\Delta g_{min} \approx 8.3 \times 10^{-6}$Å. The value which would be predicted from equation (9) is, for this case, 2.8×10^{-6}Å.

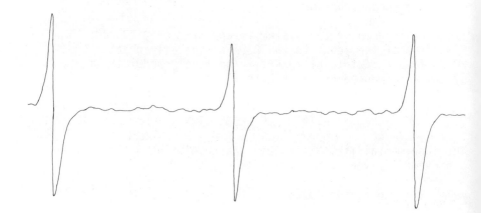

Figure 4.   Surface Vibration Detection 150 Mhz.

EXPERIMENTAL RESULTS

All the holography experiments were carried out at a frequency of 150 MHz. Figure 5(a) shows an acoustic hologram
of a micromesh structure. The periodicity of the micromesh
is 100 μm and the width of each individual bar is 25 μm.
Figure 5(b) shows a reconstruction of this hologram using a
convergent laser beam. The 100 μm periodicity is clearly
resolved. The resolution is close to the theoretical limit
of 80 μm. More recently, the system has been improved by
increasing the angular aperture and reducing the recording
spot size to 10 μm. Figure 6(a) shows an acoustic hologram
of an onion skin, and Figure 6(b) shows its reconstruction.
The cells of the skin are clearly resolved. The thickness
of the cell walls were estimated to vary from 10 μm to 30 μm.
The estimated resolution reached is close to the theoretical
limit of 30 μm.

Work to extend this technique to a frequency of 500 MHz is in
progress. The expected theoretical resolution at this frequency
will be approximately 10 μm. The need to record many more
picture points will probably require the use of a different
scanning system in order to obtain reasonable scan times.

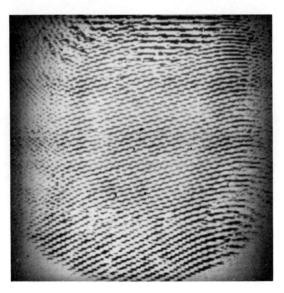

Figure 5(a).

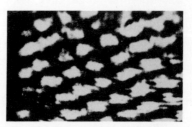

Figure 5(b).

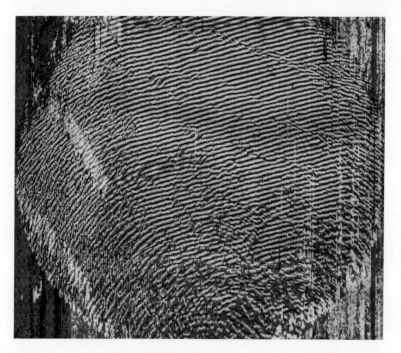

Figure 6(a).

Figure 6(b).

ACKNOWLEDGEMENTS

The author wishes to thank Professor Eric Ash for valuable
advice and many stimulating discussions. Thanks are also due
to Mr. G. Nicholls for his contributions to the electronic
design and the University of London for the award of a Research
Scholarship.

REFERENCES

1. Thomson J.K., Wickramasinghe H.K. and Ash E.A.
    'A Fabry-Perot Acoustic Surface Vibration Detector -
    Application to Acoustic Holography'. J.Phys.D.: Appl.
    Phys. Vol. 6, 1973, pp. 677-691.

2. Born M. and Wolf E. 'Principles of Optics' (New York,
    Pergamon), 1959, pp. 322-327.

3.   Metherell A.F. 'The Relative Importance of Phase and
     Amplitude in Acoustical Holography', Acoustical Holo-
     graphy, Vol. 1 (New York, London, Plenum), 1969,
     pp. 203-220.

4.   Wickramasinghe H.K. 'Acoustic Holography in Solids with
     Special Reference to Acoustic Microscopy', Ultrasonics,
     Vol. 11, No. 4, July 1973, pp. 146-147.

ULTRASONIC HOLOGRAPHY THROUGH METAL BARRIERS

H. Toffer[*,+] B.P. Hildebrand,[++] R.W. Albrecht[*]

[+]United Nuclear Industries, Richland,WA 99352
[++]Pacific Northwest Laboratories, Richland, WA 99352
[*]University of Washington, Seattle, WA 98105

ABSTRACT

Scanned ultrasonic holography was used to retrieve information about objects located behind massive metal barriers. The retrieved information was contained in the characteristic interference patterns recorded as the hologram and the visual image reconstructed from it. Particular problems associated with sound wave transmission through metal slabs, such as multiple reflections and mode conversion, were studied using holograms of point and plane wave sound sources and correlated with analytic techniques. Arising from these studies were procedures for imaging complex target configurations using a focused transducer operating in a pulse-echo mode at normal incidence with an electronically shifted reference beam. Diffusely and specularly reflecting objects were imaged through 1-inch, 2-inch, and 4-inch thick flat, and a 2-inch thick curved cylindrical surface aluminum slabs. The limitations of ultrasonic holography through material barriers and the effect of such barriers on the imaging equations were developed analytically and tested against measurements. Possible applications of the technique in the nuclear industry were considered. The potential imaging of irradiated fuel elements in metal covered storage cubicles was explored in detail.

INTRODUCTION

Retrieval of information about objects located in or shielded by opaque materials is a difficult but essential operation in the hazardous materials industry. For example, in the nuclear industry such objects may be located in intense radiation environments precluding direct observations by personnel, and limiting the usefulness of more conventional inspection techniques. Generally, the opaque materials are thick metal walls restricting access to inspection only from one surface. Sound waves can propogate through material media, such as liquids and solids, without serious loss of signal amplitude. Utilizing this unique property, ultrasonic holography was explored as a means of obtaining reflective images of objects located behind massive metal walls.

ELEMENTARY SOURCE STUDIES

Problems associated with imaging through metal barriers were studied by recording holograms of elementary sources of ultrasound. Such holograms of plane and spherical waves are readily recognizable and can be correlated with calculated results. Although ultrasonic waves can penetrate several feet into homogeneous materials without appreciable loss of energy, propagation across material interfaces can lead to significant signal amplitude reductions. Anticipating such losses for multiple interface crossings, the scanned transducer technique (possible sensitivity 10^{-12} W/cm^2)[1] for recording the holograms was selected. The experiments were carried out using the Battelle Northwest Laboratories X-Y Scanner and associated signal processing equipment.[2] The basic experimental arrangement, as displayed in Figure 1, has been extensively described in several references.[1,3]

The objectives of the elementary source studies through metal barriers were: determining the effects on holograms due to various wave incidence angles at the slab interface; experimenting with time gating and pulsing techniques for discriminating against unwanted signals; and observing the effects of slab thickness and slab surface curvature on the interference patterns.

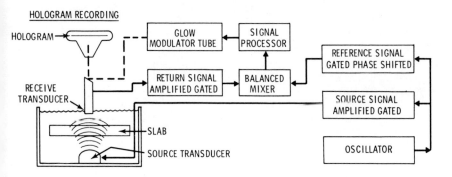

FIGURE 1.   DIAGRAM OF HOLOGRAM RECORDING TECHNIQUE AND
SIGNAL PROCESSING MODULES.

A normally incident beam of ultrasonic waves at a
liquid-solid interface will be reflected and transmitted.
The acoustic impedances, $Z_1 = \rho_1 v_1$ and $Z_2 = \rho_2 v_2$, used in
equations (1a) and (1b) will determine the magnitude of the
reflection and transmission coefficients for normal
incidence.[4],[5]

Displacement Potentials

$$A_r = \frac{(Z_1 - Z_2)}{(Z_1 + Z_2)} \; ; \; A_t = \frac{\rho_1 (2 Z_2)}{\rho_2 (Z_1 + Z_2)} \qquad (1a)$$

Energy or Intensity

$$E_r = \frac{(Z_1 - Z_2)^2}{(Z_1 + Z_2)^2} \qquad E_t = \frac{4 Z_1 Z_2}{(Z_1 + Z_2)^2} \qquad (1b)$$

For a 0-degree incidence angle, reflections between the
slab surface and the transducer will set up reverberations
limiting the availability of the transducer for receiving
target signals. To produce an off-axis hologram with a
normal incidence beam will require an electronically phase-
shifted reference beam.

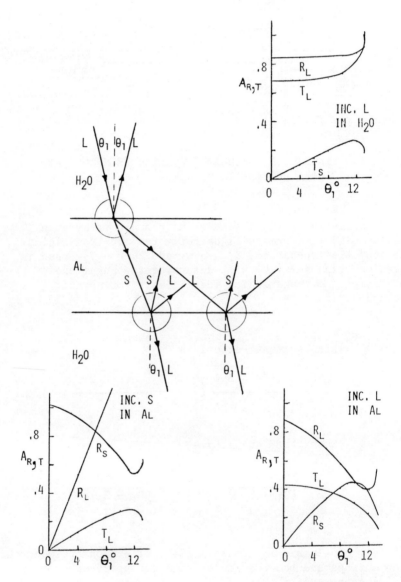

FIGURE 2.   ANGULAR DEPENDENCE OF REFLECTION, (A_R) AND
TRANSMISSION (A_T) COEFFICIENTS FOR LONGITUDINAL
(L) AND SHEAR (S) WAVES AT Al - H_2O INTERFACES.

The reflection problem can be reduced by using obliquely incident sound waves at the material interface. However, if the adjacent media are dissimilar materials, such as a liquid and solid capable of supporting shear stresses, reflection, refraction, and mode conversion of the sound waves can occur. Such interactions between obliquely incident sound waves on an aluminum slab immersed in water are shown in Figure 2. The possible interface interactions for incident longitudinal, L, and shear, S, waves are shown by the circled areas. Displacement potential reflection and transmission coefficients, A_r and A_t, were calculated according to reference 4 for incidence angles, θ_1, from normal incidence to the critical angle for L waves at a water to aluminum interface.

The signal amplitude transmitted through a slab is dependent on the angle θ_1. As θ_1 increases over the range considered, the signal amplitude received on the opposite side of the slab decreases. Using the reflection and the transmission coefficients from Figure 2, the signal amplitudes of sound waves for double transmission through the slab were calculated. Double transmission corresponds to ultrasonic waves emitted by a transducer passing through the slab, being reflected off a target plane in water, and propagating back through the slab along the same ray paths. If ultrasonic waves are incident at 10-degrees amplitude, the calculated receive signal amplitude along the longitudinal wave ray path would be 0.048 and along the corresponding shear wave ray path 0.0035. In addition to the primary mode converted rays, as seen in Figure 2, multiple reflections will occur in the slab, as shown in Figure 3. A large fraction of the incident signal amplitudes will be carried by such multiply reflected and mode converted rays.

The acoustic ray paths in Figure 3 are associated with multiple reflections and mode conversions for a 10-degree incident longitudinal, L, wave in water. The angle of incidence θ_1, and angles of refraction θ_S and θ_L are related to the corresponding velocities v_1, v_S and v_L through the following relations:

$$\frac{v_1}{\sin \theta_1} = \frac{v_L}{\sin \theta_L} = \frac{v_S}{\sin \theta_S} . \qquad (2)$$

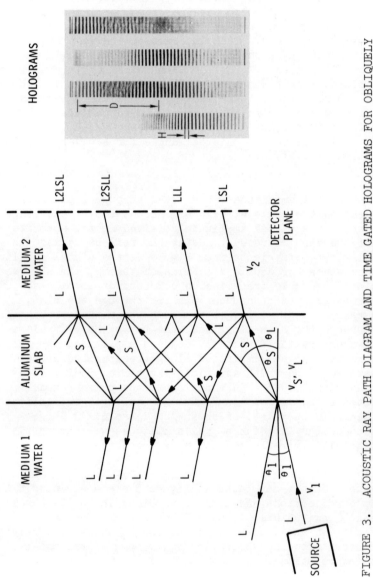

FIGURE 3.   ACOUSTIC RAY PATH DIAGRAM AND TIME GATED HOLOGRAMS FOR OBLIQUELY INCIDENT PLANE WAVES THROUGH A 2-INCH THICK Al SLAB.

The received beam patterns for a 2-inch diameter plane
wave transducer operating at 2.9 MHz through a 2-inch thick
aluminum slab were recorded in a 5 x 5-inch scan aperture.
The calculated pattern distributions for a 10-μsec wide
pulse are displayed in the displacement - time - amplitude
plot in Figure 4. The calculations were correlated with
experiments. Holograms of the transmitted beams and an electronic reference beam normal to the slab interface were recorded.
Reconstruction of these superimposed interference patterns
produced a multiplicity of source images, illustrating one
of the fundamental difficulties with imaging through metal
slabs with obliquely incident sound waves.

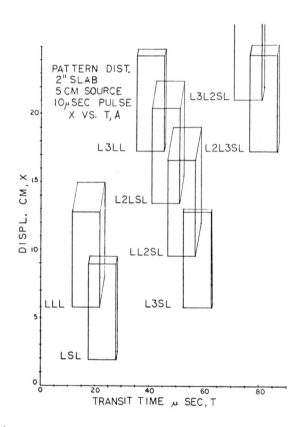

FIGURE 4.   CALCULATED RELATIONSHIPS BETWEEN SIGNAL TRANSIT
            TIME, PATTERN DISPLACEMENT, AND RELATIVE SIGNAL
            AMPLITUDES.   AMPLITUDES ARE NORMAL TO THE
            DISPLACEMENT - TIME AXIS.

Since ultrasonic waves move along the ray paths with different velocities, transit times will vary. Operating the transducer in a pulsed mode (10-μsec pulses, one every 800-μsec) and using time delays and receive signal gating, it was possible to resolve the superimposed interference patterns into component associated with particular ray paths. Strips, in Figure 3, cut from four holograms demonstrate this technique. The first strip represents the reference case, a hologram of the plane wave source transducer at 10-degree incidence angle with the 2-inch aluminum slab absent. The other three strips were recorded for the same geometric arrangement through the 2-inch slab. The second strip was obtained using a 10-μsec pulse and a 50-μsec-wide gate. The hologram contains the composite interference pattern due to ray paths LSL, LLL, L2SLL, and L2LSL. Shifting the time delay and narrowing the receive signal gate width, the composite interference pattern was resolved into the LLL component in the third strip, and into the L2SLL pattern in strip number four. Such measurements repeated with different thickness slabs had no effect on H but increased the pattern separation D with increasing slab thickness. The use of receive signal processing in this manner eliminates the multiple pattern problem for a very limited number of situations. A multi-faceted object, or different combinations of slab thickness and transducer diameter, and ultrasonic pulse width could render this technique useless for imaging objects through metal slabs; especially if both the receiver and sender are on the same side of the metal barrier.

The time gating techniques were applied to obtaining images of point sources of ultrasound through metal slabs. In Figure 5, the interference patterns produced by spherical waves with the metal barrier absent and with the 2-inch thick aluminum slab in place, are shown. The effect of time gating is apparent for the last two holograms. A wide gate allows a multitude of signals associated with multiple reflections and mode conversions to pass, resulting in a composite zone plate. The narrower gate used to produce the last hologram eliminated the undesirable signals. Holograms 1 and 3 are comparable in clarity; however, the line spacing in the last hologram is wider due to the presence of the slab. The effect of the aluminum slab is to increase the focal length of the zone plate. The image of the point source will appear displaced from the image produced by hologram 1. The measurements were repeated for various thickness slabs noting again a reduction in line density as the plate thickness was increased.

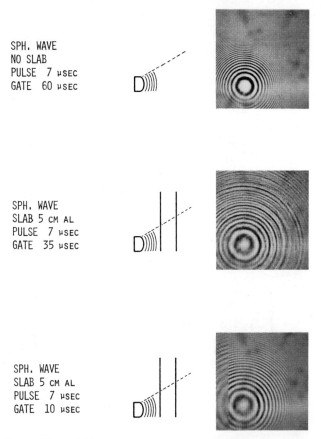

FIGURE 5.   HOLOGRAMS OF SPHERICAL WAVES THROUGH WATER
            WITHOUT AND WITH 2-INCH SLAB INTERPOSED. EFFECT
            OF TIME GATING IS DEMONSTRATED.

    In addition to recording holograms of point sources
through flat metal barriers, a curved cylindrical slab of
aluminum was investigated. The 2-inch-thick slab had a
24-inch radius of curvature. As shown in Figure 6,
holograms of point sources were obtained for a convex and
a concave orientation. The eccentricity of the elliptical
zone plates can be related to the curvature of the slab and

the distance of the source from the slab surface. The
curvature distortions can have a significant effect on
imaging through cylindrical metal walls, especially if the
radii of curvature are less than in this example. Special
signal processing methods or optical correction techniques
can be used to overcome such effects.

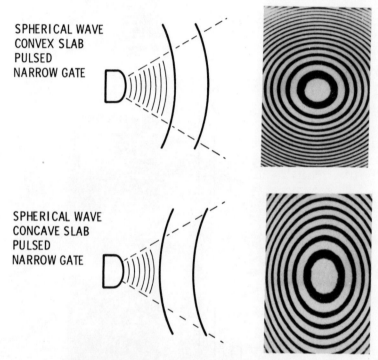

SPHERICAL WAVE
CONVEX SLAB
PULSED
NARROW GATE

SPHERICAL WAVE
CONCAVE SLAB
PULSED
NARROW GATE

FIGURE 6.   HOLOGRAMS OF SPHERICAL WAVES THROUGH CURVED
            CYLINDRICAL Al SLABS.

     Although the multiple interference patterns due to mode
conversion and multiple reflections in metal barriers present
difficulties to holographic imaging, the holograms can provide very useful information about fluid media, metal slabs,
and geometric arrangements. The distance, D, between
patterns in Figure 3 is a function of the slab thickness and
its material properties through the reflection angles θ_s and
θ_L. The spatial period, H, of the interference lines is
defined by:

$$H = \frac{V_2}{2f} \left[\sin\left(\frac{\theta_1}{2}\right) \right]^{-1} \tag{3}$$

V_2 = velocity of sound in medium 2

f = frequency of the sound waves

θ_1 = the angle of incidence in the detector plane

Small changes in θ_1 can be observed as changes in spatial periods.

An additional measurable quantity is the spatial phase variation between line patterns. Using the information contained in holograms with equations (2) and (3) and some known parameters, material constants of the layered media, geometric dimensions, and component orientations can be readily determined. Therefore, holograms of elementary sources can be considered a valuable technique for retrieving information through metal barriers.[6]

IMAGING OF TARGET CONFIGURATIONS

Mode conversion with multiple reflections and severe signal amplitude losses at interfaces limited the usefulness of oblique illumination for imaging through metal barriers. Normal incidence with an electronically shifted reference beam was selected,[2] despite some very undesirable reflection problems. A focused 1.0-inch diameter PZT transducer with a 4-inch focal length was operated at 3-MHz in a pulse-echo mode. The multiple reflections produced by a single pulse are illustrated in Figure 7. A pulse originating from the surface A will be multiply reflected between the two slab surfaces, C and D, and cause reverberations between A and C. Any information returning from the target F has to be received between the various reflected signals. Use of time gating; proper pulse width selection; and positioning of target, slab, and transducer are essential to imaging of target

configurations through metal slabs. Figure 8 illustrates
the various requirements. The center traces on CRT display
pictures show the ungated receive signals; the lower trace
is of the gated object signal. The first pulse-picture in
the sequence is for an object in water located 21.25 inches
from the focal point of the transducer with no slab present.
The pulse width is 5-µsec or about 15 cycles per pulse. The
insertion of a 1-inch thick Al slab reduces the object
signal round trip time to 840-µsec, introduces reverberations
at 500-µsec and 1000-µsec; creates a multiple reflection
bank between 500-µsec and 560-µsec.

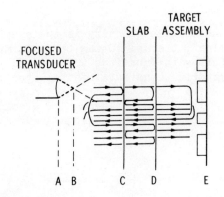

FIGURE 7.   NORMAL INCIDENCE REFLECTION PATHS BETWEEN
            TRANSDUCER, SLAB, AND TARGET.

     As thicker aluminum slabs are inserted, the distance
between A and C in Figure 8 decreases limiting available
imaging space between reverberations, and creating more
extensive reflection banks. The reflection bank for the
4-inch thick slab extends over 150-µsec.

SLAB NONE
PULSE 5 μSEC
SIGNAL TIME 870 μSEC
F. P. TO OBJ. 54 CM

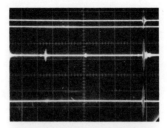

SLAB 2.55 CM AL
PULSE 5 μSEC
SIGNAL TIME 840 μSEC
F. P. TO OBJ. 54 CM

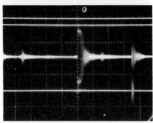

SLAB 5.1 CM AL CURVED
PULSE 5 μSEC
SIGNAL TIME 810 μSEC
F. P. TO OBJ. 54 CM

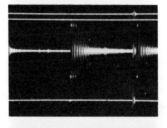

SLAB 10.2 CM AL
PULSE 5 μSEC
SIGNAL TIME 650 μSEC
F. P. TO OBJ. 45 CM

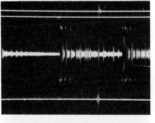

SLAB 10.2 CM AL
PULSE 25 μSEC
SIGNAL TIME 770 μSEC
F. P. TO OBJ. 54 CM

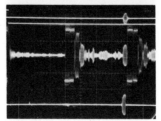

FIGURE 8. CRT DISPLAY PICTURES OF PULSE PATTERNS
ASSOCIATED WITH IMAGING THROUGH VARIOUS
THICKNESS METAL SLABS.

As the slab thickness increases, spacing between multiple reflections increases. For sufficiently thick slabs, object signals returning between reflections can be imaged.[7] The last picture shows the effect of increasing the pulse width from 5-μsec to 25-μsec. The reflection banks are very pronounced with no resolution of individual reflection signals. The multiple reflections between the slab surfaces produced a series of illuminations of the target. The resulting multiply peaked object signals can be seen in pictures 2 and 4. Selection of one peak from the return target signals resulted in images with high resolution. Allowing several image peaks to pass through the time gate destroyed the image quality.

From these examples it is apparent that to image objects through metal barriers requires: ultrasonic pulses of short duration, a damped transducer to reduce the number of reverberations, proper signal gating capabilities, flexibility to adjust the distance between the sound source and the slab surface, and a pulse repetition rate compatible with an adequate sampling frequency.

These requirements were combined with the geometric arrangement shown in Figure 7 to image a series of target configurations. The focused transducer was operated at 3 MHz in a pulse-echo mode, an off axis reference beam was simulated electronically. The 5 x 5-inch aperture was scanned at 2.4 inches/sec in 0.018-inch steps. The pulse width varied between 5 - 25-μsec and pulse repetition rate ranged from 1.8 KHz to 3.0 KHz. Target illumination was approximately 1.0-mW/cm^2.

The first target, shown in Figure 9, was a diffusely reflecting test pattern made of 1.0-inch wide and 0.125-inch thick styro-foam squares and triangles. The target was located 19 inches from the focal point of the transducer. In Figure 9, a series of reconstructions are shown of the target obtained through 1, 2, and 4-inch thick flat surface aluminum slabs, and through the curved 2-inch thick surface aluminum slab in a convex and concave orientation. For reference, a hologram and its reconstruction obtained of the target with no slab present is included. All images contain a recognizable representation of the target. The reconstruc-

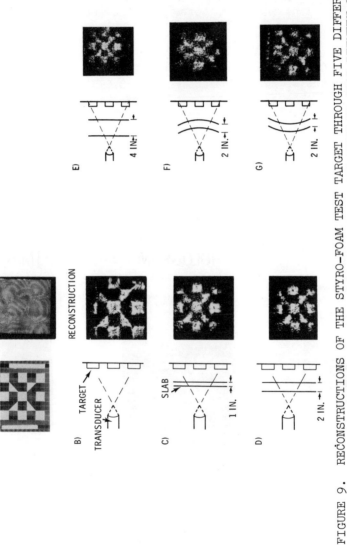

FIGURE 9. RECONSTRUCTIONS OF THE STYRO-FOAM TEST TARGET THROUGH FIVE DIFFERENT ALUMINUM SLABS. THE HOLOGRAM FOR THE CASE WITH NO SLAB BETWEEN SOURCE AND TARGET IS INCLUDED.

tions obtained with the curved slabs showed image stretching
and compression. Lateral and depth resolution and magnification were calculated for the test patterns according to
the equations listed in Appendix A. The maximum attainable
lateral resolutions of 0.073 inches for the particular transducer was the limiting value. The difference between calculated lateral magnifications and measured dimensions in the
reconstructions agreed to better than 10 percent.

A second test pattern was a resolution test chart milled
out of a 0.062-inch thick aluminum sheet (see Figure 10). The
strip widths of the test pattern were 0.275, 0.092, 0.055,
and 0.039 inches; or in terms of ultrasonic wave length in
water, 15, 5, 3, and 2λ. The target was imaged through
the same series of slabs. The observed resolution, in terms
of distinguishable line pattern, amounted to 0.092 for the
2-inch and 4-inch thick slabs. For no slab present and a
1-inch slab, the resolution approached 0.055 inch. Both
values are very close to the resolution limit of .073-inches
for the system.

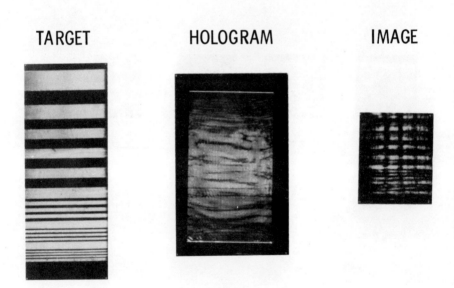

TARGET   HOLOGRAM   IMAGE

FIGURE 10.   ALUMINUM RESOLUTION TARGET, WITH THE HOLOGRAM
              AND RECONSTRUCTION OBTAINED THROUGH THE 2-INCH
              CURVED SLAB.

The third target was an assembly of specularly reflecting zirconium rings. The profile of the rings and their dimensions are shown in Figure 11.

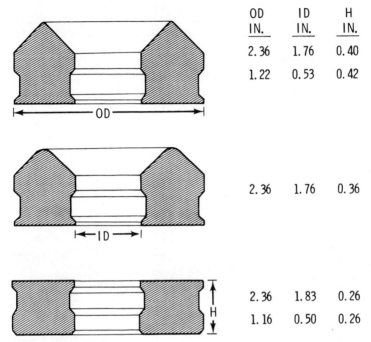

OD IN.	ID IN.	H IN.
2.36	1.76	0.40
1.22	0.53	0.42
2.36	1.76	0.36
2.36	1.83	0.26
1.16	0.50	0.26

FIGURE 11.   CROSS SECTIONAL VIEWS AND DIMENSIONS OF Zr RINGS.

The rings were mounted on low reflectivity screens in a two-dimensional and a three-dimensional arrangement. The three different surface profiles; flat, chevron, and rounded chevron, were selected to study the capabilities of ultrasonic imaging of such objects through metal slabs. The transducer used for imaging these targets was a focused transducer, 1.5-inches in diameter with a 4-inch focal length operated at 5.1 MHz. The scan aperture was increased to 6 x 6-inches. The target to focal point distance for the rings located all in one plane was 16.75-inches, and for the three-layered ring assembly the distances were 11.5, 13.8, and 16.1-inches.

The two-dimensional target arrangement is seen in Figure 12, part A. Holograms and reconstructions are shown

for the target imaged without and with the 1.0-inch thick Al
slab present. Both images contain good reconstructions of
the flat surface rings. Rings with chevron surfaces imaged
very poorly, or produced a false image. Apparently the
incident waves reflected off the slanted surface were reflected such as to simulate a point source. An interesting
observation can be made concerning the interference lines
superimposed on the ring outlines in the holograms. The
orientation of the lines can be correlated with the small
angular displacement suffered by all periferal rings due to
the sagging of the screen on which they were mounted. A
physical measurement of the angular displacement amounted to
1.0 degree.

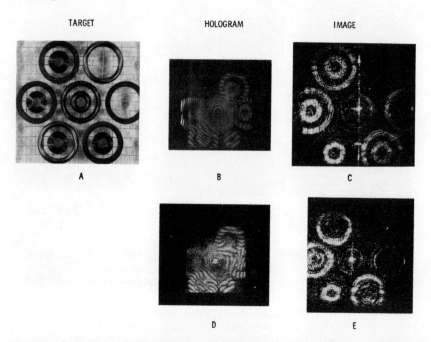

FIGURE 12.   TWO-DIMENSIONAL RING TARGET.   HOLOGRAM B AND
             IMAGE C ARE FOR NO SLAB PRESENT.   HOLOGRAM D
             AND IMAGE E WERE OBTAINED THROUGH A 1-INCH
             Al SLAB.

     Figure 13 contains a picture of the three-dimensional
ring target arrangement and reconstructions obtained for the
center row of rings in focus with the 1.0-inch Al slab absent

and present. In both images, the flat surface rings can be
readily identified, while chevron surface rings produced very
poor or false images. From the hologram for the three-layered
target, each row of rings could be reconstructed. The pulse
width used to record these targets was 25-μsec. When the
pulse width was narrowed to 5-μsec, all ring outlines were
imaged, but the information on ring dimensions deteriorated.

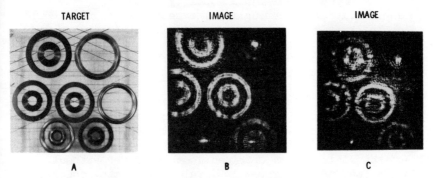

FIGURE 13.   THREE-DIMENSIONAL RING TARGET CONFIGURATION.
             IMAGES WERE OBTAINED FOR MIDDLE ROW OF RINGS
             IN FOCUS.   IMAGE B IS FOR NO SLAB PRESENT.
             IMAGE C WAS OBTAINED THROUGH 1-INCH THICK
             Al SLAB.

     Imaging of various target configurations through
different thickness metal slabs is possible. Very little
image quality deterioration was observed as progressively
thicker slabs were inserted between the transducer and the
target. Multiple reflection problems were overcome by proper
time gating methods and adjustments in the transducer to slab
distances.

APPLICATIONS

     The most challenging application of imaging through metal
barriers in the nuclear reactor plant would be information
retrieval through the 4-inch to 12-inch thick steel nuclear
reactor pressure vessel walls.[8] By locating the interrogating
instrumentation outside the vessel: the need for access ports
through the walls is eliminated, the equipment would be in a
less severe radiation environment, and serviceability of the

system would be easier than for in-core detection devices.

From the results in the preceeding sections, such an application is feasible provided the reactor vessel is accessible and the walls are made out of a single layer of low dispersion steel. Signal processing techniques, as described, will be required. Further signal reception could be enhancement by the use of special high-quality acoustical windows with anti-reflective coatings.[1] The retrieved signals could provide information on changes in media characteristics and optical images of objects such as fuel element bundles, control rods, and structural components.

More immediate applications for through-barrier imaging would be in the hydraulics systems associated with a nuclear reactor, such as: monitoring the operation of valves; strainer inspections; and checking stagnant piping legs for debris collection. Information obtained in this manner without having to open up the system is invaluable to the economic and safe operation of a power plant.

Another potential application related to fuel element storage in the Hanford N-Reactor irradiated fuel storage basin was examined in detail. Discharged zirconium-clad uranium tube-in-tube fuel assemblies are stored under 15 to 20 feet of water in this facility. The fuel remains in this basin until fission products have decayed to levels acceptable to chemical reprocessing. The fuel in the basin is packaged into cylindrical canisters and then stored in cubicles, as shown in Figure 14. Seven fuel assemblies are placed into a canister. The fuel elements in current use at the N-Reactor are either 26-inches or 21-inches long. The 21-inch long fuel elements contain higher enrichment uranium and, therefore, require more stringent nuclear criticality safety controls. For that reason, it is important to be able to account for and identify the different fuel elements in various locations of the basin, such as in the fuel element storage cubicles. These cubicles are covered with 3/8-inch thick steel covers. With the techniques discussed in this paper, it should be possible to inspect the contents of the cubicle without removing the cover plates. Such inspections should provide information about: the cubicle fuel loading status; the length of fuel elements in a canister; fuel identification markings;

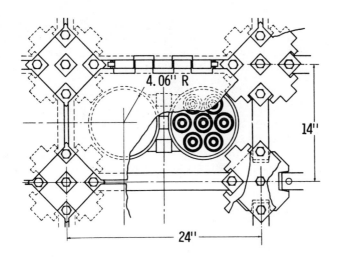

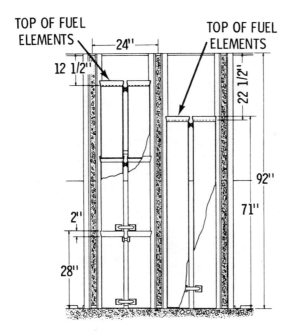

FIGURE 14. TOP VIEW AND CUTAWAY VIEW OF A FUEL STORAGE
CUBICLE SHOWING A CANISTER FILLED WITH TUBE-
IN-TUBE FUEL ASSEMBLIES.

any displacement of fuel in a canister due to damage to the cylindrical container; and the distance between the tops of the fuel assemblies and the cover plate. The last item is of importance to nuclear criticality safety considerations.[9]

The application of ultrasonic holography to the inspection of fuel element content of cubicles was explored with the zirconium ring targets, Figures 12 and 13. These rings correspond to actual end sections of N-Reactor fuel elements, and their arrangement simulates fuel element loadings in half a cubicle. The three-layered target (2-inches between layers) would correspond to fuel elements of different length in a cubicle. The reconstructions of the targets in Figures 12 and 13 show that such configurations can be imaged successfully through thick metal plates, and exemplifies that ultrasonic holography through metal barriers can have very useful applications in the nuclear industry.

CONCLUSIONS

Ultrasonic holography through metal barriers was demonstrated as a viable technique for information retrieval. The information consisted of reconstructed images of objects and of data obtained from the interference patterns composing a hologram. All the experimental work was performed with a very slow recording technique and very basic signal processing methods. Use of faster compact image recording devices, such as transducer arrays, coupled with computer-controlled signal processing techniques are needed for more wide-spread in-service application of ultrasonic holography in the nuclear industry.

ACKNOWLEDGEMENTS

The authors appreciate the generous use of the Battelle Northwest Memorial Institute Pacific Northwest Laboratory experimental facilities for performing the measurements.

REFERENCES

1. Hildebrand, B.P., Brenden, B.B., Introduction to Acoustical Holography, Plenum Press, May 1972.

2. Collins, H.D., Various Holographic Scanning Configurations for Under-Sodium Viewing, BNWL-1558, March 1971.

3. Collins, H.D., Error Analysis in Scanned Holography, Doctor's Dissertation, Oregon State University, June 1970.

4. Becker, F.L., Fitch, C.E., Richardson, R.L., Untrasonic Reflection and Transmission Factors for Materials with Attenuation, BNWL-1283, March 1970.

5. Blitz, T, Fundamentals of Ultrasonics, London: Butterworths, 1967.

6. Toffer, H., Albrecht, R.W., Hildebrand, E.P., Applications of Ultrasonic Holography in the Nuclear Industry, Proceedings of 20th Conference on Remote System Technology, September 1972.

7. Collins, H.D., Gribble, R.P., Acoustic Holographic Scanning Techniques for Imaging Flaws in Thick Metal Sections, Presented at Seminar-In-Depth on Imaging Techniques for Testing and Inspection, Los Angeles, California, February 14-15, 1972.

8. Hildebrand, B.P., Acoustic Holography for Nuclear Instrumentation, Transactions of the American Nuclear Society, Vol. 11, No. 2, 633-634, November 1968.

9. Toffer, H., Fuel Storage Basin Criticality Safety Analysis Report and Technical Basis, DUN-7824, December 1971.

APPENDIX A

RESOLUTION AND MAGNIFICANT RELATIONSHIPS

The formalisms developed by D. H. Collins[7] and B. P. Hildebrand[1] for simultaneous point source receiver scanning were adapted to through-barrier imaging. Approximate relationships defining resolution and magnification observed in the reconstructions are given below:

Lateral magnification (M_L)

$$M_L = 2 \frac{\lambda_1}{\lambda_s} \frac{r_b}{r_1} m$$

Radial magnification (M_R)

$$M_R = m \frac{r_b}{r_1} M_L$$

Maximum attainable lateral resolution $(\Delta X \text{ max})$

$$\Delta X_{max} \simeq \frac{\lambda s f}{a}$$

Holographic lateral resolution (ΔX)

$$\Delta X \simeq \frac{\lambda_s r_1}{2L}$$

Holographic depth resolution (Δr_1)

$$\Delta r_1 \simeq 1.8 \lambda_s \left(\frac{r_1}{L} \right)^2$$

Object distance from point source for through-slab imaging (r_1)

$$r_1 = r_{w1} + \frac{\lambda_m}{\lambda_s} S + r_{w2}$$

where

λ_1 = reconstruction wave length

λ_s = ultrasonic wave length in water

λ_m = ultrasonic wave length in metal

r_b = hologram to image distance

m = hologram magnification

f = focal length of the transducer

a = diameter of the transducer

L = aperture dimension

S = slab thickness

r_{w1} = distance between focal point and slab top surface

r_{w2} = distance between target and slab bottom surface

EXPERIMENTAL RESULTS FROM AN UNDERWATER ACOUSTICAL
HOLOGRAPHIC SYSTEM

Michael Wollman and Glen Wade

Department of Electrical Engineering
University of California, Santa Barbara
Santa Barbara, California 93106

ABSTRACT

This paper describes the system design, fabrication,
and preliminary results obtained from an experimental
laboratory model of an underwater holographic system whose
operation was previously simulated by computer.[1] The model
has a linear-array receiver with 100 channels and a cylindrical insonifying transducer. The receiver array and the
insonifying transducer scan simultaneously. The holographic
information is processed in parallel by 100 distinct electronic channels and is then displayed by a scanning linear
array of 100 light-emitting diodes.

The resulting images demonstrate that a properly
designed cylindrical insonifying transducer can be used to
produce holograms while scanning which are of the same
quality as those produced by a stationary transmitter.
With this method, no special techniques, i.e. no techniques
other than those conventionally used, are required to reconstruct the image.

I. INTRODUCTION

Since acoustical holography was first demonstrated
using raster scanning of single transducers, a process
which was necessarily slow, a number of researchers have

proposed and developed systems employing techniques for more rapid scanning.[2,3,4,5,6] A logical extension of a raster hologram synthesized by n scans of a single transducer is one synthesized by a single scan of a linear array of n transducers. For this case the technique of simultaneous scanning of the receiver and transmitter cannot be applied as easily as with single point transducers. This is because an astigmatism is present which places the length and width dimensions of the reconstructed image in different focal planes. This astigmatism can be corrected by using cylindrical lenses.[5,7] Nevertheless, the correction required is dependent upon the object distance. A considerable amount of time may then be needed to bring an image into focus using optical reconstruction, or alternatively, excessive computer time may be required for computer reconstruction.

As an alternative to these somewhat complicated reconstruction schemes, the system described here utilizes simultaneous scanning with a linear receiving array which requires no special technique for reconstructing the image. Image reconstruction can be done in the ordinary way, since there is no astigmatism present with this approach. This is accomplished by using a cylindrical insonifying transducer which is oriented perpendicularly to the linear array as shown in Figure 1. The conceptual operation of the system and a computer simulation is described in Volume 3 of this series and elsewhere.[1,6] Therefore, we shall not repeat it here.

The system, although intended for underwater viewing, was first built as a scaled-down laboratory prototype to experimentally test the concepts and the results of the computer simulation. This paper describes the design philosophy of the laboratory model and the tradeoffs available. It also presents some preliminary results.

II. MODEL DESIGN

The approach taken for designing the laboratory feasibility model was somewhat different from that which would be taken for an ocean prototype. The emphasis was totally on the hardware, leaving consideration of computer processing for a later time. Of course any ocean system, because of its larger size and complexity, would probably make use

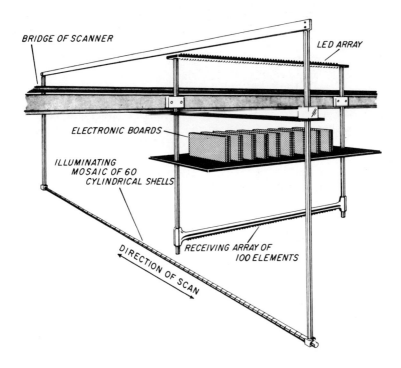

Fig. 1 - Artist's rendering of the laboratory system.
The electronic boards rest above the water
level, while the transmitting cylinder is
slightly below and off to the left of the
linear receiving array.

of extensive computer processing. Correction for both
phase distortions and position errors would undoubtedly be
done by a computer devoted to the system. In addition,
computer image reconstruction may also be used.

Since expense was the primary consideration in the
design, the parts' costs were minimized as much as possible.
This indicated a system with completely parallel analog
processing for all receiver channels.

We initially considered scaling down the wavelength
(from that of the ocean counterpart) in the exact ratio of
the physical dimensions of the systems. This would have

required operation in the megahertz frequency range. However, the ensuing problems of transducer fabrication and the added expense of the electronics operating at such a high frequency were formidable. Therefore we decided to operate at 250 KHz.

The system is pulsed, depth gated, uses phase detection, and has an electronic reference which can be either on-axis or offset by a single fixed angle. These features are generally advantageous in any scanned acoustical holographic system. An ocean system, however, might benefit from using an acoustic reference to cancel phase aberrations stemming from propagation in the water.

We selected a 100-element receiver system bearing in mind that an ocean system would probably have at least 500 elements. Each receiving transducer in the linear array has a separate electronic processing channel. The output device from each channel is a light-emitting diode. In the laboratory model the linear light-emitting diode array has the same spacing as the transducer array and scans simultaneously with the rest of the system. The output of the light-emitting diode array is photographed using time exposure, and the hologram size is reduced as was done in earlier systems.

III. SYSTEM DESCRIPTION

Our underwater facility consists of a 800-gallon water tank and a precision scanner allowing a maximum holographic aperture of 150 cm x 100 cm. This is more than adequate for making holograms using acoustic radiation with a wavelength of 6 mm.

Transmitting Transducer

The cylindrical insonifying transducer mosaic was expected to be difficult to design. Apart from the mechanical problems of supporting the cylinder and maintaining the necessary dimensional tolerances, we were quite concerned about the possibility of unwanted mode coupling and interactions between adjacent transducer shells in the mosaic.[1,8] It turned out that these problems were not as difficult to solve as we originally thought. The inter-

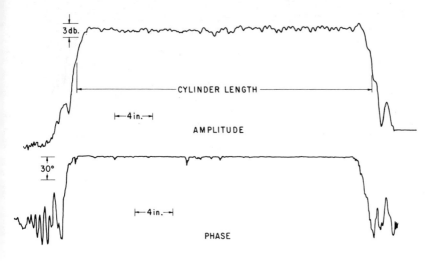

Fig. 2 - Plots of the magnitude and phase of the
        acoustic field of the cylindrical transducer.
        The measurements were made along a line
        parallel to the axis of the cylinder at a
        distance of 6 inches.

action effect is apparently only present when individual
elements in the mosaic are less than a wavelength long.
After testing several prototypes, we have settled on a
transducer which is 31 inches long, 1 inch in diameter and
has 60 1/2-inch cylindrical shells placed end-to-end and
connected electrically in parallel.

The transducer, made of a PZT ceramic material, is
operated in a thickness or radial mode slightly below
resonance to minimize the rapid phase variation with frequency which occurs at resonance. The outside of the
cylinder, which is electrically grounded, is immersed uncoated into the water.

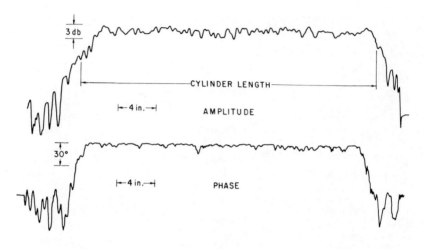

Fig. 3 - Field plots of the magnitude and phase of
         the cylinder at a distance of 18 inches.
         This corresponds approximately to the
         object spacing.

A precise measuring system was built for plotting the
amplitude and phase of the acoustic field of the cylindrical transducer under pulsed operation. Figures 2 and 3
show the field plots obtained along a line parallel to the
axis of the cylinder and at distances of 6 inches and 18
inches respectively. The latter corresponds approximately
to the normal object distance from the transducer. These
curves compare well with the theoretical ones although the
excursions are larger here. The greatest experimental contribution to the phase variations is mechanical misalignment. The amplitude variations are largely caused by differing responses of the individual transducer elements.

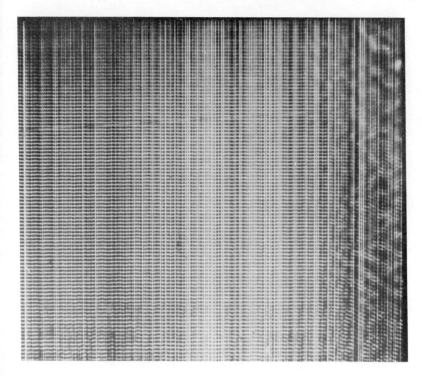

Fig. 4 - Hologram of the cylindrical transducer under
excitation. The cylinder is located below
the photograph and is oriented horizontally.
The vertical lines are irregularly spaced
scan lines, while the horizontal lines are
the wavefronts from the cylinder. The degraded lines on the right show the rapid
variations in the phase at the end of the
cylinder as predicted by theory.

The actual wavefronts were later made visible by recording a hologram with the insonifying cylinder in a
fixed position and with no object in the tank. A conventional raster scan was used with one receiver instead of
scanning with the linear array of receivers, and an electronic plane-wave reference was employed. Figure 4 is a
reproduction of such a hologram. The fringes are spaced
about a wavelength apart and therefore can be thought of
as representing approximately the wavefronts of the acoustic

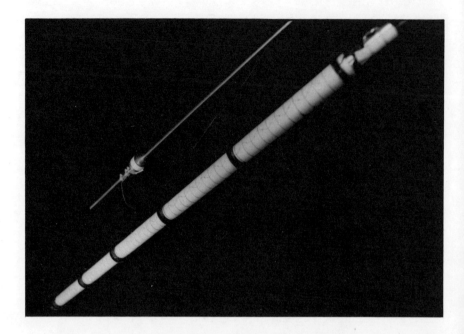

Fig. 5 - Sixty-element transmitting transducer and one-
element prototype.

radiation from the cylindrical transducer. The cylindrical
transducer, which is oriented horizontally with respect to
the page, is located about 100 wavelengths away from the
nearest fringe at the bottom of the photograph. This value
is greater than the source-object distance under normal
circumstances. Figure 5 shows the cylinder next to a one-
element prototype.

Receiving Transducer Array

The receiving transducers are also made of a PZT
material and are operated in the thickness mode below all
resonances. The thickness resonant frequency is 1 MHz and
all other resonances are above 500 KHz. Their size is 0.120
inch square, and they have a center-to-center spacing of
0.160 inch. This spacing corresponds to a distance of about
2/3 wavelength in water. The use of a dicing technique[9,10]

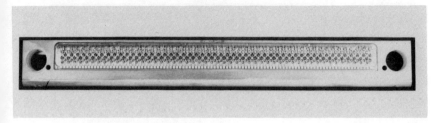

Fig. 6 - View of the receiving transducer array and
first electronic stage while partially
encapsulated.

(of partially sawing through segments on a strip of ceramic
material) to isolate transducer elements was not employed here
but probably will be employed for systems involving many
more channels where it would be impractical to use discrete
transducers. The first stage of the electronics is encapsulated at the transducer array.

Figures 6 and 7 show respectively the complete transducer array and a detailed view before encapsulation.

Electronics

As stated earlier we tried to build the feasibility
model as easily and inexpensively as possible. This naturally meant making a tradeoff between parts' costs and labor
costs. The receiving electronics system is therefore a
combination of inexpensive integrated circuits and discrete
devices. For a prototype system of more than 100 elements
a custom-hybrid or thick-film circuit would certainly be
used. Figure 8 is a block diagram of each receiving electronics channel. Because of the size of the electronics
boards we found it necessary to separate them by about 18
inches from the transducer array. This was detrimental
in that the individual cables to the electronics boards
were plagued by electrical pickup of the electronic reference. This pickup constitutes a "noise source" which is
difficult to remove. In fact this "noise source" is the
dominant one, having a typical value of 500 microvolts
referred to the input. The dynamic range of the system for
a S/N of unity is then 500 microvolts to 5 milivolts or

Fig. 7 - Detailed view of receiving array.

about 20 db. Beyond 5 milivolts the amplifiers saturate,
and a phasigram (phase-only hologram) results. Within
this limited dynamic range adjacent channel crosstalk is
negligible.

System timing is controlled by a Dranetz 206 timing
unit connected to our reference tone burst-and-driver unit.
Figure 9 illustrates the range of timing values used in
our experiments. We are able to trim off the transient
trailing edge of the received signal quite easily. The
electronic reference is offset from the normal by an angle
of 41° by simply inverting the reference signal to adjacent
receiver channels.

Display and Reconstruction

No attempt was made to build a real-time system. Nevertheless, with a 10 second scan and by using rapid-develop
film, it should be possible to reconstruct an image within
a minute.

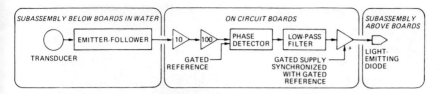

Fig. 8 - Block diagram of receiver channel electronics.

Light-emitting diodes were chosen for the output
because of their linear light-intensity response with current. The center-to-center spacing of the LED array is
identical to that of the transducers. This arrangement was
used to provide the correct aspect ratio in the one-to-one
mapping of the hologram aperture. For convenience, the
output light is reflected to a camera by a large front-
silvered mirror above the water tank. The film used is Kodak
Shellburst which has extended red sensitivity down to the
far-red spectrum of the LEDs.

The LEDs were first roughly matched using a photometer.
Then an equalizing mask was made by lightly exposing a strip
of photographic film placed over the LED array which has
constant current excitation. This film was then developed
and carefully repositioned in the same location above the
LED array. Although there is some attenuation, the response
with the mask is quite uniform.

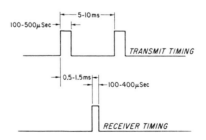

Fig. 9 - Timing Diagram.

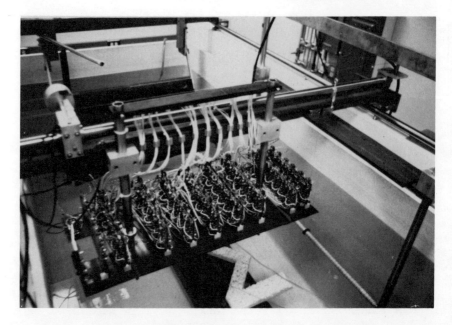

Fig. 10 - Photograph of system.

Reconstruction is done in the ordinary manner with converging laser light after the hologram is shrunk by a factor
of 200.

IV.   RESULTS

In one experiment, we used as the object a styrofoam
letter M with 1/4 inch styrofoam balls glued to the surface.
The balls make the object a diffuse reflector to the 6 mm
acoustic wave. We first made a hologram of the object with
the cylindrical insonifying transducer in a fixed position,
while raster-scanning a single-point receiving transducer.
Clusters of the styrofoam balls are visible in the reconstructed image showing that the theoretical resolution of
about 3/2 λ is approached. The stroke width of the letter
M is about 12 λ. A photograph of the reconstructed image
is shown in Figure 11.

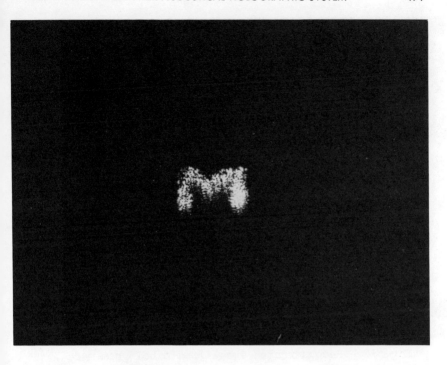

Fig. 11 - Reconstructed image from a scanned hologram
made with the cylindrical transmitter fixed.

The object was then rotated 90° (to optimize the illumination) and a scanned hologram was made with the cylinder
scanning simultaneously with the rest of the system. Figure
12 is a photograph of the reconstructed image in this case.
Note the perspective foreshortening of the vertical strokes
of the letter because the receiving array did not scan
directly over the object. This arrangement was used in
order to produce an offset object beam. The two images of
Figures 11 and 12 do not look identical because slightly
different object and cylindrical transducer positions were
used in the 2 experiments. In the first experiment, the
resolution is somewhat better because the hologram aperture
is effectively larger. Except for this fact the two images
would be of approximately equal quality.

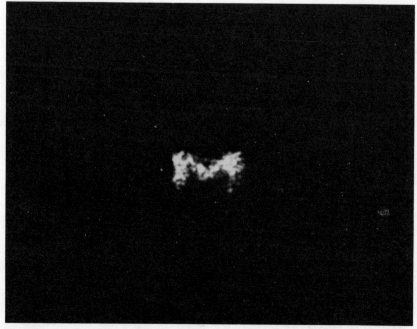

Fig. 12 - Reconstructed image from a hologram made
         with simultaneous scanning of the cylindrical
         transmitter and receiver.

V.   CONCLUSIONS AND FUTURE WORK

    We have demonstrated that our scanned holographic
system will work in a laboratory environment. We have also
shown that a rather large and complex system can be built
economically.

    Some of the limitations of the system could be removed
by using a more integrated approach to the system design.
Either a monolithic circuit or a hybrid would reduce the
electronics of each channel to one active package which
would enable us to place the entire processing unit underwater adjacent to the transducers. Perhaps some tradeoff
between multiplexing and parallel processing should be
made, especially in an ocean prototype.

The next goal is to design a prototype which will work in the field (perhaps in a lake or lagoon). This new system should permit tradeoffs between hardware and software techniques to allow for multiplexing and computer processing. It should also have a means for correcting position errors in the scanning process. Work on this task has now begun.

Acknowledgement

We gratefully thank Mr. Ernest Richards for his interesting and creative ideas as well as his practical assistance on this project.

REFERENCES

1. G. Wade, M. Wollman, and K. Wang, "A Holographic System for Use in the Ocean," Acoustical Holography Vol. 3, Plenum Press, New York, pp. 225-245 (1971).

2. B. P. Hildebrand and K. A. Haines, "Holography by Scanning," J. Opt. Soc. Am., 59(1):1 (1969).

3. N. H. Farhat, W. R. Guard, and A. H. Farhat, "Spiral Scanning in Longwave Holography," Acoustical Holography Vol. 4, Plenum Press, New York, pp. 267-297 (1972).

4. B. P. Hildebrand and B. B. Brenden, An Introduction to Acoustical Holography, Plenum Press, New York, pp. 67-96 (1972)

5. Ibid., pp. 187-191.

6. G. Wade, M. Wollman, and R. Smith, "Acoustic Holographic System for Underwater Search," Proceeding of the IEEE, 57:2051-2052 (1969).

7. T. Iwasaki and Y. Aoki, "Optical Processing of Anamorphic Holograms Constructed in an Ultrasonic Holography System with a Moving Source and Electronic Reference," Presented at the Fifth International Symposium on Acoustical Holography and Imaging, Palo Alto, California: July 1973.

8. D. L. Carson, "Diagnosis and Cure of Erratic Velocity
    Distributions in Sonar Projector Arrays," J. Acoustic
    Soc. Am., 34(9) (1962).

9. R. L. Cook, "Experimental Investigations of Acoustic
    Imaging Sensors," IEEE Transactions on Sonics and
    Ultrasonics, SU-19(4):444-447 (1972).

10. V. G. Prokhorov, "Piezoelectric Matrices for the
    Reception of Acoustic Images and Holograms," Soviet
    Physics-ACOUSTICS, 18(3):408-410 (1973).

ACOUSTICAL HOLOGRAPHIC TRANSVERSE WAVE SCANNING TECHNIQUE
FOR IMAGING FLAWS IN THICK-WALLED PRESSURE VESSELS

H. Dale Collins and Byron B. Brenden

HOLOSONICS, INC.
2950 George Washington Way
Richland, Washington 99352

ABSTRACT

Transverse (i.e., shear) wave scanning configuration
used in acoustical holography for imaging internal flaws
or voids in thick-walled pressure vessels are discussed.
Simultaneous source-receiver scanning configuration was
employed. Because of the severe off-axis nature of the
shear waves, large third order aberrations are present in
the image. Methods of reducing and effectively eliminating
these aberrations are discussed.

The severe resolution limitations of the traditional
nondestructive testing techniques to image radial planar
flaws deep in reactor pressure vessels has been alleviated
by the use of T-wave scanned holography techniques with
aberration compensation procedures. The image resolution
capability exceeds all other NDT methods at depths greater
than a few cm.

One of the major advantages of using transverse waves
in nondestructive testing is the gain in resolution due to
the fact that shear waves are approximately one-half the
wavelength of longitudinal waves in given solid at a
given frequency. Furthermore, many internal cracks, voids,
etc. have geometrical orientations which require slant
angle insonification and viewing for detection. The vertical
or radial flaws can then be viewed from either side at

approximately 45 degrees. The apparent flaw height or projection in the image is usually always less than the true height as a result of the viewing angle.

The flaw images appear two-dimensional as a result of the large difference between the construction and reconstruction wavelengths. However, the hologram does contain the complete depth information and this is easily extracted by focusing on different planes in the image field. Different depths within the image field are brought into focus by adjusting the effective position of the reconstruction light source through the use of a lens. The lateral and longitudinal dimensions of the flaws are then calculated using the conventional image location and magnification equations.

The resolution and aberration correction capabilities have been demonstrated by experiment.

CONCLUSIONS

Transverse wave acoustical holography technique employing simultaneous source-receiving scanning configuration has successfully imaged planar radial type defects in thick metal sections and pressure vessels. The use of conical illumination provides the necessary conditions for imaging radial type defects, cracks, etc. with extremely small acoustic longitudinal cross sections. The transverse wave acoustic cross sections (i.e., 40° view) of the internal flaws are usually adequate to ensure off-axis holographic images with sufficient resolution to define their geometrical configurations. The image resolution is approximately twice that of the equivalent longitudinal wave configuration.

The aberration correction technique of satisfying the angular conditions (between the object and reconstruction source) effectively reduces the astigmation and enhances the resolution in the optical image.

INTRODUCTION

This paper discusses the theory and application of

transverse wave scanned acoustic holography techniques for imaging planar flaws with extremely small longitudinal acoustic cross-sections in thick-walled vessels (i.e., 30 cm). The use of transverse wave illumination complements the existing traditional longitudinal wave holographic imaging systems used in nondestructive testing. Acoustic holography imaging techniques have previously been extremely successful employing L-wave illumination to image flaws, etc., in thick metal sections.[1,2,3,4,5,6,7] The use of L-waves restricts the illumination angles to less than fifteen degrees with respect to the metal surface. This severely limits the angular viewing aperture and precludes imaging vertical planar flaws. These flaws usually have sufficient acoustic cross-sections for imaging if insonified from various slant angles greater than fifteen degrees. A spherical source is used to provide conical illumination from twenty-four to fifty-eight degrees in the metal. The planar flaw can then be viewed from either side at approximately a forty degree central slant angle. The apparent flaw height or projection in the reconstruction is always less than the true height unless the flaw orientation is perpendicular to the central illumination angles. The true planar flaw height can be evaluated if multiple images are obtained at different viewing positions. The extreme off-axis illumination requirements of transverse wave holography introduces severe third order aberrations (i.e., astigmation, etc.) in the reconstructions. The correction technique reduces the astigmatism by inclination of the hologram with respect to the reconstruction source at the effective construction illumination angle. Thus, the hologram is tilted at approximately forty-five degrees in the reconstruction process.

Imaging with transverse waves in thick metal sections such as nuclear reactor pressure vessels, etc. employing simultaneous source-receiver scanning configuration has yielded excellent images of internal planar flaws. The obtainable transverse wave image resolution was approximately twice that of the equivalent L-wave configuration.

DESCRIPTION OF HOLOGRAPHIC IMAGING SYSTEM

The "HolScan Imaging System" used in the construction of transverse wave holograms for the inspection of nuclear reactor pressure vessels is shown in Figure 1.

Figure 1.   HolScan Imaging System On Reactor Vessel Wall

        The rectilinear scanner with the associated optics
is mounted vertically on a simulated vessel wall. Eight
electromagnets or vacuum feet are used to adhere the scanner
to the reactor vessel. Figure 2 is a closeup of the scanning
mechanism showing the transparent plastic coupler mounted
against the vessel.

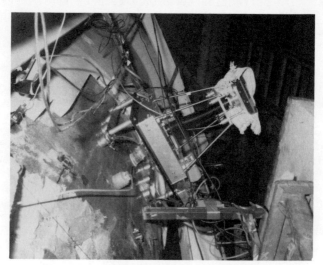

Figure 2.   HolScan Scanner With Transverse Wave Coupled
            On Reactor Vessel Wall

The coupler has three adjusting screws to position the focused transducer. The coupler has a styrofoam liquid seal and is spring loaded to provide liquid acoustic coupling to the vessel. A reservoir and pump is used in conjunction with the coupler to maintain adequate liquid level in the coupler. The amount of leakage is naturally proportional to the vessel surface roughness.

The scanner controls, r-f gating controls, signal processor, etc., are located in the console as shown in Figure 1. The control console is usually located remotely from the scanning unit when imaging on reactor pressure vessels.

The "holographic transverse wave imaging system" is essentially the L-wave configuration with the exception of the on axis (i.e., zero degrees inclination) acoustic coupling device. The transverse wave configuration employs a seventeen degree inclination coupler that houses the focused transducer and provides acoustic coupling between the liquid-steel interface. The inclination angle of the 2.54 cm diameter, 10 cm focal length transducer provides shear wave illumination over a 34 degree cone in the steel. The focused transducer simulates a point source and receiver at the focal point. Simultaneous focused source-receiver scanning has many unique advantages in comparison with the various other configurations. (See references 3 and 5.)

Figure 3 is the block diagram of the S-wave holographic imaging system showing the various components. The L-wave holographic system is adequately described in references 4 and 5, and is identical to the S-wave system with the exception of the seventeen degree coupler. The pulse generator operates at approximately 1KHz and gates the oscillator producing μ-sec sinusoidal 3MHz pulses which after amplification and impedance matching drive the source transducer. The transverse waves illuminate the internal flaws in the steel at various slant angles and the reflected energy is then detected by the focused point receiver. The area scanned is usually 10cm by 10cm aperture, with a line density of 2.5 lines/mm and the hologram contains approximately 250 lines of information. The received echos from the internal flaws are selectively time-gated to enhance the signal-to-voice ratio and then phase detected with the reference signal in a balanced mixer. The output of the mixer is time averaged, amplified

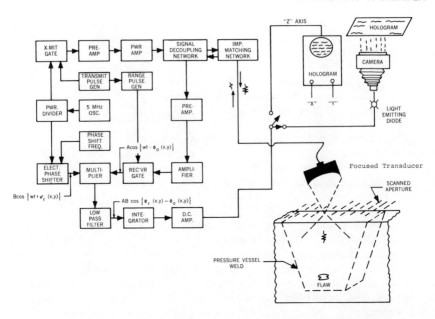

Figure 3.   Simplified Block Diagram Of The HolScan Holographic System For Imaging Flaws In Thick-
Walled Pressure Vessels

and the large voltage peaks clipped before the signal is
used to modulate the intensity of the glow modulator tube
or the "z" axis on an oscilloscope. The glow modulator
tube system employs a fiber-optic light-pipe that provides
the optical path to the camera mounted above the scanning
plane. The light-pipe is attached to the receiving transducer and coupled to the tube which eliminates the requirement of x-y position signals. The oscilloscope system
requires x-y position signals and the hologram is constructed
on the tube face. Either system requires the hologram to
be recorded on transparency film.

HOLOGRAM RECONSTRUCTION PROCESS

The reconstruction of the acoustic hologram to produce
the optical image of the flaw requires a simple optical
computer as shown in Figure 4.

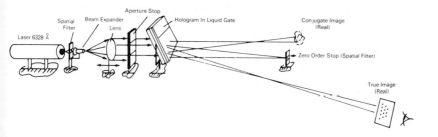

Figure 4.   Schematic Of The Acoustic Hologram Reconstruction
System

A laser provides the source of coherent light to illuminate and reconstruct the hologram for viewing. The spatial
filter shapes or filters the beam to ensure the light source
approaches a point source. The mechanical or electronically
timed shutter provides the necessary light exposure when
photographing the flaw images for permanent records. The
adjustable mechanical aperture (i.e., rectangular) provides
the required light over the entire area of the hologram.
The lens position is variable and moving the lens brings
the true (real) image of the flaw into focus on the viewing
screen as shown in Figure 4. Different lens positions
correspond to different flaw depths in the vessel wall and
by simply moving the lens back and forth simulates looking
through the interior of the vessel. It gives one the
sensation of moving through space and viewing various flaws,
voids, etc., when the lens is continually in motion. The
hologram is shown tilted with respect to the reconstruction
source axis for aberration correction.

The hologram is inserted in a liquid gate containing
a solution with an index of refraction approximating that
of the hologram film (i.e., Polaroid 46L.) The liquid
gate essentially eliminates the undesired effects of film
thickness variations (i.e., phase errors.) The solution
surrounds the film between two optical flats making the
film appear as thick as the width of the gate to the coherent light. The optical smooth surfaces now represent
the film surfaces, thus eliminating the thickness variations.

The image screen (ground glass) is located usually at a specified distance from the hologram. The flaw images are then viewed directly on a television monitor and permanent records obtained merely by replacing the monitor screen with a camera.

ABERRATION CORRECTION ANALYSIS

One of the primary difficulties in the practical application of transverse wave acoustical holography has been the presence of astigmatism in the optical image. This off-axis aberration severely reduces the resolution and distorts the image. Aberrations in acoustical holography have been previously investigated by Leith and Vest.[8] Their techniques were to vary the hologram geometry and thus reduce the aberrations.

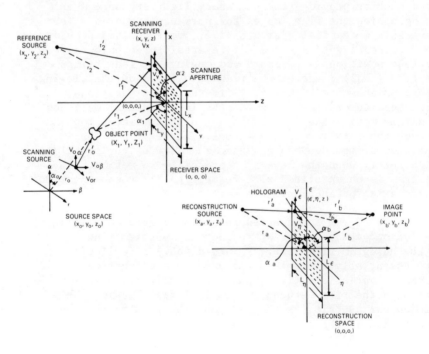

Figure 5.   Geometry For Scanned Acoustic Holography

This analysis is directed toward the reduction or minimization of a single aberration (i.e., astigmatism). The astigmatism coefficients are reduced to zero and the required conditions experimentally verified.

The general scanning source receiver hologram construction and reconstruction geometrics are shown in Figure 5. We have used Hildebrand and Haines' notation and expressions for the Gaussian image points and aberrations.[9] They have modified Champagne's analysis to include simultaneous source-receiver scanning in acoustic holography and then matched the phase of the reconstruction source after it passed through the hologram with that of the hypothetical image point. The normal procedure is the power series expansion of the distance terms in the phases. The first order terms yield the Gaussian image location equations:

$$\frac{1}{r_b} = \pm \frac{\lambda_2'}{\lambda_1 m^2}\left[\frac{1}{r_1} + \frac{1}{r_o}\right] - \frac{1}{r_a} \tag{1}$$

$$\frac{x_b}{r_b} = \pm \frac{\lambda_2}{\lambda_1 m}\left[\frac{x_1}{r_1} + \frac{(x_1 - x_o)}{r_o}\right] - \frac{x_a}{r_a} \tag{2}$$

$$\frac{y_b}{r_b} = \pm \frac{\lambda_2}{\lambda_1 m}\left[\frac{y_1}{r_1} + \frac{(y_1 - y_a)}{r_o}\right] - \frac{y_a}{r_a} \tag{3}$$

where $r_2 = \infty$ (plane wave reference beam) and the magnification $m = m_x = m_y = m_z$.

The third order terms yield the complete aberration expressions for both the true and conjugate images.

Spherical aberration (coeffients of x^4, y^4)

$$S = \pm \frac{\lambda_2}{\lambda_1 m^4}\left[\frac{1}{r_1^3} + \frac{1}{r_o^3}\right] - \frac{1}{r_a^3} - \frac{1}{r_b^3} \tag{4}$$

Coma (coefficients of x^3, y^3)

$$C_x = \pm \frac{\lambda_2}{\lambda_1 m^3}\left[\frac{x_1}{r_1^3} + \frac{x_o}{r_o^3}\right] - \frac{x_a}{r_a^3} - \frac{x_b}{r_b^3} \tag{5}$$

$$C_y = \pm \frac{\lambda_2}{\lambda_1 m^3} \left[\frac{y_1}{r_1^3} + \frac{y_o}{r_o^3} \right] + \frac{y_a}{r_a^3} + \frac{y_b}{r_b^3} \qquad (6)$$

Astigmation (coefficients of x^2, y^2)

$$A_x = \pm \frac{\lambda_2}{\lambda_1 m^2} \left[\frac{x_1^2}{r_1^3} + \frac{x_o^2}{r_o^3} \right] - \frac{x_a^2}{r_a^3} - \frac{x_b^2}{r_b^3} \qquad (7)$$

$$A_y = \pm \frac{\lambda_2}{\lambda_1 m^2} \left[\frac{y_1^2}{r_1^3} + \frac{y_o^2}{r_o^3} \right] - \frac{y_a^2}{r_a^3} - \frac{y_b^2}{r_b^2} \qquad (8)$$

Field Curvature (coefficients of $(xy)^2$)

$$A_{xy} = \pm \frac{\lambda_2}{\lambda_1 m^2} \left[\frac{x_1 y_1}{r_1^3} + \frac{x_o y_o}{r_o^3} \right] - \frac{x_a y_a}{r_a^3} - \frac{x_b y_b}{r_b^3} \qquad (9)$$

Investigation of equations (4-6) reveals under various conditions some or all of the aberrations disappear. All of them disappear when one uses the same wavelength to construct and reconstruct the hologram (i.e., $\lambda_1 = \lambda_2$, $r_a = \infty$, $r_2 = \infty$, $m = 1$, $\cos\alpha_a = \pm \cos\alpha_2$). The conditions that we have in transverse wave holography employing simultaneous source-receiving scanning are $r_2 = \infty$, $\lambda_1 \gg \lambda_2$, $r_o = r_1$, and

$$\cos\alpha_b = \pm \frac{2\lambda_2}{\lambda_1 m} \cos\alpha_1 - \cos\alpha_a . \qquad (2a)$$

The astigmation appears to dominate the image distortion in all of our previous experiments. The most simple approach is to identify the conditions that reduce the astigmation coefficient to zero.

If the object points are off-axis in the "x" direction, the coefficient A_x must be reduced. Equation 7 can be rewritten in the following form:

$$A_x = \pm \frac{2}{m^2} \frac{\lambda_2}{\lambda_1} \frac{\cos^2\alpha_a}{r_a} - \frac{\cos^2\alpha_a}{r_a} - \frac{\cos^2\alpha_b}{r_b} \qquad (10)$$

where $\alpha_o = \alpha_1$, $r_2 = \infty$, $r_o = r_1$. For this specific case the image location equation is

$$\frac{1}{r_b} = \pm \frac{\lambda_2}{\lambda_1 m^2} \left(\frac{2}{r_1}\right) - \frac{1}{r_a} \tag{11}$$

and $\cos\alpha_b + \cos\alpha_a = \pm \frac{2\lambda_2}{\lambda_1 m} \cos\alpha_1$. $\tag{12}$

Typical values of the experimental variables are $\lambda_2/\lambda_1 = 6.1 \times 10^{-4}$, $r_1 = r = 0.125$ meter, $r_2 = \infty$, $r_b = 10$ meters, $m = 0.25$, $\alpha_1 \simeq 45°$ and $\alpha_2 = 90°$ from which it may be calculated that

$$\cos\alpha_a + \cos\alpha_b = \pm 0.00345 \tag{13}$$

and

$$\frac{1}{r_a} + \frac{1}{r_b} = \pm 0.1562 \text{ (meter)}^{-1} \tag{14}$$

In all examples, use of the minus sign yields the true image whereas use of the plus sign yields the conjugate image. On the basis of Equation 13

$$\cos\alpha_b \simeq \pm \cos\alpha_a . \tag{15}$$

From Equation 14

$$r_a = 17.794, \text{ or } - 3.904 \tag{16}$$

where -3.904 meters is the required value to form the true image at a distance of 10 meters from the hologram.

With these conditions established we may write

$$A_x \simeq \pm \frac{\lambda_2}{m^2 \lambda_1} \left(\frac{2}{r_1}\right) \cos^2\alpha_1 - \left(\frac{1}{r_a} + \frac{1}{r_b}\right) \cos^2\alpha_a \tag{17}$$

or

$$A_x \simeq \pm \frac{2\lambda_2}{m^2 \lambda_1 r_1} \left(\cos^2\alpha_a - \cos^2\alpha_1\right) . \tag{18}$$

Equation 18 shows that astigmatism will be eliminated if
the hologram is tilted to an angle α_a with respect to the
reconstruction source such that

$$\cos\alpha_a \approx \pm \cos\alpha_1 \tag{19}$$

Of course, for a given value α_a only one direction α_1 in
the object field is fully corrected for astigmatism but
as in all corrections of this sort, sufficient correction
extends over an appreciable range of angles.

EXPERIMENTAL RESULTS OF ABERRATION CORRECTION

Figure 6 shows the simulated construction geometry of
simultaneous source-receiver scanned transverse wave holography. A focused transducer is shown acting as both the
source and receiver at the focal point. The transducer is
tilted at the central transverse wave angle in metal (i.e.,
forty degrees). The object is a series of cylindrical rods
simulating point reflectors forming a "Y" pattern. The
point reflectors on each leg of the "Y" are spaced 1,2 and
4mm apart (edge to edge). The 2mm separation is the theoretical image resolution using a 10 cm focal length, 2.54 cm
diameter transducer, 3MHz illumination and the 12.7 cm object
to hologram distance.

Figure 7a shows the optical image of the "Y" pattern
with the hologram perpendicular to the reconstruction source
axis (i.e., $\alpha_a = 90°$). The correct angle should be equal
to approximately forty-five degrees--the angle between the
object and the effective aperture center (i.e., α_1). The
"Y" pattern exhibits gross distortion in the optical image.
The pattern is elongated or stretched in the "X" direction.

Figure 7b shows the optical image with the hologram
tilted thirty degrees with respect to the vertical (i.e.,
$\alpha_a = 30°$). The image appears less distorted in the "X"
direction than (a). Figure 7c shows the "Y" pattern image
with the hologram tilted at forty degrees, the optimum
theoretical correction angle predicted in the analysis.
The image has excellent resolution compared with the two
previous images verifying the predicted correction procedure.

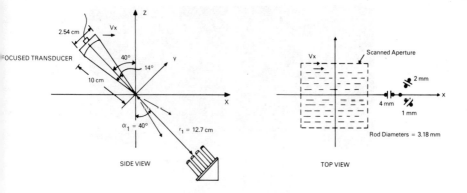

Figure 6.   Simulated Transverse Wave Hologram Construction Geometry In Water

TRANSVERSE WAVE EXPERIMENTAL RESULTS

The first experiments consisted of verifying the various
basic image parameters using simultaneous (inclined) focused
source-receiver scanned shear wave holography technique.
The preliminary experiments were on aluminum blocks to
ascertain the flaw resolution, range and projected height.
The results were extremely encouraging and the final in
situ work was completed on the EBOR vessel at the National
Reactor Test Site, Idaho Falls.

A simulated radial flaw was constructed on the interior
side of the 15 cm thick EBOR vessel. The flaw was a circular
saw cut perpendicular to the interior vessel surface. The
vertical or radical flaw was then pressure cycled to induce
crack growth. Transverse wave holographic images of the initial
saw cut and the progressive crack growth were obtained and
the results were conclusive that "T-wave" holography is a
viable tool in pressure vessel inspection.

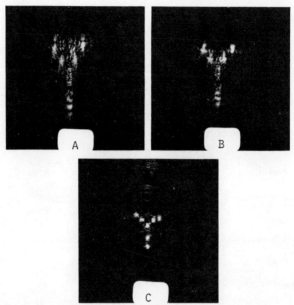

Figure 7.   Hologram Reconstruction Of The "Y" Pattern; (A)
$\alpha_a = 90°$, (B) $\alpha_a = 30°$ and (C) $\alpha_a = \alpha_1 = 40°$.

The results are presented in sequential form by first
showing a sketch of the flaw, hologram, construction geometry,
the acoustic hologram, the reconstructed image and the photograph of the flaw itself.

TRANSVERSE WAVE RESOLUTION MEASUREMENTS

The image resolution as a function of object distance,
frequency and effective aperture is an extremely important
parameter to verify experimentally. Figure 8 shows the array
of seven flat-topped holes drilled with their tops 17 cm
on the diagonal from the block surface. The holes are
6.4 mm in diameter (i.e., 250 mils) and placed in a "Y"
pattern. In the first row the hole spacing is 2 mm (edge-
to-edge), second row 3 mm, third row 4 mm.

The maximum obtainable theoretical lateral and longitudinal resolution using a focused transducer is given
by the following equations:

$$\text{Lateral Resolution} \simeq 1.22 \frac{\lambda f}{a} \qquad\qquad (20)$$

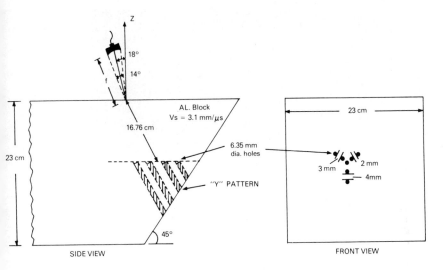

Figure 8.   Schematic of Aluminum Block Constructed For
Shear Wave Lateral Resolution Tests

$$\text{Longitudinal Resolution} \simeq \frac{\lambda}{2} \left(\frac{f}{a} \right)^2 . \qquad (21)$$

where λ = acoustical wavelength in the medium
f = focal length in the medium
a = diameter of the transducer or lens

The maximum lateral and longitudinal resolutions for our
transducer operating at 3MHz are 2 mm and 16 mm respectively.
This means we can resolve flaws separated 2 mm laterally
and 1.6 cm in depth.   The lateral resolution is valid to a
depth of approximately ten aperture lengths.   If the flaw
depths exceed these values then the holographic expressions
of resolution dominate, and must be used to predict flaw
resolution.   The crossover point is easily determined by
equating the classical lens resolution equations with the
holographic equations and solving for the depth (r_1)

$$r_1 \simeq \frac{2.44fL}{a} \qquad (22)$$

where r_1 = flaw to hologram distance

and L   = aperture length .

The holographic resolutions are expressed in the following form:

Lateral Resolution $\simeq \dfrac{\lambda r_1}{2L}$ (23)

Longitudinal Resolution = $\dfrac{\lambda}{2}\left(\dfrac{r_1}{L}\right)^2$.   (24)

Using a 10 cm focal length, 2.54 cm diameter transducer and
effective aperture length of 5 cm the cross over point is
approximately 50 cm.

The results of the transverse wave resolution tests are
shown in Figure 9, A, B and C. The third row of holes from
the left are separated 2mm edge-to-edge and are resolved
in the reconstructed image (see Figure 8.) This experiment
essentially verifies that the system has the capability
of resolving flaws separated 2 mm laterally at a depth of
approximately 17 cm. The longitudinal resolution requirements are less critical because the pulse-echo transient
time provides extremely accurate depth information.

SIMULATED RADIAL EBOR FLAW

This experiment was performed to simulate the EBOR
flaw and verify the true flaw height when imaging with
approximately 45 degree transverse waves. Radial flaws when
viewed at angle will always appear diminished in height
(see Figure 10.) The true height can be resolved if 1)
the object's geometrical orientation is known or 2) the
object is viewed from various locations. Naturally the
defect geometry is usually unknown and the flaw must be
viewed from various positions to accurately measure the
true height. The identical argument holds for viewing the
true width of the flaw. If views are obtained from both
sides (i.e., 180 degrees rotation of the scanned aperture)
on a simple planar radial flaw then the true height is
easily calculated using the following equation (see Figure
10 for the symbols.)

True Flaw Height (h) $\simeq \sqrt{h_a^2 + h_b^2}$ (25)

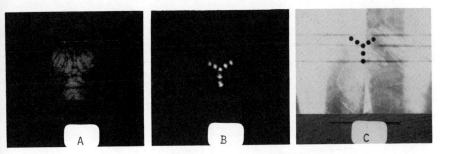

Figure 9.   Holography Transverse Wave Resolution Test; (A)
Acoustic Hologram of the Holes Pattern, (B)
Reconstructed Image and (C) Photograph of the
Hole pattern from the Bottom Side of the Block.

Figure 11 is the transverse wave hologram construction
geometry of the simulated "EBOR" radial saw cut.

The simulated circular cut was imaged using 2.25MHz
insonification and a 11 cm focal length transducer.   The
reconstructed image of the saw cut projection is shown in
Figure 12B.   Figure 12A and C are the acoustic hologram
and the photograph of the saw cut.   The apparent height and

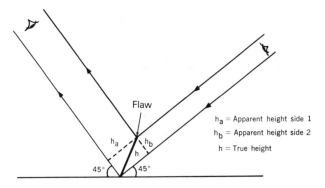

h_a = Apparent height side 1
h_b = Apparent height side 2
h = True height

Figure 10.   Apparent and True Flaw Height of a Radial Planar
Flaw

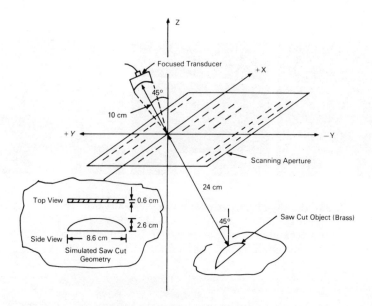

Figure 11.   Hologram Construction Geometry of the Simulated
             EBOR Flaw

length are easily calculated by measuring them directly
from the image and dividing by the magnification. The apparent height is 1.7 cm and the length 8.38 cm. The true

          (a)   (b)   (c)

Figure 12.   Simulated EBOR Flaw; (A) Acoustic Hologram of
             the Flaw, (B) the Reconstructed Image, (C)
             the Photograph of the Curved Flaw

height is then calculated using the previous equation;

$$h = \sqrt{h_a^2 + h_b^2} = \sqrt{2h_a} = 2.4 \text{ cm} \tag{26}$$

where $h_a = h_b$ and the vertical flaw is 45° with respect to
the acoustic shear wave illumination. The actual height of
the flaw at the apex is 2.54 cm. The resulting error in the
measurement of height is 5.5% or 1.4 mm. The magnitude of
the error is very reasonable considering the image height
resolution at 2.25 MHz is approximately 2.5 mm which again
is very reasonable when compared with the system resolution.
The image quality--shape, size, etc., is excellent considering the extremely long flaw distance (i.e., 24 cm.)

INTERIOR SAW CUT CRACK IN THE EBOR VESSEL

A circular saw (10 cm radius) was used to construct a
narrow slotted artificial flaw in the inner surface of the
vessel wall. The circular slot was approximately 3 cm in
height at the apex and 9.27 cm at the base or inner wall
surface. A rectangular pressure boss was welded over the
exterior slot to provide hydraulic coupling, enabling the
slot to be pressure cycled. The slot was then cycled with
extremely high pressure bursts of oil to promote crack
growth over the entire periphery of the circular slot.
Before initiation of the pressure cycling, shear wave holograms were constructed of the saw cut. Figure 13 shows
the acoustic hologram, reconstructed image and the photograph of the slot itself from the vessel inner surface.

Figure 13 is the reconstructed holographic image of
the saw cut projection (i.e., view from approximately a
45 degree angle.) The viewing angle has the effect of ob-
scuring the saw cut curvature. The projection or apparent
maximum height is easily calculated by measuring the image
at the apex and dividing by the magnification. The true
height was then calculated to be 2.84 cm using the central
shear wave refraction angle of 40 degrees. The actual saw
cut measured 2.99 cm at the apex. The error in the reconstructed image height is approximatley 5% or approximately
the holographic resolution. The image length at the base
of the saw cut is 8.8 cm and the actual length 9.27 cm. An
error of 5% or 4.7 mm in the length is again expected
considering the resolution limit at 2.5 MHz.

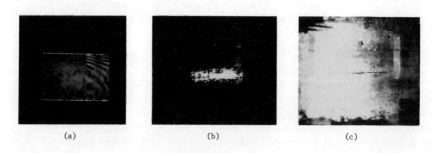

(a)   (b)   (c)

Figure 13. Saw Cut Flaw Constructed in the EBOR Vessel;
            (A) Acoustic Hologram, (B) the Reconstructed
            Image 0.16 magnification and (C) the End View
            of the Flaw from the Interior of the Vessel

The quantitative holographic image information defining
the geometrical size, shape, etc. of the saw cut is very
accurate and demonstrates the extremely unique capabilities
of Acoustical Holography in pressure vessel inspection.

INTERIOR SAW CUT IN THE "EBOR" VESSEL AFTER PRESSURE CYCLING

The interior saw cut was subjected to extremely intense
high pressure (30,000 psi to 50,000 psi) bursts of oil over

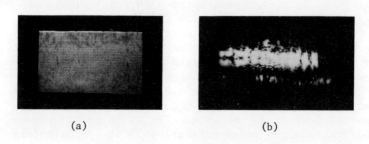

(a)   (b)

Figure 14. Internal Saw Cut in the "EBOR" Vessel after
            Pressure Cycling; (A) Acoustic Hologram and
            (B) the Reconstructed Image

28,000 cycles.   The main objective was to promote or induce a crack growth along the periphery of the curved saw cut. The test was conducted to provide acoustic emission data as the crack propagated from the slot.   The test also provided a seminatural flaw in a 15 cm thick vessel wall for image evaluation with shear wave acoustical holography.   Holograms were constructed of the internal saw cut after the pressure cycling was completed and the results are shown in Figure 14.   Hologram construction geometry is identical to the previous setup.

The reconstructed image shown in Figure 14b shows the saw cut has definitely cracked around the entire periphery. The base after cracking increased in length to 13 cm.   The height is now 3.8 cm and fissures or small cracks radiate outward from the periphery as shown by the apparent halo above the solid rectangular crack and saw cut region.   The small cracks extend above the 3.8 cm height or solid region to approximately 7 cm.   Naturally, the verification of the radial appearing images above the solid region as being small cracks can only be verified if this volume is extracted and sectioned.

The base length was easily measured from the interior of the vessel and this measurement verified.

REFERENCES

1) Aldridge, E.E., Clare, A.B. and Shepherd, D.A., Exploring the Use of Ultrasonic Holography in Nondestructive Testing, Ultrasonics for Industry Conference Papers, 1969.
2) Aldridge, E.E., Clare, A.B., and Shepherd, D.A., Ultrasonic Holography, Atomic Energy Research Establishment, Harwell, 1969.
3) Hildebrand, B.P. and Haines, Kenneth, Holography by Scanning, J. Opt. Soc. Am., Vol. 59, p. 1, 19, 1969.
4) Collins, H.D. and Gribble, R.P., Acoustic Holographic Scanning Techniques for Imaging Flaws in Thick Metal Sections, Seminar-in-Depth on Imaging Techniques for Testing and Inspection (SPIE), February 14-15, 1972, Los Angeles, California.

5) *Collins, H. Dale and Hildebrand, B.P., Evaluation of Acoustical Holography for Inspection of Pressure Vessel Sections, Paper presented to the joint ASME (USA) Institution of Mechanical Engineers, London, England, May, 1972.*

6) *Collins, H. Dale, Investigation of Acoustic Holography for Visualization of Flaws in Thick-Walled Pressure Vessle, Project VI, In Service Inspection Program For Nuclear Reactor Vessels, SWRI Biannual Progress Report No. 5, EEI Project RP-79, May 28, 1971.*

7) *Collins, H. Dale, Investigation of Acoustic Holography for Visualization of Flaws in Thick-Walled Pressure Vessels, Project VI, In Service Inspection Program for Nuclear Reactor Vessels, SWRI Biannual Progress Report No. 6, Vol. II, EEI Project RP-79, January 7, 1972.*

8) *Leith, E.N. and Vest, C.M., Aberrations in Acoustical Holography, Investigation of Holographic Testing Techniques, Semiannual Report, June - Nov. 1972, Radar & Optics Laboratory, Willow Run Laboratories, P. 37--54.*

A SURVEY OF SOUND PROPAGATION IN SOILS

Thomas G. Winter

Department of Physics, University of Tulsa

600 South College, Tulsa, Oklahoma, 74104

INTRODUCTION

Acoustical scientists could make an enormous contribution to society if they could develop techniques to
image objects buried near the surface of the earth. It
would be possible to locate and define the surface of the
water table or bedrock. Builders could locate boulders
and other obstructions in construction sites, geologists
could trace the details of subsurface geology,
archeologists could locate tombs and ruins. The list of
applications is almost endless.

The development of acoustical holography would seem
to provide the technology necessary to image over large
surface areas. The question remains, however, are the
propagation characteristics of soils adequate to make
acoustical imaging feasible with wave-lengths in the
one meter to one centimeter range.

With this in mind, a search of the literature on
acoustical propagation characteristics of soils was
initiated. The literature has turned out to be much more
extensive than expected. In addition to acoustical and
geophysical journals, there are a large number of papers
in Civil, Mechanical and Petroleum Engineering journals.
There have also been a large number of government
sponsored studies for purposes of acoustic intrusion
detection, mine and tunnel detection, etc., which are not

in the open literature. This paper is a summary of the
results of this literature search. It contains no new
information.

For the purposes of this paper, the term soil means
the unconsolidated clay, silt, sand and gravel which lies
over the solid rock of the earth's crust. It is useful
to distinguish between two types of soil, sand and clay.
Sand particles are bulky and compact and range from 0.05
to 2 mm. in diameter. Adhesive forces are small and
sand resists shear stress even when wet. Clay particles
are flat and plate-like with the largest diameter less
than 0.002 mm. In the presence of water, strong electrical forces exist between particles and at some critical
water contents, the shear modulus drops precipitiously.
With variation in water content, a clay soil will crack,
swell and slip while a sand soil will not. A mixed soil
will behave like a clay when there is more than 20% clay
by weight.

It appears that the principal obstacles to any kind
of sophisticated imaging in soils are the severe velocity
gradients and multipath conditions encountered in soils.
These will be treated first and most extensively.
Following this, there will be a discussion of acoustic
absorption, seismic background noise and the acoustic
impedance of soils.

VELOCITY OF SOUND IN SOILS

The principal factors affecting the velocity in a
soil are the confining pressure and the water content (1).
One mathematical model of a soil is an assembly of very
small uniform spheres (2,3). Figure 1 shows the compressional wave velocity through an assembly of spheres,
each sphere having the density and elastic properties of
granite as calculated by Gassman (2). On the right is the
velocity within 10 meters of the surface and on the left
the behavior at greater depths. There are three important points to note. First, the velocity in dry spheres
increases very rapidly with depth, especially within the
first few meters, and second, the horizontal velocity is
different from the vertical. The reason for both of these
facts is that the bulk modulus of an assembly of spheres
increases rapidly with stress and the vertical stress from
the overburden is greater than the horizontal stress.

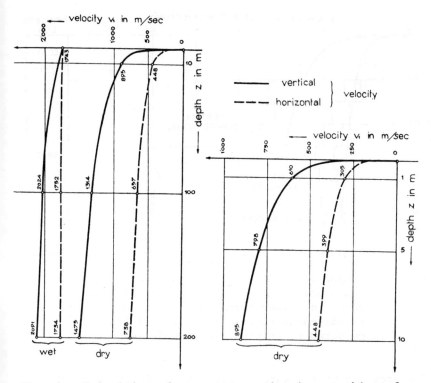

Fig. 1. Velocities of wave propagation in a packing of
spheres from Ref. (2).

The third point to note is that the velocity in water
saturated spheres is roughly 15% greater than that in
water alone and is virtually independent of depth.

Many of the theoretical predictions for an assembly
of spheres are born out in real soils. Figure 2 presents
the results of a field study by White & Sengbush (4).

Explosions were set off at various depths in holes
drilled in a sand formation. Receivers placed above the
shot location in the same hole or in adjacent holes measured the travel time of vertical and horizontal waves.
The smooth curve is Gassman's theory, the vertical lines
are experiment. The variation of velocity with depth and
the large increase of velocity below the water table at
fifty feet are confirmed here and in other studies.

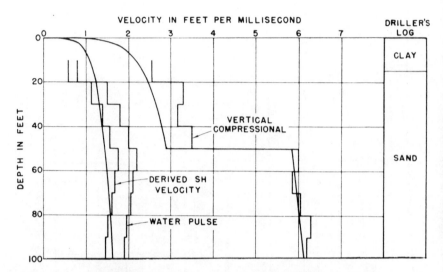

Fig. 2.   Comparison of theoretical and experimental
          sound velocities of compressional and shear
          waves in a sand formation. The smooth curve
          is Gassman's theory, the vertical lines are
          measured velocities. From Ref. (4).

The presence of the water table has no effect on shear
velocity. This is expected because a compressional wave
involves a volume change and water is much less com-
pressible than the sand skeleton alone. The shear wave
involves a shearing action which is not resisted by the
water. The general conclusion from many experiments is
that the shear velocity in dry sand is about half that
of the compressional velocity and both increase with pressure. In saturated sand, the shear velocity increases
with depth just as in dry sand but the compressional
velocity is constant at about 6,000 fps independent of
depth.

     In this formation, the horizontal and vertical
velocities were identical contrary to prediction. In
both chalk and shale formations, the same authors
observed that the horizontal compressional velocity is
greater than the vertical (see Figure 3). In any
layered soil one may expect that the horizontal and
vertical velocities will differ, partly because of the
way the layers were deposited and partly because the

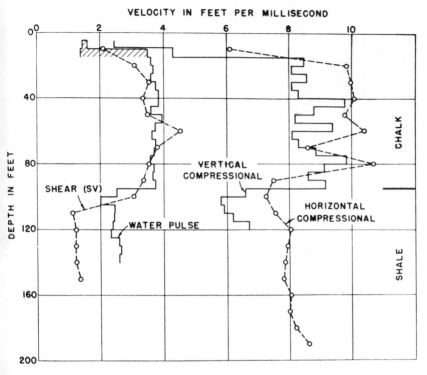

Fig. 3.   Sound velocities as a function of depth in a
          chalk and shale formation. The water table
          is at ten feet. From Ref. (4).

horizontal and vertical stresses are not necessarily the
same.

In the laboratory soil composition can be controlled,
and the effect of soil composition can be investigated.
Figure 4 shows that the compressional wave velocity
through clay-sand mixtures is independent of soil composition. It also shows that the velocity increases slowly
with moisture content until near saturation and then
it increases very rapidly to the saturated velocity of
6,000 fps.

Figure 5 is a similar graph for shear waves which
shows that velocity decreases with increasing moisture
content and increasing sand content. The explanation of
the difference in the effect of water on the two types

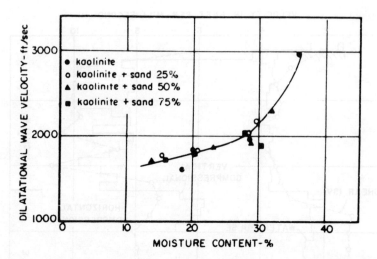

Fig. 4.   Influence of moisture content on dilitational
          wave velocity in mixtures of clay and sand at
          confining stress of 40 psi.   From Ref. (1).

of waves may be that the water increases the density of
the soil in both cases leading to a decrease in velocity,
but in the case of the compressional wave, the effect on
the bulk modulus is greater than the effect on the density.

Figure 6 shows that on a small scale, the variation of velocity with depth is not at all uniform.   In
this experiment a gated 800 Hz pulse generator was located
about six feet beneath the surface and the time of
travel of a pulse to a receiver above the generator was
recorded.   After each transmission the receiver was
located six inches deeper in the ground.   The difference
in travel time for each transmission yielded the velocity
through the preceding six inch layer.   This layering
was found to exist in both Texas and Virginia soils.

All experiments indicate that there is no velocity
dispersion below 50 kHz.   The velocity of both shear and
compressional waves decreases slightly with increasing
amplitude, but this effect becomes negligible at high
confining pressures.

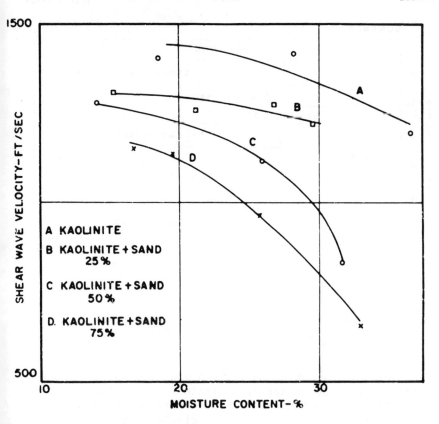

Fig 5.   Influence of moisture content on shear wave
          velocity in mixtures of clay and sand at
          confining stress of 40 psi.   From Ref. (1).

        This behavior of the acoustic velocity affects any
imaging process in at least three ways.   First, it affects
the amount of energy available to illuminate the reflector.
If the velocity gradient with depth were constant, the

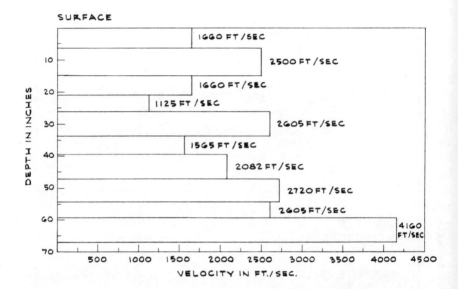

Fig. 6. Velocity vs. depth showing incremental
         velocity layers. From Ref. (5).

path of any ray making an angle θ with the normal to the
surface would be an arc of a circle (6) as shown in Figure
7. Velocity gradients in the earth are not constant so
the ray path is not the arc of a circle, but the angle
of ray entry can still be related to the deepest penetration of that ray by the following reasoning. Divide the
soil into n horizontal layers, each thin enough that the
ray path within that layer may be considered a straight
line.

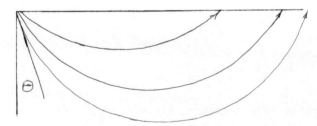

Fig. 7. Ray paths of waves in a soil with the sound
         velocity proportional to depth.

Snell's law holds for each layer:

$$\frac{\text{Sin } \theta_1}{c_1} = \frac{\text{Sin } \theta_2}{c_2} = \ldots = \frac{\text{Sin } \theta_n}{c_n}$$

Where θ_n is the angle the ray in layer n makes with the normal and c_n is the velocity of sound in layer n. The quotient in the first layer is equal to the quotient in the nth layer. At the depth of greatest penetration, Sin θ_n = 1 and the initial angle of that ray is given by Sin θ_1 = c_1/c_n. Because of the very large difference between surface velocities and velocities at depth, only those rays making a very small angle with the vertical will penetrate very deeply.

The second effect of the velocity characteristics of soils on imaging is refraction effects. Any imaging method must incorporate a receiving array many wavelengths in diameter and a foreknowledge of the ray path between the object and each receiver in the array. The variation of velocity has been shown to be large (often 10:1), irregular, and unpredictable. Due to refraction, the ray paths will also be irregular and unpredictable. Any imaging system using wavelengths less than a meter which assumes an isotropic medium will produce hopelessly blured images.

The third problem is that as the wavelengths get smaller, all of the local variations in velocity described in connection with Figure 6 become reflectors. It is well known that if continuous waves are incident normally on a reflecting layer imbeded in a medium, the reflecting layer is invisible if the layer thickness is much less than a quarter wavelength, and the reflections are largest when the layer thickness is odd integer multiples of one quarter wavelength. The same type of behavior is observed with pulses, discontinuities which are much smaller than the pulse width are invisible and for certain ratios of discontinuity width to pulse width, the energy reflected becomes a maximum (7). As the wavelengths become small enough to give adequate resolution, soils become filled with scatterers and reflectors.

MULTIPATH IN SOILS

Any imaging system requires that the sound source be separated from the receivers. But in a layered soil, there are many sound paths between any two points in addition to the path due to reflection from the object. The result is that there will appear to be as many reflectors as there are sound paths.

It is the common experience of anyone who has done seismic field work that a single seismic impulse will produce a train of impulses arriving by different paths which may last for thirty seconds. No reference has been found which describes multipath propagation in shallow layers for short wavelengths but such propagation may be infered from common experience at greater depths and longer wavelengths. Figure 8 shows some of the paths by which acoustic energy may travel from the source to the receiver. Path 1 is the surface Rayleigh wave. Path 2 is the wave refracted by the velocity gradient in the soil. Path 3 is the desired direct reflected wave. Path 4 is a wave propagated through a higher velocity layer and reradiated to the receiver. This is the mode of propagation used in the seismic technique known as "refraction shooting." Path 5 represents a wave reflected several times within the layer before it escapes to the receiver. At each reflection, some compressional wave energy will be converted to shear waves and vice versa, and since shear waves travel at about half the velocity of compressional waves, any path containing a segment of shear wave propagation represents a new path. The resulting wave train is so complex that the contributions of the various paths can be understood only in the simplest cases.

One simple case which has been completely analyzed (8) is a model study in which the earth was simulated by 1/16-inch sheet metal. It turns out that so long as the wavelength is much greater than the thickness, both shear waves and compressional waves propagate in sheet metal much as they would in the bulk material. A 9 cm. wide sheet of brass, a low velocity material, was brazed to an "infinitely" deep sheet of high velocity stainless steel with dimensions shown in Figure 9. A compressional pulse with the waveshape shown in Figure 10 was transmitted from the source and the waveforms at

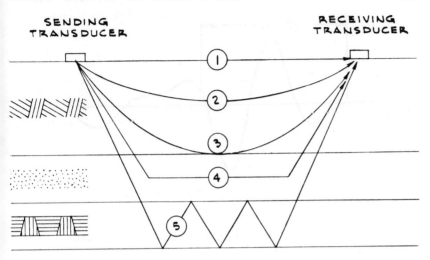

Fig. 8.   Some of the paths by which sound can travel
from the sending transducer to the receiving
transducer in a layered soil.

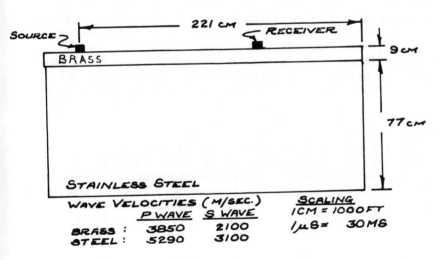

Fig. 9.   Schematic diagram of two-dimensional analog
model.   From Ref. (8).

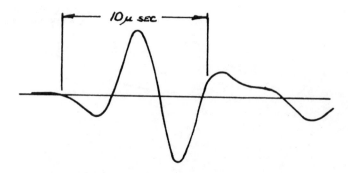

Fig. 10.   Source pulse.   From Ref. (8).

various distances down the sheet from the source were
recorded.

A 100 microsecond portion of the received wave train
at distances from 115 cm to 125 cm from the source is
shown in Figure 11. The single impulse has grown into a
train of pulses, some of which have suffered a 180° phase
inversion. The paths of some of these pulses are shown
in Figure 12. A compressional wave, (P wave), incident on
the brass-steel boundary will produce a compressional reflection, designated P_{2x}, in Figure 12(a). This is the
large amplitude wave in Figure 11. The compressional wave
will also produce a reflected shear wave, (S wave), and
this is designated PS_x. A compressional wave in the brass
will also set up a compressional wave in the steel which
will travel with the speed of sound in steel and also re-
radiate back into the brass. This mode is called P_2 and
is illustrated in Figure 12(b). If it reradiates as a
shear wave, it is designated PS and if it was a shear
wave in the brass before it reached the steel and converted to a P wave, it is designated SP. If there was an
additional reflection at the air-brass interface as
shown in Figure 12(c&d), the mode is designated P_4 or
P_{4x}.

The actual existance of these various paths was
demonstrated by calculating a theoretical seismogram
from the superposition of the first seventeen predicted
arrivals and comparing the theoretical waveform with the
observed waveform. The agreement was very good as may
be seen from Figure 13 which shows the theoretical

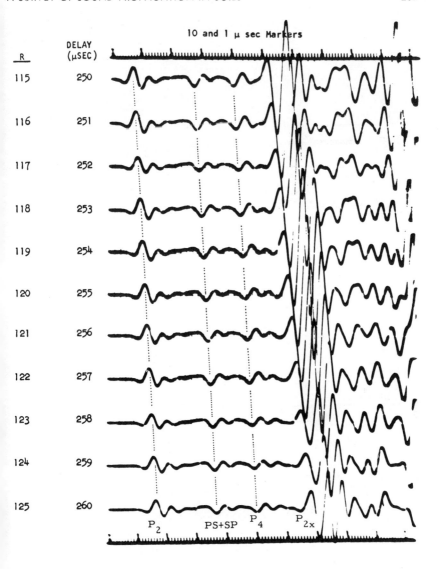

Fig. 11. Typical model seismograms recorded with top
transducer. Vertical motion. From Ref. (8).

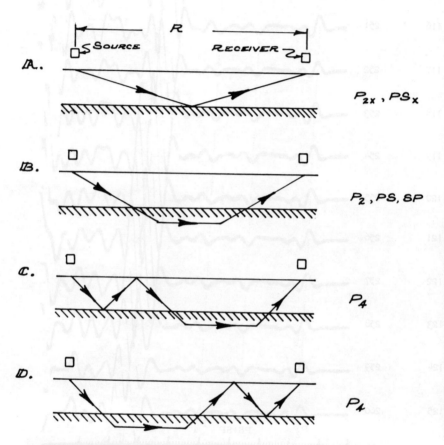

Fig. 12.   Some acoustic paths from the source to the
           receiver in Fig. 9.   From Ref. (8).

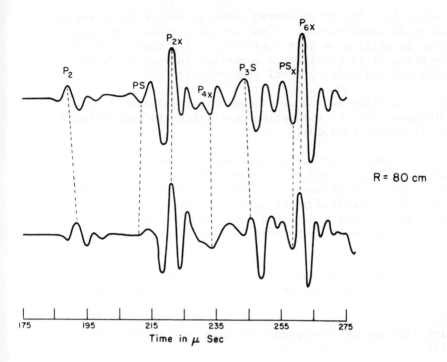

Fig. 13.   Comparison of theoretical and experimental
           seismogram.   The theoretical seismogram is
           at the top.   From Ref. (8).

seismogram at top and the observed seismogram below at a
distance of 80 cm from the source.

This was a particularly simple model because there
were only two layers, the interface was flat and parallel
to the surface, and the transit time through the upper
layer was long compared to the pulse width.   Any imaging
system must have the ability to pick the P_{2x} pulse out of
the wave train.   In this case, P_{2x} was always the first
large amplitude pulse to arrive, but it is not clear that
this would always be the case in real soils.

ABSORPTION

Absorption studies in the earth are difficult because the multipath, scattering, refraction and

reflection effects discussed above introduce large errors
in field measurements. Also, not many studies of absorption in soils have been published. With these limitations
in mind, it is possible to summarize current knowledge
with the following general statements:

1)   Absorption in a homogeneous soil is exponential with
distance, and the absorption coefficient is proportional
to frequency (5,9,10).

2)   The absorption coefficient is 2-4 times greater in
water saturated sand than dry sand (11). This result may
be in error as it has been observed that sound absorption
in freshly prepared water saturated soil is much higher
than in soil which has been saturated for many days because of gas bubbles present in the freshly prepared
soil (12).

3)   Absorption decreases with increasing confining pressure (11). Loose unconfined soil is opaque to sound (13).

4)   The absorption coefficient increases with amplitude in
dry sand and is independent of amplitude in saturated sand
(11).

5)   Typical values of absorption coefficient in dB/ft reported in field studies are 0.00012f in shale (9) and
1.2+0.002f in a dry silty clay (5) where f is the frequency
in Hz.

SEISMIC BACKGROUND NOISE

     The mechanical noise level due to wind, traffic and
earthquakes, etc., is about forty dB above the electronic
noise level of a geophone, so seismic noise sets the limit
of sensitivity unless signal averaging is used. Seismic
noise levels above 100 Hz have not been published. Figure 14 is typical of the studies of seismic noise below
100 Hz (14,15). This figure plots particle velocity for
a one cycle bandwidth as a function of frequency at
several locations in California. Note that the noise
falls off rapidly with frequency, and that wave action at
the beach generates noise several orders of magnitude
greater than inland sites.

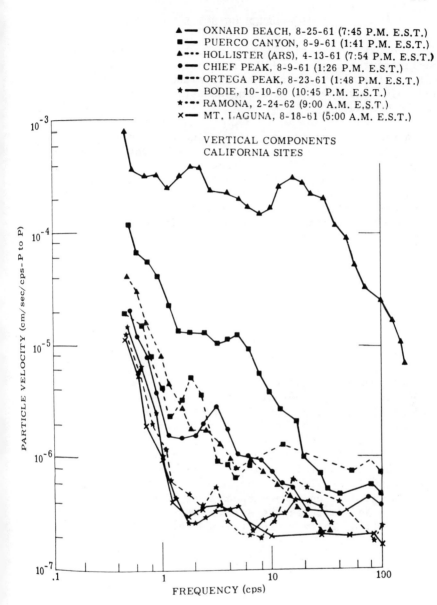

▲— OXNARD BEACH, 8-25-61 (7:45 P.M. E.S.T.)
■— PUERCO CANYON, 8-9-61 (1:41 P.M. E.S.T.)
▲··· HOLLISTER (ARS), 4-13-61 (7:54 P.M. E.S.T.)
●— CHIEF PEAK, 8-9-61 (1:26 P.M. E.S.T.)
■··· ORTEGA PEAK, 8-23-61 (1:48 P.M. E.S.T.)
★— BODIE, 10-10-60 (10:45 P.M. E.S.T.)
★··· RAMONA, 2-24-62 (9:00 A.M. E.S.T.)
✕— MT. LAGUNA, 8-18-61 (5:00 A.M. E.S.T.)

VERTICAL COMPONENTS
CALIFORNIA SITES

Fig. 14.   Earth noise spectra for eight locations in
            California. From Ref. (14).

IMPEDANCE OF SOILS

The characteristic impedance is the product of the density of the medium and the speed of sound in that medium. For a dry surface soil, typical quantities are 2400 kg/m^3 and 500 m/s yielding an estimate of 1.2x10^6 MKS Rayls.

A piston vibrating on the surface of a soil sees a driving point impedance which may be represented by a compliance, a mass and a resistance (15,16). At low frequencies the compliance dominates and the impedance becomes purely real at a few hundred Hertz. In one experiment, (15), is was found that the driving point impedance of a certain soil (expressed in terms of acoustic impedance in MKS Rayls at a frequency f) was well represented by the empirical expressions:

$$z = 1.2 \times 10^6 + j \left(200 \, \pi \, f - \frac{4.85 \times 10^8}{2 \, \pi \, f} \right) \text{ MKS Rayls}$$

SUMMARY

Soils present a hostile environment for imaging with wavelengths of the order of a meter or less. This is due primarily to the large velocity gradients near the surface and to multipath propagation produced by the layered nature of soils. These problems become less important at wavelengths large compared to the discontinuity, but at present no method of imaging small objects with large wavelengths is available. The application of static pressures to the soil under a surface transducer will decrease absorption and decrease the surface velocity gradient which in turn will decrease refraction and reflection of energy.

ACKNOWLEDGMENTS

This work was supported by the U.S. Army Research Office Durham. The author wishes to express his gratitude to the personnel of the U.S. Army Corps of Engineers Waterway Experiment Station and the Mobility Equipment Research and Development Center for pointing out the Army studies listed in the references.

REFERENCES

1. R. V. Whitman, "Response of Soils to Dynamic Loading," AD-708-625*, May 1970.

2. F. Gassmann, Geophysics 16, 673 (1951).

3. J. Duffy and R. D. Mindlin, J. Appl. Mech. 24, 585 (1957).

4. J. E. White and R. L. Sengbush, Geophysics 18, 54 (1953).

5. L. S. Fountain and T. E. Owen, "Investigation of Seismic Parameters Related to Shallow Tunnel Detection," AD-820-238*, July 1967.

6. J. S. Saby and W. L. Nyborg, J. Acoust. Soc. Am. 15, 316 (1946).

7. R. L. Sengbush, P. L. Lawrence, F. J. McDonal, Geophysics 26, 138 (1961).

8. S. J. Laster, M. M. Backus and R. Schell, "Seismic Refraction Prospecting," A. W. Musgrave Ed. Society of Exploration Geophysicists, Box 1067, Tulsa, Oklahoma, 1967, p. 15.

9. F. J. McDonal, F. A. Angona, R. L. Mills, R. L. Sengbush, R. G. VanNostrand and J. E. White, Geophysics 23, 421 (1958).

10. L. Knopoff, "Physical Acoustics III B," Warren P. Mason Ed. Academic Press, 1965, p. 287.

11. F. E. Richart, Jr., J. R. Hall, Jr., and J. Lysmer. AD 286075.*

---

*Documents referenced by AD numbers available from Defense Documentation Center, Cameron Station, Alexandria, Virginia, 22314.

12. W. L. Myborg, I. Rudnick and H. K. Schilling, J.
     Acoust. Soc. Am. <u>22</u>, 422 (1950).

13. H. F. Eden and P. Felsenthal, J. Acoust. Soc. Am.
     <u>53</u>, 464 (1973).

14. G. E. Frantti, Geophysics <u>28</u>, 547 (1963).

15. B. Isacks and J. Oliver, Bull. Seism. Soc. Am.
     <u>54</u>, 1941 (1964).

16. W. R. Runyon and R. E. Anderson, J. Acoust. Soc.
     Am. <u>28</u>, 73 (1956).

17. W. R. Runyan and J. F. Mifsud, AD 97517.*

APPLICATIONS OF ACOUSTIC SURFACE WAVE VISUALIZATION TO NONDESTRUCTIVE TESTING*

G. A. Alers, R. B. Thompson and B. R. Tittmann

Science Center, Rockwell International

Thousand Oaks, California 91360

ABSTRACT

Acoustic surface waves offer many potential advantages as a tool in nondestructive testing. The concentration of the energy of these waves near the surface of a solid makes possible efficient interaction with subsurface defects which might otherwise be quite difficult to detect. Also the high propagation velocity of surface waves suggests that they may be used to achieve rapid inspection rates on large surface areas. This paper describes a simplified version of the visualization system of Adler, Korpel, and Desmares which has been used to investigate these applications. Photographs to be presented illustrate the deformation of the crests and troughs of surface waves which occur when they encounter thickness variations and holes in metal sheets and plates, cracks in surfaces, and metallurgical variations in nominally defect free material.

*This work was supported by the Rockwell International IR&D Interdivisional Technology Program under the sponsorship of the Nondestructive Testing Technical Panel.

INTRODUCTION

The use of ultrasonic waves to interrogate the interior
of structural or biological materials has been a standard
procedure for many years. Present techniques usually involve
bulk acoustic waves which can interact with defects or internal structure in either a reflection or a transmission mode.
A number of the papers in this symposium deal with methods
to present the information so obtained in a visual display
to allow the human operator to use his own pattern recognition capabilities in interpreting the data.

Use of ultrasonic surface waves in nondestructive
testing has not been nearly so widespread. This paper
reports the results of feasibility studies to determine
whether an inspection system based on the display of surface
waves can in fact demonstrate the anticipated advantages of
high sensitivity to near surface defects and rapid inspection rates on large surface areas. A modification of the
visualization scheme of Adler, Korpel, and Desmares[1] has
been chosen for these studies. The technique produces a
stationary display of the crests and troughs of the waves
and thereby provides a characterization of those physical
and chemical properties affecting the elastic wave velocity
near the surface. Abrupt discontinuities in properties are
indicated by scattering of the surface waves, while more
gradual changes in properties may be detected by changes in
wavelength and the accompanying bending of wavefronts. Our
implementation of the system is described in the next section. A number of experimentally obtained images are then
presented which demonstrate the ability of the system to
detect thickness variations and holes in metal sheets and
plates, cracks in surfaces, and metallurgical variations in
nominally defect free material.

TECHNIQUE AND APPARATUS

The technique for visualizing the surface waves is essentially the same as that proposed by Adler, Korpel and
Desmares[1] and presently applied at their laboratory towards
the development of acoustic cameras[2] and microscopes.[3] The
basic principles of operation are illustrated in Fig. 1.

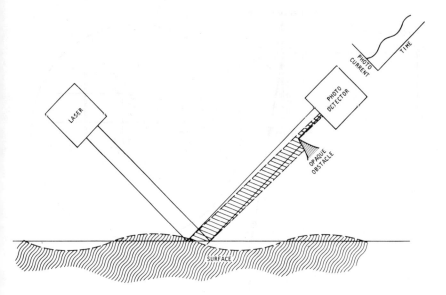

Fig. 1 Principles of Surface Wave Detection.

When an ultrasonic surface wave propagates on a specularly reflecting surface, an incident light beam will be
reflected at an angle which differs from its quiescent value
by twice the instantaneous tilt of the surface. The phase
and the amplitude of the tilt, and hence of the surface wave,
are determined by partially blocking the beam with a knife
edge and collecting the remaining light in a photodiode.

To generate a display of the surface wavefronts over a
finite area, additional optical scanning and electronic
signal processing techniques must be used. Figure 2 shows
the optical components of our system. The beam from a 50 mW
He-Ne laser is directed onto the part to be inspected via a
pair of scanning mirrors and a lens. The mirrors allow the
beam to be scanned over the surface of the part so that the
local vibration phase and amplitude may be sampled on a
point-by-point basis. In order to recover this information,
it is necessary that the reflected beam be partially blocked
by the knife edge before entering the photodiode. Ideally,
this can be accomplished for all scan angles by a single
knife edge and detector if the knife edge and scanners are

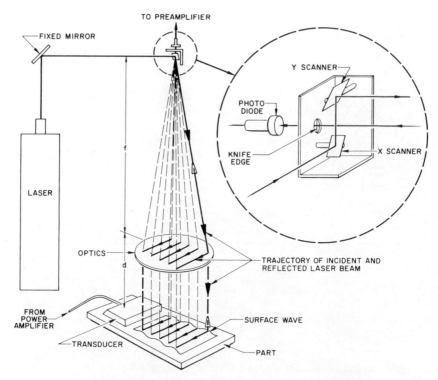

Fig. 2 Optical Components of Visualization System.
The exploded view shows the details of the scanner-
knife edge combination.

appropriately placed in the focal plane of a lens[1,4] as indicated in Fig. 2. In practice, the returning beam sometimes
misses the knife edge due to irregularities in the surfaces
of the part or lens. Best results are obtained when all
elements are placed as close to the optical axis of the lens
as possible, as shown in the exploded view of the scanner-
knife edge combination.

The distance d between the lens and part may assume a
variety of values, depending upon the application. At low
frequencies for which the laser spot size is small with respect to the surface wavelength, it is often convenient to

use small distances. At higher frequency, it is necessary
to focus the laser beam to ensure that it acts as a point
detector of the surface vibration. It is then desirable to
choose d equal to the focal length of the lens.

The electronics used to visualize the surface waves are
shown in Fig. 3. Schematic waveforms at various points are
included to define the function of the various components.
The essence of the system is the derivation of a video pulse
whose height is proportional to the product of the amplitude
of the surface vibration at the point of interest times the
cosine of its phase relative to a continuous wave reference.
A voltage comparator intensifies the beam of a cathod ray
tube when positive going pulses are present (- $\pi/2$ < phase
< $\pi/2$). When the x and y positions of the cathode ray beam
are scanned in synchronization with the laser probe, an image

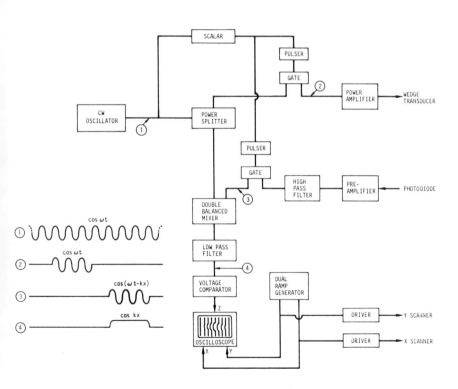

Fig. 3 Electronic Components of Visualization System.

is produced which may be thought of as stroboscopically displaying the crests and troughs of traveling waves as light and dark lines. The response time of the mechanical mirrors limits the speed of our system to about one frame per minute. the image may be readily viewed on a storage oscilloscope or directly recorded on film using a standard oscilloscope and a time exposure.

This approach differs from previous work in the field[1-4] by the use of simpler but slower scanning and display elements. The result is a system which can be constructed from readily available or relatively inexpensive components and which is easily used to evaluate the application of surface wave visualization to nondestructive testing.

RESULTS

This system was used to display acoustic surface waves, and their interaction with defects, on a number of different structural materials. The study encompassed frequencies ranging from 200 kHz to 10 MHz and samples such as steel sheet, titanium block, and honeycomb panels. The major limitation was the requirement that the surface be relatively flat and specular over the area of interest.

Figure 4 shows the results obtained at 3 MHz for a number of defects in a 0.019 in. (0.048 cm) thick steel sheet. In this case the elastic wave was the first symmetric (longitudinal) Lamb mode traveling with a phase velocity of 5.1×10^5 cm/sec. Lamb waves, which fill the plate with energy, were chosen to demonstrate the ability of the technique to detect defects on the side of the sheet opposite the transducer and optical probe and hence not directly visible at the surface. Part a) illustrates the results when no defects were present. The white and dark lines may be thought of as delineating the crests and troughs of the wave, and these are relatively straight as would be expected in a uniform plate. Part b) shows the same plate, but now with a small 3/8 in. (.953 cm) diameter 17% thickness reduction etched in the bottom side. The circularly spreading waves scattered by the defect are now clearly seen in addition to the incident straight-crested waves. Similar results were obtained when a 1/4 in. (.635 cm) diameter washer was bonded to

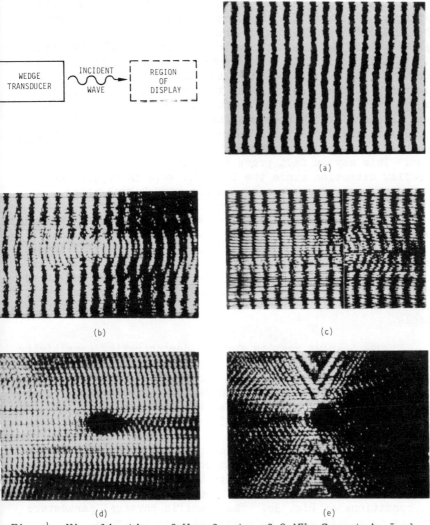

Fig. 4   Visualization of Wavefronts of 3 MHz Symmetric Lamb
Waves (λ = 0.067 in.) scattered from various defects and
structures bonded to the underside of .019 in. steel sheet.

   a)   no defects
   b)   .003 in. deep 3/8 in. diameter etched hole on bottom
   c)   1/4 in. diameter washer bonded to bottom
   d)   3/8 in. diameter drilled through hole
   e)   same as d) but with detector preferentially sensitive
        to waves scattered at 90°.

the bottom of the sheet as shown in Part c). Further interesting results are shown in Part d) which illustrates both
the reflection of sound from the edge of a 3/8 in. (.953 cm)
diameter hole and the diffraction of sound around the back
of the hole to fill in the geometric shadow. Note also that
the radiation pattern of the transducer is visible in this
picture. Part e) is identical to Part d) in all respects
except that the knife edge is rotated 90° about the light
beam as an axis. This causes the system to be most sensitive
to elastic waves traveling perpendicular to the incident wave,
and these components of the scattered wave are now clearly
seen. This may in fact prove to be the more sensitive method
for flaw detection since the incident wave is suppressed and
the scattered waves are seen against a dark background.

In the experiments described above, the scanning mirrors
were not available and the sample was mechanically scanned
underneath the stationary light beam by x - y stepping motors.
As a consequence, the time for each exposure was 45 minutes.
However, the results to follow were all taken in times on the
order of a minute using the scanning mirrors previous described to rapidly move the laser beam over the stationary
sample.

Figure 5 shows results further demonstrating sensitivity
to subsurface defects. In this experiment, 200 kHz antisymmetric (flexural) Lamb waves were excited in a 1/4 in.
(.635 cm) thick glass plate. In Part a), the straight crested
waves radiated by a transducer at the left are seen. A 2 in.
(5.08 cm) diameter 1/4 in. (.635 cm) thick steel washer was
then bonded to the back surface of the plate. The distorted
waveform which resulted is shown in Part b). It is quite
interesting to note that the bonded washer caused primarily
a distortion of the wavefronts in this case, while a similar
bonded washer caused primarily a scattering of energy under
the conditions of Fig. 3c. The specific physical parameters
causing this difference have not yet been isolated.

In these experiments, the flexural wavelength is approximately 1/2 in. (1.27 cm) and the area of the plate displayed
is about 16 in.^{2} (103 cm^2). This ability to inspect large
areas in short periods of time is an important feature of
the visualization technique.

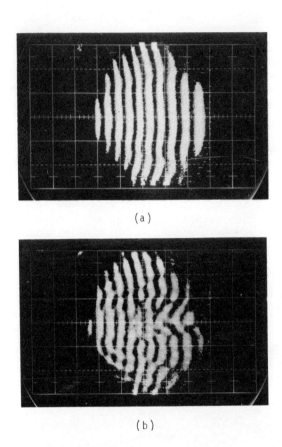

(a)

(b)

Fig. 5 Visualization of Wavefronts of 200 kHz Antisymmetric
Lamb Waves (λ = 0.5 in.) on 1/4 in. Glass Plate.
Area displayed has approximately 4 in. diameter.

    a) no defect
    b) 2 in. diameter, 1/4 in. thick steel washer bonded to
        bottom

Figure 6 is a display showing 760 kHz waves propagating on the surface of an adhesively bonded honeycomb sandwich panel. The shape of the area [5 in.^{2} (32 cm^2)] over which the waves are seen is a result of both the finite width of the transducer [2 in. (5 cm)] and the heavy attenuation of the waves due to scattering from the core. No discrete defects are located in the field of view. However, both the attenuation and the surface wave velocity, which can be deduced from the known frequency and the wavelength as measured on the display, are important parameters in studying the characteristics of the adhesive bond between the face sheet and core.[5] This points out the potential of the surface wave display technique in the characterization of intrinsic properties of monolithic and heterogeneous materials as well as in the location of discrete flaws and defects.

An important problem in evaluating structural materials is the detection of small surface cracks such as appear during fatigue or corrosion. Such defects can act as sources of concentrated stress during loading and lead to failures. Figure 7 illustrates the interaction of a Rayleigh wave with an eloxed slot which simulates a crack on a stainless steel surface. This slot was 0.009 in. (0.023 cm) wide by 1/8 in. (0.318 cm) long at the surface and had the shape of a segment of a circle penetrating to a maximum depth of 0.012 in. (0.030 cm). Since the Rayleigh wave also has a surface wave-

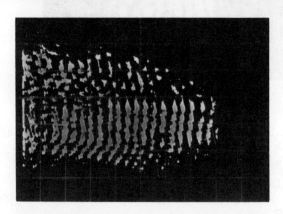

Fig. 6 Visualization of Wavefronts of 760 kHz Wave Propagating on the Surface of a Honeycomb Sandwich Panel.

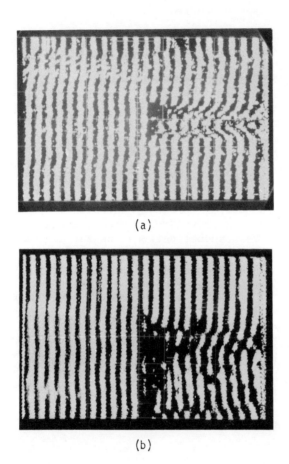

(a)

(b)

Fig. 7 Visualization of Wavefronts of 10 MHz Rayleigh Waves
(λ = 0.012 in.) incident upon a 1/8 in. long, .012 in. deep
eloxed slit in steel.

    a)   global view (vertical scale compressed)
    b)   local view (approximately 1:1)

length of 0.012 in. (0.030 cm) this represents a severe
discontinuity and is readily observed. The waveform in the
geometric shadow of the crack appears to be a result of both
diffraction around the edges and underneath the bottom of
the crack.

Figure 8 illustrates the variation of intrinsic properties in a nominally homogeneous sample of rolled titanium
plate[6], as viewed with 10 MHz Rayleigh waves propagating on
the edge of the plate. The upper part of the figure shows
the wavefronts over an area of approximately 0.09 in.^{2}
(0.6 cm^2). The wavefronts would be expected to be reasonably
straight, as in Figs. 4 and 5a. In fact, severe interference
effects and bending of the wavefronts are observed. This is
believed to arise from the layer-like micro-structure fluctuations which can be seen in the optical micrograph of the
same area shown in the lower half of the figure. These fluctuations in structure are probably introduced by inhomogeneous deformation during the rolling of the plate material.
A one-to-one correspondence between the surface wave display
and the microstructure is not obvious, presumably because
of the fact that the phase of the surface wave is not a local
variable, but rather depends upon the properties of all of
the material through which it has passed. Nevertheless, there
is a marked similarity between the boundaries of the stri-
ations and the breaks in the wavefronts, hence the display
appears to give a qualitative measure of the uniformity of
the material. Furthermore, it provides a graphic demonstra-
tion of one of the problems in ultrasonic pulse echo inspection of titanium. The same elastic inhomogeneities seen here
can produce a multitude of small reflections of a bulk wave
which can give the appearance of a high noise level and mask
returns from defects.

CONCLUSION

A modification of the system of Adler, Korpel, and
Desmares[1] has been used to visualize acoustic surface waves
on structural materials. The results demonstrate the power
of the technique in locating surface and subsurface defects
and in studying the intrinsic properties of structural
materials. An important feature of the technique is the
high speed with which large areas can be inspected. The primary limitation is the requirement of flat, specularly reflecting surfaces. The development of a scanned detection

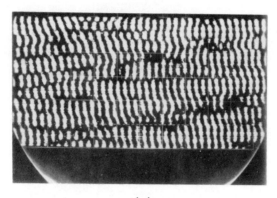

(a)

(b)

Fig. 8 Visualization of Wavefronts of 10 MHz Rayleigh Waves (λ = 0.012 in.) on Titanium.

   a) bending of wavefronts due to local elastic in-homogeneities

   b) optical micrograph of same area show fluctuations in metallurgical structure.

system which overcomes this difficulty would open the way to even more widespread application and work toward this end is in progress.

ACKNOWLEDGMENTS

Special thanks are due to N. A. Massie and M. A. Tennison for their important contributions to this work.

REFERENCES

1. R. Adler, A. Korpel and P. Desmares, "An Instrument for Making Surface Waves Visible", IEEE Trans. Sonics Ultra-sonics SU-15, 157-161 (1968).

2. R. L. Whitman, A. Korpel, M. Ahmed, "Novel Technique for Real-Time Depth-Gated Acoustic Image Holography", Appl. Phys. Lett. 20, 370-371, (1972).

3. A. Korpel, L. W. Kessler, P. R. Palermo, "Acoustic Micro-scope Operating at 100 MHz", Nature (GB) 232, 110-111 (9 July 1971).

4. R. L. Whitman and A. Korpel "Probing of Acoustic Surface Perturbations by Coherent Light", Applied Optics 8, 1567-1576 (1969).

5. R. B. Thompson, G. A. Alers and D. O. Thompson, "Non-destructive Measurement of Adhesive Bond Strength in Honeycomb Panels", Materials Evaluation XXXI, 33A(A), 1973.

6. Titanium 6Al-4V.

ACOUSTICAL HOLOGRAPHY - A COMPARISON WITH PHASED ARRAY SONAR

P. N. Keating

Bendix Research Laboratories

Southfield, Michigan 48076

ABSTRACT

The primary difference between phased-array sonar and acoustical holography lies only in the order in which the temporal and spatial processing operations are carried out. It is shown that this difference indirectly leads to an advantage for holography in terms of either signal-to-noise performance or reduced processing complexity in many types of application. The holographic approach has an important signal-to-noise advantage over scanned phased-array systems because of parallel processing. Compared with multibeam sonar, the holographic spatial processing of data obtained from uniform arrays can be carried out with fewer operations than phased-array parallel beam-forming via adders and tapped shift registers because of the Cooley-Tukey algorithm. A significant reduction in processing operations due to data reduction in holography, especially for active systems, is also possible.

INTRODUCTION

The concept of holography can be summarized by the phrase "wavefront reconstruction," which implies storage of information regarding the wavefront prior to subsequent reconstruction. In this respect, it is rather different from real-time directional processing, such as the time-delay beam-forming typically carried out in phased-array sonar systems. However, in many other respects, far-field acoustical holography is very similar to phased-array sonar. Both these similarities and the major differences are the subject of this paper, together with a discussion of the practical implications of the similarities and differences.

We begin by examining the basic similarity between the two approaches. We denote the electrical signal at the output of the kth hydrophone in a linear array as $G_k(t)$. Both the phased-array system and the holographic system perform, in essence, two operations on the set $\{G_k\}$:

(a) Linear combinations of the $G_k(t)$ which are equivalent to a Fourier transform are formed. In the phased array system (PAS), this is the phasing or beam-forming operation which directs the array pattern in differing directions. In the holographic system (HS), this is the reconstruction operation.[3]

(b) The time-varying signal is temporally processed (e.g., filtered) to improve the signal-to-noise ratio.

The primary difference between the PAS and HS approaches is the order in which the two operations are carried out.

In the phased-array system, the first operation is (a) above, which we shall represent by the transform

$$F_\ell(t) = \sum_k{}' G_k(t) e^{2\pi i k\ell/N} \tag{1}$$

where $F_\ell(t)$ is the signal in the ℓth beam ($\ell = -j, -j+1, \dots +j$) and $N = 2j+1$ is the number of elements in the uniform (i.e., equally-spaced) array. The expressions given in this section apply to a uniformly-spaced line array but the generalization to two-dimensional (e.g., square, rectangular) uniform arrays is straightforward.[4] The main lobe of the ℓth beam pattern lies in a direction θ given by

$$\sin\theta = \frac{2\pi \ell c}{Na} = \frac{\ell\lambda}{Na} \qquad (2)$$

where λ, ω, and c are the acoustical wavelength, frequency and velocity, respectively, and a is the hydrophone spacing. In order to make an explicit comparison, we shall represent the temporal processing [(2), above] as a filtering operation in which the signal is mixed with a local oscillator signal and integrated. In other words, the ℓth beam output in the passband centered on ω is

$$F_\ell^{PAS}(\omega) = T_1^{-1} \int_t^{t+T_1} dt\, e^{-i\omega t} \sum_k G_k(t) e^{2\pi i k\ell/N} \qquad (3)$$

In the holographic case, the filtering is carried out first and the linear combination formed last, during reconstruction. Thus,

$$F_\ell^{HS}(\omega) = T_2^{-1} \sum_k e^{2\pi i k\ell/N} \int_t^{t+T_2} dt\, e^{-i\omega t} G_k(t) \qquad (4)$$

It is clear from Eqs. (3) and (4), that the two approaches do have a strong basic similarity. The order of the operations in itself is not important, and the two approaches can differ only in the way they are carried out.

COMPARISON OF THE TWO APPROACHES

If we compare a linear holographic system with a scanning linear sonar, then an important signal-to-noise difference arises. The object field must be scanned successively through N different directions with the single-beam phased-array sonar whereas, during data acquisition, the holographic system is receiving data from all directions. In other words, a scanned beam-former only samples a section of the ω-k spectrum[5] at one time. As a consequence of this difference, if T is the time allowed for one frame of the total object field, then

$$T_1 = T/N$$

$$T_2 = T.$$

Thus, the sample length T_1 appropriate to the scanned phased-array system is N times shorter than that (T_2) appropriate to the holographic system; thus, the signal-to-noise ratio is $N^{\frac{1}{2}}$ times better for the latter if the signal is coherent and the noise is Gaussian.

The last result is due, of course, to the fact that the holographic system utilizes parallel acquisition and processing [or serial processing in an essentially noise-free environment (e.g., a digital computer)], whereas the scanned PAS uses sequential acquisition and processing of the noisy data. However, phased-array systems with parallel processing (multibeam systems) are in existence, a good example being Anderson's DIMUS system.[6,7] The improved signal-to-noise performance of the multibeam system does, of course, require a considerable increase in hardware. The basic performance of this type of system should be similar to that of the holographic system if binary words of the same length are obtained from digitization in both cases.[8] Nevertheless, the holographic spatial processing can be carried out in a multipurpose computer, and the hardware of the multibeam PAS is then replaced merely by software (assuming the computer has the necessary extra capacity available). Thus, there would be an advantageous hardware-to-software trade-off in going to the holographic approach.

However, it is more appropriate to compare apples with apples. Even if we make a direct hardware-hardware comparison, i.e., if we assume hardwired processing, there appears to be a definite reduction in processing necessary in the holographic approach compared to that necessary in a DIMUS-type system, as described by the following. The time-evolving output of each hydrophone in the DIMUS system is sampled and digitized and passed down a tapped shift-register delay line. The output from each tap is fed to one of the N inputs on one of the N adders such that the output from each adder is a sampled version of one of the N beam signals $F_\ell(t)$. Thus, N^2 binary operations (additions) are carried out, in addition to the shift-register delay operations. In the holographic system, it is, of

course, necessary to carry out the same type of processing.
However, a hardwired Fast Fourier Transform processor using
the Cooley-Tukey algorithm[9] can be employed in the holographic case; the N^2 operations are thereby replaced by
$N \log_2 N$ for a uniformly-spaced array. As is well-known,[9,10]
this is a considerable reduction in processing. In addition, the data storage occurring with holography allows
processing to occur within the time between transmit pulses.
Hence, the processor and A/D converter can, in effect, be
time-shared to reduce hardware costs further.

Another possible area in which the amount of processing
might be reduced in the holographic case lies in the data
reduction which can occur in that approach. For ease of
illustration, we shall here treat the case where the temporal processing is simple filtering, as in Eqs. (3) and
(4), although a similar argument also applies if more
sophisticated processing is used. Digital phased-array
methods involve carrying out the N^2 operations at each
sampling interval in time and thus $N^2 T/\tau$ operations are
required before the filtering is carried out, where τ is
the sampling interval. In holography, the temporal processing is completed before the spatial processing, and the
data are compressed to only two values (in effect, amplitude
and phase) in each channel per frequency band. If the data
in all of the T/τ frequency bands available are required,
then a similar amount of processing is required in both
cases - other than the Cooley-Tukey gain mentioned above.
However, in active systems, and often in passive systems,
only a few frequency bands contain data of interest and
another reduction in processing will be the result. One
might, in the passive case, argue from the other direction,
i.e., that sometimes fewer search directions than frequency
bands contain data of interest. This is certainly not
likely to be true of active systems, however.

There are a number of other differences between the
two different methods which are worth mentioning. For
example, the holographic approach is much better for operation at shorter distances, i.e., nearer than the far-field
zone. Fresnel-zone operation involves a trivial extension
in the holographic case where software is utilized, but
causes considerable difficulties for a phased-array system.
Another advantage for the holographic system is the fact
that the array shading and processing methods can be
rapidly changed to allow increased flexibility. On the

other hand, however, a conventional range-gated holographic
system fails to detect part of the return pulse in many of
the channels when the target is off-axis, whereas a time-
delay phased system does not encounter this problem.

SUMMARY

In summary, there is a basic similarity between the
phased-array and holographic approaches, but there are
certain significant advantages in the latter. The holographic approach has an important signal-to-noise advantage
over scanned phased-array systems because of parallel processing. Compared with multibeam sonar, the holographic
spatial processing of data obtained from uniform arrays can
be carried out with fewer operations than phased-array
parallel beam-forming via adders and tapped shift registers
because of the Cooley-Tukey algorithm. A significant reduction in processing operations due to data reduction in
holography, especially for active systems, is also possible.

ACKNOWLEDGEMENTS

The author is indebted to Dr. R. K. Mueller, G. Goodrich,
G. Zilinskas, and R. F. Koppelmann for useful discussions.

REFERENCES

1.   See, for example, H. G. Frey, "High Resolutions Sonar
     Technology," Vol. II, p. 43, Classified Report No.
     69-R-NRC:MAC:2027.

2.   See, for example, H. R. Farrah, E. Marom, and R. K.
     Mueller, p. 173, Acoustical Holography, Vol. 2, Ed.
     by A. F. Metherell and L. Larmore (Plenum Press, New
     York, 1970); R. K. Mueller, Proc. IEEE, 59, 1319 (1971).

3.   For the purposes of this discussion, we define phased-
     array systems as those in which the spatial processing
     (e.g., beam-forming) is carried out before the temporal
     processing, and holographic systems as those in which
     the reverse is true, the reconstruction being carried
     out last.

4. See, for example, P. N. Keating, R. F. Koppelmann, R. K. Mueller, and R. F. Steinberg (to be published).

5. M. O. Fein and E. S. Eby, Naval Underwater Sound Laboratory Technical Memo No. 2242-173-69 (July 1969).

6. V. C. Anderson, J. Acoust. Soc. Am., 32, 867 (1960).

7. P. Rudnick, J. Acoust. Soc. Am., 32, 871 (1960).

8. In actual fact, the original DIMUS approach used only one-bit words whereas current holographic systems use 8-12 bits for each word. However, one-bit words can be used for holography [see, for example, W. J. Dallas and A. W. Lohmann, Acoustical Holography, Vol. 4, p. 463, Ed., G. Wade (Plenum Press, New York, 1972)] with the same drop in performance as in DIMUS. Alternatively, a comparison with current holographic systems should involve an "improved" DIMUS with 8-12 bit digitization and the corresponding reduction in quantization "noise".

9. J. W. Cooley and J. W. Tukey, Math. Comput., 19, 297 (1965).

10. B. Gold and C. M. Rader, Digital Processing of Signals (McGraw-Hill, New York, 1969).

THRESHOLD CONTRAST FOR THREE REAL-TIME ACOUSTIC-IMAGING SYSTEMS

Keith Wang
University of Houston
Houston, Texas
Glen Wade
University of California
Santa Barbara, California

ABSTRACT

Three approaches to real-time acoustic imaging are presently being worked on in various laboratories. These approaches involve the use of static-ripple diffraction, dynamic-ripple diffraction and Bragg diffraction. Experimental results indicate that good images can be obtained from each and that the systems currently appear to be quite competitive. This paper compares the ultimate potential performance of systems of each of the three types in terms of threshold contrast and sensitivity. The comparisons not only give an indication of the inherent capabilities and limitations of the systems but also of how far the present systems are from achieving ideal behavior.

The analysis is based on idealized models of each type and also of a non-existing, hypothetical system with highly desirable theoretical characteristics. This latter system would be difficult to build as a practical instrument but is useful in analyses since it can serve as a theoretical standard of excellence against which to measure the performance of the other systems.

The systems are shown to vary in different ways with various parameters such as acoustic frequency of operation, resolution-cell area and power in the laser beam from which the optical image is eventually derived. However, for a set of consistent operating conditions selected to be compatible with the application of medical diagnosis, a sample calculation shows that the calculated capabilities of the

three systems are of the same order. Thus the decision of
employing one particular system in preference to another
in this application would have to be based on practical
considerations and a knowledge of how close the system
comes to achieving the ideal performance calculated.

INTRODUCTION

The analysis[1] presented in last symposium is extended
to three real-time systems that have obtained good acoustic
images. These three systems utilize approaches to acoustic
imaging via static-ripple diffraction[2], dynamic-ripple
diffraction[3], and Bragg-diffraction[4]. The ideal potential
performance of the three systems are characterized and
compared in terms of acoustic threshold contrast. The
derivation[5] is based on idealized models of each type where
only the most fundamental noise[1] limits the detection. In
all three approaches, optical detection is involved; hence
the most fundamental image noise is quantum noise in the
photon beam. Water is assumed to be the medium and diffraction limited resolution is assumed of interest.

THRESHOLD CONTRASTS FOR THE THREE SYSTEMS

The Bragg-Diffraction Imaging System

The manifestation of quantum noise in the Bragg-diffraction imaging system may be classified as originating from the
signal fluctuation, Brillouin scattering, or zero-order
flare light. The threshold contrast expressions are derived
with the assumption of predominance of each of the quantum
noise contributions. When quantum noise due to the signal
fluctuation (QNS) limits the performance, the threshold
contrast is

$$C_{aQNS} = 0.6 \times 10^7 \; k_o \sqrt{\frac{\lambda\alpha}{P_I}} \sqrt{\frac{hF}{qP\tau_f}} \tag{1}$$

where

P_I = total laser power per unit height of the light wedge in watt/cm

α = half-wedge angle of light wedge in radians

λ = laser wavelength in vacuum in cm

F = acoustic frequency in Hz

q = quantum efficiency of the pickup device

τ_f = frame time in sec.

P = acoustic power transmitted by the more transparent object element in watts

k_o = output threshold signal-to-noise ratio

h = Planck's constant

Equation (1) sets the upper limit of ideal performance. It is applicable for high transmitted sound levels. For ordinary room temperature operations at low* transmitted sound intensities, the Brillouin-scattered photon density rate is greater than the Bragg-diffracted photon density rate. When the quantum noise due to Brillouin scattering (QNB) predominates, the threshold contrast is

$$C_{aQNB} = 0.6 \times 10^7 \; k_o \sqrt{\frac{\lambda \alpha}{P_I}} \sqrt{\frac{hF}{qP\tau_f}} \; \sqrt{\frac{5.32 \; F\alpha LKT}{I_s \Lambda^2}}$$

---

* When $I < \dfrac{5.32\alpha \; F \; L \; KT}{\Lambda^2}$ watt/cm^2 where L is the height of the light wedge in cm, K Boltzmann constant, T absolute temperature, Λ acoustic wavelength in cm.

where I_s is the intensity transmitted by the more transparent element in watt/cm^2.

Unless some scheme* is used to reject or reduce the zero-order** light, the effect of zero-order diffraction predominantly impairs the image of objects with low acoustic transmission.

The threshold contrast due to the quantum noise in the zero-order flare light (QNZ) for an optimum operation (compatible with the scheme in reference 4) is

$$C_{aQNZ} = 0.91 \times 10^{-7} \, k_o \, \frac{\lambda^{\frac{3}{2}}}{\Lambda} \sqrt{\frac{1}{P_o q \tau_f}} \frac{L\alpha}{P} \tag{3}$$

where P_o is the total laser power in watts.

The Dynamic-Ripple Imaging System

In the dynamic-ripple imaging system, the unperturbed projection of the scanner spot is essentially responsible for producing the dc component in the photodiode. The threshold contrast from shot noise consideration is

$$C_a = 0.58 \times 10^5 \, k_o \, \frac{\lambda}{\pi\Lambda} \sqrt{\frac{h\nu m^2}{q P_o I_s \tau_f}} \tag{4}$$

where m^2 is the number of resolution elements.

---

* For instance by the use of heterodyning or polarization discrimination techniques to reject the zero-order light or by apodization to reduce the zero-order light.

** By zero-order diffraction we mean the light at the original laser frequency diffracted into the image region by the finite cylindrical lens aperture.

The Static-Ripple Imaging System

In the static-ripple imaging system, the threshold contrast determined by both quantum noise in the signal and quantum noise in the zero-order reflected light is

$$C_a = k_o \sqrt{\frac{h\nu}{q\tau_f \gamma \sigma P_o}} \sqrt{\frac{1.95 \times 10^{-12} + 4.77 \times 10^{-3} \frac{\Lambda^4}{m^2 \lambda^2} I_r I_s}{4.77 \times 10^{-3} \frac{\Lambda^4}{m^2 \lambda^2} I_r I_s}} \qquad (5)$$

where I_r is the reference wave intensity in watt/cm^2, γ the duty factor, σ the power reflection coefficient. The angular separation between the reference wave and the central object component is assumed to be 60°.

SUMMARY

The acoustic threshold contrasts concentrates on the ideal performance of systems. The derivation ignores component imperfections and includes only the most fundamental noise which, even in principle, cannot be avoided. The noise that ultimately limits the system sensitivity is characteristic of each individual system scheme. In the analysis[1] of basic scanning modes with real-time potential, the positively-scanning transmitter (PST) arrangement was found to be the best theoretical candidate with distinctive advantages. Although the PST system is not in existence and may be difficult to build, it can be used as a reference for comparison or as a theoretical standard of excellence against which to measure the performance of other systems.

With the assumption of the most important noise source for each system as listed in Table I, Table II gives the threshold contrast expressions. In the first expression for the PST system, K is the Boltzmann constant, T absolute temperature, η transducer conversion efficiency, P the sound power in watts transmitted by the more transparent of the two adjacent objective resolution elements.

TABLE I

Assumed most important fundamental noise in each system.

Hypothetical positively-scanning transmitter (PST) system

..... thermal noise

Bragg-diffraction imaging (BDI) system

..... quantum noise in the zero order flare light

Dynamic-ripple imaging (DRI) system

..... quantum noise in the unperturbed projection of the
scanner spot

Static-ripple imaging (SRI) system

..... quantum noise in the first order and zero order
diffraction

TABLE II

Threshold contrasts for the hypothetical and the three existing real-time systems.

$$C_{aPST} = k_o \sqrt{\frac{2kTm^2}{\eta P \tau_f}}$$

$$C_{aBDI} = 0.91 \times 10^{-7} \, k_o \, \frac{\lambda^{\frac{3}{2}}}{\Lambda} \sqrt{\frac{1}{P_o q \tau_f}} \frac{L\alpha}{P}$$

$$C_{aDRI} = 0.58 \times 10^5 \, k_o \, \frac{\lambda}{\pi \Lambda} \sqrt{\frac{h\nu m^2}{q P_o I_s \tau_f}}$$

$$C_{aSRI} = k_o \sqrt{\frac{h\nu}{q \tau_f \gamma \sigma P_o}} \sqrt{\frac{1.95\times10^{-12} + 4.77\times10^{-3} \, \dfrac{\Lambda^4}{m^2 \lambda^2} I_r I_s}{4.77\times10^{-3} \, \dfrac{\Lambda^4}{m^2 \lambda^2} I_r I_s}}$$

The threshold contrasts vary in different ways with various parameters such as acoustic frequency, acoustic intensity, resolution cell area and laser beam power. A measure of relative sensitivity of the systems can be obtained by forming ratios of their threshold contrasts. The minimum detectable insonification can be obtained by setting the threshold contrasts to unity.

With an important application---biomedical diagnosis in mind, a set of compatible operating conditions is chosen: room temperature, $F = 3MHz$, diffraction limited resolution of $5/2 \ \Lambda^2$, $P_o = 2$ watts, $\lambda = 0.488\mu m$, $\tau_f = 1/30$ sec, $\eta = 0.8$, $q = 0.1$ for BDI and SRI, $q = 0.5$ for DRI, $\alpha = 0.1$ rad, $L = 10$ cm, $I_r = 1$ watt/cm^2, $\gamma = 0.01$, $\sigma = 0.15$, $m^2 = 1.6 \times 10^4$, $k_o = 5$. This choice warrants the applicability of Tables I and II.

The threshold acoustic contrasts are compared with C_{aPST} as reference. At the low end of detectable transmitted intensity,

$$C_{aBDI} = 18 \, C_{aPST}$$

$$C_{aDRI} = 9 \, C_{aPST}$$

$$C_{aSRI} = 11 \, C_{aPST}$$

The minimum insonification needed to produce an effective image of two elements having the best possible condition of contrast is given below for each of the four systems.

$$I_{min \atop PST} = 1.9 \times 10^{-11} \, watt/cm^2$$

$$I_{min \atop BDI} = 6.1 \times 10^{-9} \, watt/cm^2$$

$$I_{min \atop DRI} = 1.6 \times 10^{-9} \, watt/cm^2$$

$$I_{min \atop SRI} = 2.6 \times 10^{-9} \, watt/cm^2$$

Under the conditions assumed and with the perfect components hypothesized, the inherent capabilities of the three existing real-time systems are of the same order. The actual capabilities of the three existing systems, however, are far from being as good as the above ideal capabilities.

It should be emphasized again that the calculations leave out many important practical effects such as fluid streaming, turbulence, Tyndall scattering, transducer wavefront aberration, saturation and noise generated by the actual components. There are many practical problems as well as attractive aspects of great importance associated with the operation of the existing systems. For a situation of this kind, where the theoretical capabilities show little difference, the decision of employing one particular system in preference to another should be based on practical considerations and its deviation from ideal performance.

The theoretical capability of the hypothetical system is considerably better than the existing systems, as we would expect from the simplicity of operation. The major difficulty to be overcome in the implementation of such a system is in electronically steering a focused acoustic beam. A possibility[5] is to use the technique of holographic optical memory (with Fourier-transform hologram of the acoustic zone plate pattern) in conjunction with a photo-conductive piezoelectric layer.

REFERENCES

1. K. Wang and G. Wade, "Threshold Contrast for Various Acoustic Imaging Systems," Acoustical Holography, Vol. 4, Plenum Press, New York, pp. 431-462 (1972). For another interesting analysis, see D. Vilkomerson, "Analysis of Various Ultrasonic Holographic Imaging Methods for Medical Diagnosis," ibid. pp. 401-429.

2. B. B. Brendon, "Real Time Acoustical Imaging by Means of Liquid Surface Holography," ibid. pp. 1-9.

3. R. L. Whitman, M. Ahmed, and A. Korpel, "A Progress Report on the Laser Scanned Acoustic Camera," ibid. pp. 11-32.

4. J. Landry, H. Keyani, and G. Wade, "Bragg-Diffraction Imaging: A Potential Technique for Medical Diagnosis and Material Inspection," ibid. pp. 127-146.

5. K. Wang. Threshold Contrast for Various Acoustic Imaging Systems, an unpublished Ph.D. dissertation, University of California, Santa Barbara, 1972.

A NEW ULTRASOUND IMAGING TECHNIQUE EMPLOYING TWO-DIMENSIONAL
ELECTRONIC BEAM STEERING

F. L. Thurstone and O. T. von Ramm

Department of Biomedical Engineering
Duke University
Durham, North Carolina 27706

In recent years, ultrasound imaging based on B-mode
echosonography has become an accepted and useful technique
in medical diagnosis. However, the application of this technique has been limited by the time required to obtain an
adequate image, the resolution that can be obtained in the
image and problems related to the dynamic range of the echo
information. A new ultrasound imaging system which removes
or substantially reduces many of these limitations has been
developed in the hope that ultrasound tomography may find
even more widespread application and greater diagnostic value.

This imaging system is based on phased array principles
rather than the usual mechanical positioning to scan the
ultrasound beam through the object volume and in this respect
shares some common features with the system developed by J.C.
Somer (1, 2). A linear array of ultrasound transducers is
electronically phased during transmission as well as during
the reception of echo information to produce real time high
resolution B-mode tomographic images in a circular sector
format at high image frame rates. A dedicated digital computer acting as a programmed controller controls the phasing
of the array during the entire scan and provides flexible
interactive control of the scan format and imaging parameters.

In the attempt to eliminate or reduce the limitations of
scanners currently in use, the design and realization of this
new imaging system was guided by several objectives. Foremost
among these was the development of a system utilizing

sufficiently high data acquisition rates so that dynamic
structures could be imaged in real time. Improved image
resolution and the ability to present a large range of echo
amplitudes in the usable brightness range of the display
constituted additional major objectives as did the design
of a clinically useful instrument. In this latter context,
a large field of view, flexibility in control of the scan
format and ease in clinical application are desirable.

A prototype of this system has been designed to operate
on-line with a minimum image frame rate of twenty frames per
second with each frame consisting of 256 individual B-scan
lines in a sector format. The lateral or azimuthal image resolution varies from less than 2 mm. at short ranges to a
little less than 4 mm. at a target range of 15 cm. Logarithmic compression of the echo information makes it possible to
present 60 db of echo information in approximately the 10 to
1 brightness range of a typical video system. In this way,
an extended range of echo amplitudes can be presented in a
gray scale image. In the present configuration, a sector
scan angle of 60 degrees is available resulting in a field of
view of 15 x 15 cm. Since the scan format is under computer
control, it can readily be modified for scan angle, number
of lines per frame, maximum range, etc. Facility in clinical
use is provided by a hand-held transducer in contact with
the skin surface so that the anatomical plane to be imaged
can be selected by manual orientation of the array.

The transducer presently employed is a 16 element linear
array 25 mm. long. Each element is 14 mm. high and approximately 0.7 mm. in width. The fundamental resonance of the
heavily damped array is 2.25 MHz.

In describing the operation of this imaging system, it
is advantageous to consider the transmitting operation and
the receiving operation separately since these two are fund-
amentally different.

The high data rates required to image dynamic structures
can be achieved by phasing the individual elements of the
array in a manner such that the effective azimuthal orientation of the insonification pulse can be altered from one
transmit operation to the next. Suppose that the elements
of the array described above are excited with electrical
pulses in a linear time sequence. Then, according to Huy-
gen's Principle, the resulting acoustic wave will propagate

at an azimuth angle from the direction normal to the array. By varying the linear time relation of the electrical pulses, it is possible to direct the acoustic beam at any angle. In this way, then, the basic scan format is produced. Each subsequent ultrasound pulse is propagated at increasing azimuth angles until a full circular sector scan image is generated. For a 256 line image of 15 cm. maximum range 50 ms are required; thus, 20 image frames are generated per second.

This principle of imposing a time sequence on the transmit pulses of the array can also be used to produce a focussed wavefront. A timing sequence which results in a circular wavefront produces a cylindrical Gaussian focus at a specific range. This focussing action can then be superimposed on the linear timing to produce a wave focussed at any azimuth angle. Although this produces an excellent focus, theoretical calculations and Schlieren studies (3) show that the insonification along the axis of the transducer is quite variable outside the so-called focal depth. Since each transmit pulse is utilized in writing one entire image line, nulls in the axial beam pattern would seriously degrade the resulting image. This problem may be alleviated by phasing the elements to produce an aspheric wavefront, a wavefront characterized by a continuum of foci over an extended distance along the axis rather than one focus at a particular range. Schlieren studies indicate that such an aspheric phasing in conjunction with some degree of apodization does produce a more uniform acoustical intensity distribution along the axis.

The actual beam steering employed by the system is a combination of the above concepts and is diagramatically illustrated in Figure 1. In this figure, excitation pulses are indicated as arriving at the transducer elements with an aspheric time relationship. The resulting acoustic wave propagates at a specific azimuth angle and produces an extended axial maximum or focal line along this direction. With the array dimensions presently employed, the azimuth angle may be directed 30 degrees on each side of the normal resulting in a 60 degree field of view.

Processing of the echo information during the receive operations shares some similarities with the transmit operations yet differs significantly in several respects. Echoes

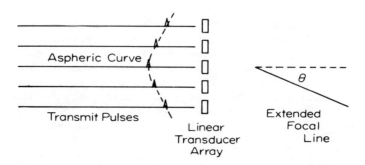

Figure 1. Transmit Beam Steering

returning from targets in the direction of the transmitted
pulse arrive at the transducer elements at different times
necessitating a phasing of the received signal so that the
effective orientation of the array during reception corresponds to the orientation during transmission. In this system accurate phasing is accomplished via the use of switchable wide band lumped-constant analog delay lines which
pass the acoustic information with different delay times
for each element. In this way, echo information from each
element of the array will arrive at a summing amplifier in
phase. After summation, the signal can be processed for
B-mode display.

The incorporation of switchable delay lines in the
electrical configuration of the receiver permits the receiver
to be focussed at a given range and allows tracking of this
focus in synchrony with the range of returning echoes. This
technique is a significant departure from previously
reported systems and is primarily responsible for the
improved resolution of this system.

Figure 2 illustrates the phase processing during reception. Echo information from targets on the right propagate
to the left, back to the transducer. Resulting electrical signals prior to summation are delayed by an amount of
time determined by the delay controller which can rapidly
alter the time delay in each transducer channel. Since
echo information results from a sounding pulse, the time of
the returning echo is directly related to the target range.
It is thus possible to focus the receive system at short
range immediately after transmitting and then to increase

the focal distance in synchrony with the increasing range
of target echoes. With such a moving focus, all targets
are in focus, and there is no depth of focus limitation on
resolution as there is in transmit or with a single focus
receiver. In the prototype system, the receiver uses ten
different foci or focal zones resulting in an almost perfect optical figure over the display range.

This moving focus receive phasing is superimposed on a
linear phase shift to produce steering in azimuth, and, thus,
a two-dimensional steering in both range and azimuth angle
results. The receive beam steering exactly follows the
transmitted direction in order to produce a synchronized
two-dimensional scan.

With the above principles in mind, attention is now
turned to a description of the operation of the entire
sector scan imaging system. For clarity, a block diagram
of only a single transducer channel is illustrated in Figure
3. The single basic component in this system is the programmed digital controller which, in this case, is a dedicated PDP-11 computer. This controller, together with an external controllable timing unit, controls and synchronizes
the entire imaging system.

Perhaps the system operation can best be understood
by following the sequence involved in the generation of a
single line of image information. The controller first
provides deflection information to the oscilloscope display
in the form of the sine and cosine of the azimuth angle to
be investigated. The controller then preloads the transmitter timing circuits to produce an extended focus along

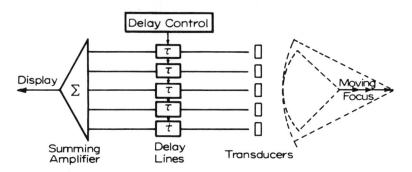

Figure 2. Receive Processing

a particular azimuth direction. The transmit and deflection
operations are then initiated. Immediately after the transmit operation, the controller sets the receiver delays
for a focus close to the transducer. As echoes are received
from more distant targets, the controller changes the
receiver delays to maintain the focus of the receive system.

Within the core memory of the digital controller is
stored all of the necessary direction coding, timing, and
receiver focussing information to produce a two-dimensional
B-scan tomogram in the plane of view of the transducer.
By reducing redundancy in this information it has been
possible to store this information in approximately 3,000
computer words of 16 bits each.

Simple software modifications available by input
switches or teletype, can control the scan format in such
parameters as angular field of view, number of lines per
frame, maximum range. etc.

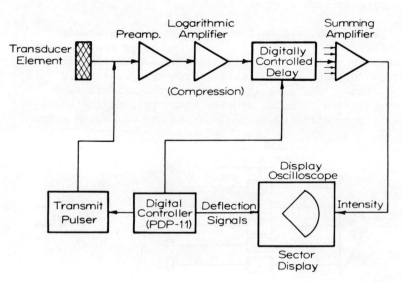

Figure 3.   Block Diagram of One Channel of the Sector Scan
            Imaging System

This system diagram also illustrates the application
of nonlinear processing of the received signals in a manner
described earlier (2.4,5). Echo information from the individual elements is first amplified in a low noise preamplifier and then amplified with logarithmic compression before
being delayed. This compression serves several purposes.
First, the large dynamic range of the echo information available in such a simple scan is compressed to more closely
match the range available in the display device. This
reduces the effects of specular reflection and produces a
more useable gray scale image presentation. Second, by
logarithmically compressing the signals before summation,
the output of the summing amplifier more closely approximates the product of the individual received signals than
their linear sum. This multiplicative processing produces
an improvement in image resolution, a reduction in sidelobe
response, and an improved depth of field. Third, compression prior to the digitally controlled delay lines improves
the signal-to-noise ratio because switching transients in
the delays are of less consequence.

Each channel of the receiver delay system is composed
of six binary coded analog delay elements. Thus, a six
bit control word from the computer can produce any incremental delay from 0 to 63 times the basic delay unit which
is 125 ns in this system.

In practical use, the diagnostician can interact with
the imaging system in three different ways. First of all,
he may interact with the digital controller to modify the
scanning format in terms of field of view, speed, etc. As
well, he may interact with the display oscilloscope to control image brightness, contrast. scale dimension, range
marks, signal inversion, etc., and most importantly, he
selects the two-dimensional anatomical plane that is being
imaged by manipulating the position of the transducer array.
Recognizing that the image presented represents a fan-shaped
azimuthal sector scan emanating from the array, it is clear
that this plane can be raised or lowered in elevation angle
simply by rotating the transducer about its long dimension. Rotating the transducer about the normal to its surface rotates the image plane about that axis. The sector
scan can be moved in azimuth by either tilting the transducer about its short dimension to the extent that contact
can be maintained or by setting the controller to provide
the desired off-axis scan.

During operation, the following scan format parameters are under immediate control. The angular limits of the scan can be set and need not be symmetric about the transducer axis; that is, the image plane can be offset in the azimuth dimension. Next, the line density may be reduced which results in a proportionately higher image frame rate. As well, the maximum range of target information can be preset which determines the transmitter repetition frequency, and, in turn, the image frame rate. Image frame rate is also directly related to the total angle of the sector scan, in particular, the greater the angle the lower the frame rate.

Primary image display parameters such as image contrast, brightness, background rejection, range compensation and display size are directly accessible to the diagnostician. In addition, range markers, image inversion, and nonlinear brightness controls are also available. In the final image, a brightness range in excess of a 10:1 ratio is available. and approximately 60 db of echo information is presented as a gray scale image in this brightness range. A second oscilloscope monitor is viewed by a vidicon TV camera, and the images recorded on video tape for later or repetitive study.

In order to evaluate the resolution capability of this system, the response to a single target consisting of a nylon monofilament thread supported with its axis along the line of elevation or normal to the tomographic image plane was measured by mechanically transporting the array in the azimuth dimension. At a target range of 70 mm. the distance between the first off axis nulls was approximately 4 mm. The half-power beamwidth is slightly less than 2 mm. and multiple targets spaced 2 mm. have been resolved at this range. It is probably incorrect to relate this response to a beamwidth since no such beam actually exists. This response is the result of an aspheric apodized transmitted beam, a Gaussian receiver focus followed by nonlinear multiplicative processing.

Other response curves taken at different target ranges are similar and show an azimuthal resolution capability which is essentially linear with increasing range. Thus at 110 mm., the resolution limit drops off to almost 3 mm., and at a range of 150 mm., it is almost 4 mm.

These resolution limits are slightly better than the so-called "Rayleigh Limit" throughout the field of view. This results from the nonlinear multiplicative processing in the receive mode together with an essentially perfect optical figure on the receive focus at all target ranges.

Similar measurements were conducted by transporting the array in the elevation dimension while the monofilament target was positioned parallel to the tomographic image plane. At 70 mm., the resolution limit was found to be 5 mm. increasing to 8 mm. at a target range of 110 mm. It is apparent that the resolution in elevation is significantly reduced from that achieved in azimuth, in part a result of the smaller aperture size in elevation, and in part a result of the lack of phase processing in this dimension.

A single image frame is shown in Figure 4. In this model system the targets were six nylon monofilaments supported normal to the image plane. The targets were on a rectangular grid with spacings of six and twelve millimeters. The mean target range was 12 cm.

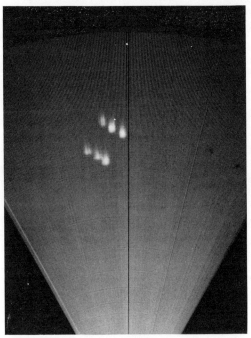

Figure 4.   Sector Scan of Model Targets

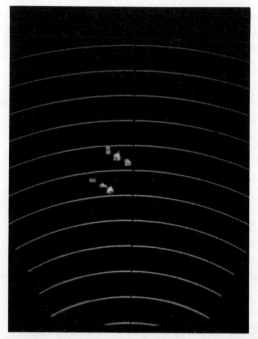

Figure 5.   Sector Scan with Range Markers

Figure 5 shows a similar scan of the same targets shown in the previous figure but in this case, the background level was reduced, and range markers have been inserted at one cm. intervals. Of course, such static images as these cannot convey the real time imaging capabilities of this system.

In summary, the development of a new imaging system which was designed for very high data acquisition rates has been described. This system was specifically designed for diagnostic use in evaluating the dynamic behaviour of the heart. Many design considerations have been predicated on this application and might need to be modified for other use. However, even in its present form, this instrument may have considerable usefulness in many areas of ultrasound diagnosis.

It is hoped that this type of dynamic two-dimensional imagery will provide a great deal more information than the one-dimensional time record presently used in echocardiography and that these techniques will find widespread application in clinical medicine.

This work was supported in part by USPHS grants HL 41,131; HL 12715; and HL 14228.

REFERENCES

1. Somer, J.C.: Electronic Sector Scanning with Ultrasonic Beams, Proceedings of the First World Congress on Ultrasound Diagnostics in Medicine, 1969, June 2-7, Vienna, Austria.

2. Somer, J.C.: New Processing Techniques for Instantaneous Cross-Sectional Echo-Pictures and for Improving Angular Resolution by Smaller Beams; Proceedings of the Fourth Congress of the International Society for Ultrasonic Diagnosis in Ophthalmology, 1971, May 6-9, Paris, France.

3. Miller, E.B., Smith, S.W., and Thurstone. F.L.: A Study of Near Field Ultrasonic Beam Patterns from a Pulsed Linear Array, Acoustical Holography and Imaging, Volume V, Plenum Press, 1973 (in press).

4. Lobdell, D.D.: A Nonlinearly Processed Array for Enhanced Azimuthal Resolution; IEEE Transactions on Sonics and Ultrasonics, SM-15, no. 4, p. 202, 1968.

5. Thurstone. F.L. and Melton, H.E.: Biomedical Ultrasonics, IEEE Transactions on Industrial Electronics and Control Instrumentation, IECI-17, No. 2, p. 167, 1970.

A STUDY OF NEAR FIELD ULTRASONIC BEAM
PATTERNS FROM A PULSED LINEAR ARRAY

E.B. Miller, S.W. Smith, F.L. Thurstone

Department of Biomedical Engineering
Duke University
Durham, North Carolina 27706

INTRODUCTION

A major advantage of pulse-echo imaging in the near-field of an acoustic array is the improved lateral resolution that focusing makes possible. Usually, equivalence is tacitly assumed between a lateral beam dimension for continuous wave excitation and a lateral pulse dimension for the same geometry. Although distinctions certainly exist between the lateral dimensions in these two cases, it is convenient in dealing with pulsed ultrasound to speak of a "beam" or "beams" of ultrasonic energy. By analogy to the continuous wave case, such a beam may be defined as the envelope of a pulse dimension. This dimension may be specified by some absolute or r.m.s. pressure magnitude criterion (e.g., the distance to a point at which the absolute pressure is down "n" dB from the maximum value in the field).

Transducers with sampled apertures (arrays) or unsampled apertures have certain common main-beam characteristics when unfocused. These characteristics are observed with both continuous wave and pulsed excitation. For example, without focusing lateral resolution is governed in the near-field by the lateral beamwidth which is controlled, in turn, by the size of an array aperture. A reduction in aperture size improves lateral resolution in the near-field but moves the near-to far-field transition closer to the array so that far-field beamwidth is increased. Thus, without focusing, aperture specification involves a compromise between decreased lateral resolution for small apertures in the far-field and decreased resolution for large apertures in the near-field. Obviously, unfocused beam geometry may differ drastically between large and small apertures so that acceptable lateral resolution over any extended range may not be possible. This is a fundamental problem of unfocused transducer design that can be resolved by focusing with a sufficiently large aperture.

The word "focus" may cause one to think only of a focal point. However, a point focus is but one of a continuum of possible focal locii that may be obtained by manipulating, for example, the phasing of the array elements.

In fact, a point focus imposes a serious limitation upon the interrogation of the near-field of an acoustic array by the pulse-echo method. This limitation may result in a data acquisition rate that is lower than necessary and stems from the fact that improved lateral resolution near the focal point is obtained at the expense of reduced lateral resolution at distances very much different than the focal distance. Thus, a separate firing of all the array elements may be required for each point interrogation.

A higher data acquisition rate is possible if the transmitted pulse is focused in a convergent manner that approximately preserves the lateral pulse dimension over the entire range of interest. This may require some increase in beamwidth over that for a point focus but involves little sacrifice of range resolution. Thus, a single firing of all the array elements can permit unambiguous interrogation of not just one point, but of all the points within the range of interest and along the path of maximum pulse intensity.

The higher order diffraction effects of arrays may differ considerably from those of transducers with continuous apertures. For arrays, these effects may be an important design consideration since they can give intensity maxima that occur at relatively large angles to the main beam. This is true for both pulsed and continuous wave excitation, whether focused or unfocused. Over much of the near field, the higher diffraction orders for an array of equally spaced slits are found at angles to the main beam which are approximately given by

$$\theta \sim \arcsin n\lambda/b \qquad n = 0,1,2 \qquad\qquad (1)$$

where λ is a nominal acoustic wavelength and b is the spacing
between two adjacent elements. Generally, one need only worry
about the first order, because the amplitude of this effect
attenuates rapidly with the order, n.

Obviously, from equation (1), the tendency for intensity
maxima to occur at discrete angles to the main beam may be
reduced by appropriate non-uniform element spacing. Frequency variation over a sampled aperture or a suitable elemental
pulse spectrum may also be used to decrease the angular density
of the diffraction orders.

Parameters that may affect beam geometry of a pulsed
array include: elemental pulse length, element firing times,
element weights, elemental pulse frequencies and element locations.

One of the goals of this paper is to demonstrate methods
that have been developed for studying the effects of some of
these parameters upon beam geometry. In so doing, it will be
possible to show, in a qualitative way, some of the effects
themselves. Another goal is to demonstrate that simple methods
exist, for analyzing and visualizing the field of a pulsed
acoustic array which are not based on the questionable assumption of continuous wave excitation.

MODELING PARAMETERS

To demonstrate the methods that have been developed, a
linear array is modeled that was designed principally for in-
vivo, real time, cardiac tomography. Although this array is
a part of a system[1] which is designed to beam-steer to extreme
angles of $\pm 30^{\circ}$ from the array symmetry axis, an on-axis focus
is considered. This halves the number of computations required
to specify the pressure field without significantly affecting
the focus at the steering extremes.

Pulse lengths, envelopes and frequencies are presumed to
be the same for all array elements. The effects upon the
resultant time-dependent pressure field of element firing times,
element weights and elemental pulse length, are used to demonstrate field mapping methods which are useful in beam optimization.

Because of the cardiac application of this array, the range of interest is 3 centimeters to 20 centimeters from the plane of the array. This is well within the far-field of the individual array elements which are 0.7 millimeter wide and 14 millimeters long.

The array may thus be modeled by sixteen infinite line sources, all but the central two of which are spaced 1.5 millimeters apart. The total aperture of this model is 24 millimeters. The sources are parallel to each other and perpendicular to the plane in which the pressure is studied.

The two dimensional spatial character of this model is not essential to the results shown here. A distribution of point sources may be used, rather than line sources, to model the three dimensional spatial character of the actual array more precisely. Such sophistication would involve more computation time but would introduce no new conceptual problems.

The elemental pulses are modeled by sine functions which start at zero and terminate at zero after "m" half-wavelengths Thus, m will be used as an index of pulse length. The assumed wavelength is 0.684 millimeter. This may be obtained from a nominal propagation speed[2,3] in soft tissue of 1540 meters per second and the frequency at which the array elements are pulsed, which is 2.25 megahertz.

THEORY

At typical diagnostic amplitudes, ultrasonic phenomena may be described by use of the wave equation,

$$\nabla^2 p - \frac{\partial^2 p}{\partial \tau^2} = 0 \qquad\qquad (2$$

where τ is the propagation speed multiplied by time. Linearization of the basic fluid equations produces the perturbation pressure, p. The dissipative effects of viscosity, thermal conductivity and relaxation are usually taken to be negligible Dispersion of propagation speeds[3] does not appear to be an imp ortant effect for typical pulse spectra in soft tissue. Thermal conduction and viscosity need only be considered in the vicinity of relatively solid boundaries[4] and thus are boundary layer phenomena.

The use of equation (2) in a boundary value problem en-
compassing all sixteen array elements is not possible since
such a problem[5] is not properly posed. However, because the
range of interest lies completely within the far-field of the
individual elements, an array of line or point sources may be
used to model the plane of the transducer elements. Boundary
effects are then approximated, because of the symmetry of the
composite pressure field.

This approach requires the solution of equation (2) for
point or line sources. A simple dimensional argument is sufficient to show that when r/λ is much greater than unity,
where r is the distance from source to field point and λ is
a typical wavelength, the boundary value problems for infinite
plane, infinite line and point sources are formally similar.
Equation (2) becomes

$$\frac{\partial^2 \phi}{\partial \tau^2} - \frac{\partial^2 \phi}{\partial r^2} = 0 \qquad (3)$$

The product $pr^{q/2}$ defines ϕ, where q is zero for an infinite planar source, unity for an infinite line source and
two for a point source. As initial conditions, $\phi(r,0)$ and
$\partial \phi/\partial \tau$ are set equal to zero for r greater than zero. For τ
greater than or equal to zero, the boundary conditions that
$\phi(\infty,\tau)$ equals zero and that $\phi(0,\tau)$ equals the forcing function
$F(\tau)$, (which may be either transient or continuous wave in
nature) apply. Hence, the solution is

$$p = \frac{F(\tau-r) \ U(\tau-r)}{r^{q/2}} \qquad (4)$$

where U is the unit step function and q will subsequently take
on the value of unity.

From equation (4) and the assumed elemental pulse, the
pressure at a field point is

$$p(r,\tau) = \sum_{k=1}^{16} \frac{A_k}{r_k^{1/2}} \ \sin \frac{2\pi}{\lambda}(\tau_k - r_k) \qquad (5)$$

where the kth term only contributes if the inequalities

$$m\lambda/2 \; > \; \tau_k - r_k > 0 \tag{6}$$

are satisfied. The parameter, m, gives elemental pulse length
in half wavelengths while

$$r^2 = x^2 + y^2 \tag{7}$$

and

$$r_k{}^2 = (x - x_k)^2 + y^2 \tag{8}$$

The lateral field-coordinate, x, is measured from the
symmetry axis of the field, while the range field-coordinate,
y, is zero at the array and increases with range. A_k is an
amplitude factor corresponding to the line source located at
x_k.

Figure 1 shows the result of applying equation (5) to the
construction of one sort of pressure field map. The field
map is a collage of individual pulse maps for uniformly incremented propagation times (ranges). These computer generated plots correlate well with schlieren photographs (Figure
2) obtained from the actual array.

Characters were overprinted on a line printer to obtain
the seven level gray-scale shown in Figure 3. The gray levels
appearing in Figure 1 are related in a logarithmic manner to
the values of absolute pressure.

Stroboscopic illumination was synchronized with array
firing to obtain the schlieren photographs. Uniformly incremented delays were used between strobe and array firing
to produce the individual pulse photographs. These are assembled in the two collages of Figure 2. Each pulse photograph is an integration over many array firings.

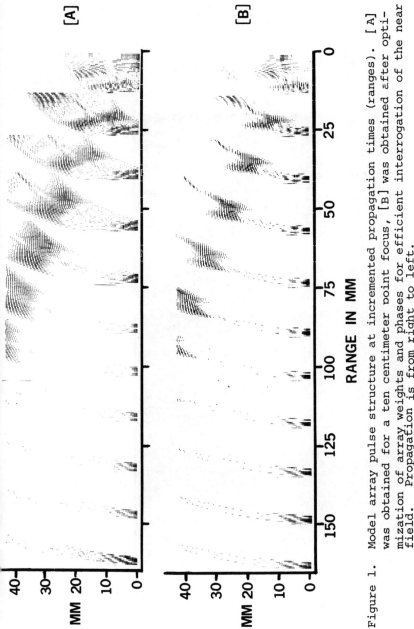

Figure 1.   Model array pulse structure at incremented propagation times (ranges). [A] was obtained for a ten centimeter point focus, [B] was obtained after optimization of array weights and phases for efficient interrogation of the near field. Propagation is from right to left.

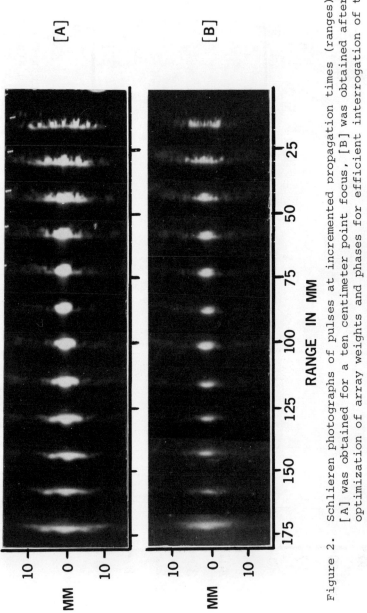

Figure 2. Schlieren photographs of pulses at incremented propagation times (ranges). [A] was obtained for a ten centimeter point focus, [B] was obtained after optimization of array weights and phases for efficient interrogation of the near field. Propagation is from right to left.

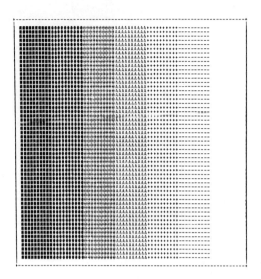

Figure 3.   Seven Level gray scale.

Figure 4 is intended to clarify the sign conventions
and parametric inter-relationships that are used in developing a Fresnel-like approximation to equation (5).   This
figure shows that if y_{l_k} and y_{t_k} are the y coordinates locating the leading and trailing edges, respectively, of the
pulse from the kth line source, then for a given x and τ

$$y_{l_k}^2 = (\tau - d_k)^2 - (x - x_k)^2 \qquad (9)$$

and

$$y_{l_k}^2 = (\tau - d_k - \frac{m\lambda}{2})^2 - (x - x_k)^2 \qquad (10)$$

The parameter d_k is the delay, expressed in units of distance,
between the firing of the initial elements and the firing of
the kth elements.

Expansion of equations (9), (10) and the definition of
r_k in power series having terms that are small in the region
of interest shows that

$$y_{1_k} \sim \tau - d_k - \frac{(x-x_k)^2}{2\tau} \tag{11}$$

$$y_{t_k} \sim y_{1_k} - \frac{m\lambda}{2} \tag{12}$$

and

$$r_k \sim y \tag{13}$$

Trignometric expansion of equation (5) after application of these approximations gives

$$p(w,z,\tau) = \frac{1}{\tau^{1/2}} \left(1 + \frac{\lambda w}{4\pi\tau}\right) \left[\sin w \sum_{k=1}^{16} A_k \cos z - \cos w \sum_{k=1}^{16} A_k \sin z\right] \tag{14}$$

where

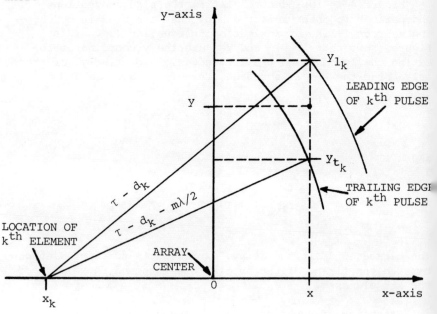

Figure 4.   Some Parameters and Sign Conventions.

$$w = \frac{2\pi}{\lambda} (\tau - y) \tag{15}$$

and

$$z = \frac{2\pi}{\lambda} \left[d_k + \frac{(x-x_k)^2}{2\tau} \right] \tag{16}$$

Terms only contribute to the sums in equation (14) if k satisfies the inequalities

$$\frac{m\lambda}{2} > y_{1_k} - y > 0 \tag{17}$$

DISCUSSION

The method of characterizing the pressure field that is shown in Figure 1 has the advantage of relatively direct comparison with schlieren photographs. However, it is difficult to interpret and requires a prohibitive amount of computer time for reasonable range resolution of field structure. Because of this, a way was sought to compress field information into a simpler format that would preserve the main points of interest.

Figure 5 will help in understanding a method which is an improvement over that of Figure 1 in the areas discussed above. This figure shows model pulse structure in the vicinity of a nine centimeter focus. That is, phasing is given by

$$d_k = \frac{a^2}{2f} \left[1 - \left(\frac{x_k}{a} \right)^2 \right] \tag{18}$$

where "f" is the focal distance (nine centimeters) and "a" is the half aperture distance. The elemental pulses were two cycles in length.

Figure 5. Pulse Structure at a Nine Centimeter Point
Focus, m=4.

The important thing to notice in Figure 5 is the existence
of significant structure out to a considerable distance from
the symmetry axis. Also, structure near the pulse envelope
is not always representative of what may be found at the same
lateral position inside the pulse. One is therefore faced
with the problem of finding an index that is sensitive to off-
axis structure at a given pulse propagation time, although
this structure may vary slightly with the range coordinate, y.

Since target range resolution requires that pulses have
a small spatial extent in the direction of travel, one might
be tempted to reduce the problem to two dimensions by averaging over the coordinate y for a given x and τ. This is, in
fact, what has been done here with some fairly minor but important modifications.

First, if one simply averages as suggested above, the
larger, more important, absolute values of pressure could be
washed out by a large number of smaller less important values.
One might therefore prefer to have an index off-axis structure
that is just sensitive to values of pressure above a certain
threshold. This can be accomplished by only summing those
values above a given threshold value and then dividing by
their number. Thus, if only one point were above the threshold for all y at fixed values of x and τ, the index would be
equal to the absolute value of pressure at that point.

Further, instead of averaging along the coordinate y, one
can average over the coordinate, w, which shows less variation
with range. Then, from equation (14) an index can be defined
which is only a function of x and τ. Thus, if

$$w_{min} = \frac{2\pi}{\lambda} (\tau - y_{1_k}) \tag{19}$$

and

$$w_{max} = \frac{2\pi}{\lambda} (\tau - y_{t_k}) \tag{20}$$

a sampling interval, c, may be defined by the expression

$$c = \frac{w_{max} - w_{min}}{N} \tag{21}$$

where $N + 1$ is the number of samples over the interval. An index, $P(x,\tau)$, of important lateral pulse structure may now be defined for any given τ by

$$P(x,\tau) = \frac{1}{M} \sum_{n=0}^{N} p(x,w,\tau) \, U(p-p_t) \, \delta(w-nc) \tag{22}$$

where

$$M \leq N + 1$$

is the number of samples above the threshold value, p_t, and δ is the Dirac delta function.

This technique has been used in producing Figures 6 through 8 which show rather clearly the effects of the variation of certain sets of parameters upon the model pressure field of the array. Figure 6 shows the effect of pulse length for a nine centimeter point focus. This effect cannot be

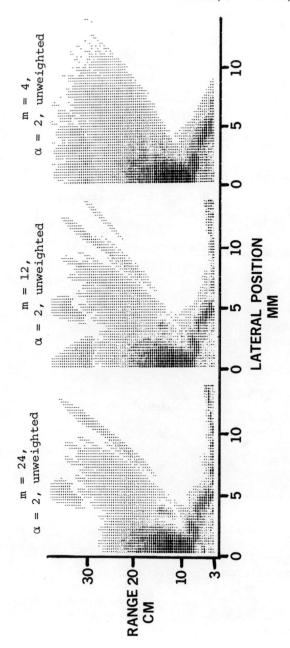

Figure 6.　The effect of pulse length upon the pressure field
structure-index.　The pulse propagation direction
is upward.

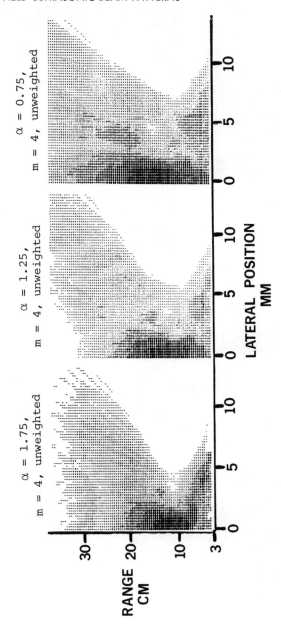

Figure 7. The effect of phasing upon the pressure field structure-index. The pulse propagation direction is upward.

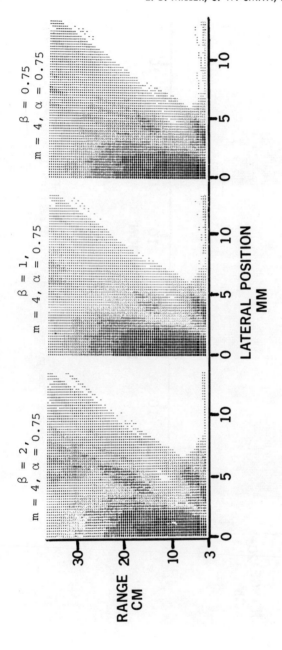

Figure 8. The effect of element weighting upon pressure field structure-index. The pulse propagation direction is upward.

described by theories based on continuous wave excitation of the elements.

Shortening the pulse length has the effect of removing structural detail from the pressure field. Since such field structure can mask or artificially emphasize target structure, shorter pulses are preferable. Thus, retaining the shortest pulse from Figure 6 (two cycles in length) one can consider the effect of variation in firing delays.

Figure 7 shows the effect of varying the power, α, in an expression that is analogous to equation (11) for a point focus,

$$d_k = \frac{a^2}{2f} \left[1 - \left| \frac{x_k}{a} \right|^\alpha \right] \tag{23}$$

From left to right, in this figure, α takes on the values 1.75, 1.25, and 0.75 for a value of f equal to nine centimeters. This sort of phasing is well known in optics. A device[6,7,8] producing such a phase relation is termed an axicon. It has the property of producing an extended focus which approaches a focal line for α equal to unity. An extended focus of this sort is apparent in Figure 7 for $\alpha = 0.75$. As was noted in the introduction, such a focus is a desirable feature since it allows a higher data acquisition rate.

However, side structure is also apparent in Figure 7. This structure becomes more prominent with decreasing α. Such structure can be reduced by appropriate element weighting. For example, in Figure 8 where m = 4, f = 9 centimeters and $\alpha = 0.75$ the function

$$A_k = B \left[1 - \left| \frac{x_k}{a + b} \right|^\beta \right] \tag{24}$$

was used. Here B is given by the condition that

$$\sum_{k=1}^{16} A_k = 100 \tag{25}$$

which also applied to previous unweighted cases shown in Figures 6 and 7. As before, "b" is the spacing between elements which is 1.5 millimeters and "a" is the half aperture distance. The reduction of the side structure is seen in moving from left to right across Figure 8. The three cases correspond from left to right to the respective β-values of 2.0, 1.0 and 0.75.

Another aspect of this optimization procedure is seen in Figure 9 where values of $P(x,\tau)/P_{max}$ along the symmetry axis are plotted for three cases. From top to bottom these correspond to the right hand gray-scale plots in Figures 6, 7 and 8, respectively. It is obvious from Figure 9 that in moving through the optimization procedure outlined by these figures, the axial distribution of $P(x,\tau)/P_{max}$ becomes more of a monotonically decreasing function over the range of interest. This is a desirable trend if one wishes to amplify returning echoes in order to correct for the increased attenuation that accompanies increased range.

There is also possible potential in the fact that once the main beam is optimized for a given value of f, adjustments in the delays, corresponding to increases or decreases in f about the optimization value, have the effect of shifting the region of uniform beam dimension either down-range or up-range, respectively, with little attendant change in beam character.

Figure 1 compares a beam optimized for f = 9 centimeters, but with f set equal to 10 centimeters, with a 10 centimeter point focus. The optimized beam is uniform over a range that is at least twice that for the point focus. This is borne out by the schlieren photographs of Figure 2. The difference in the two beam geometries would be even more dramatic if the point focus were chosen closer to the array.

The lateral beam dimension has been compressed in Figure 1 in order to show first order diffraction in the two cases. This appears in the form of a more diffuse beam that makes an angle of about 27° with the symmetry axis. First order

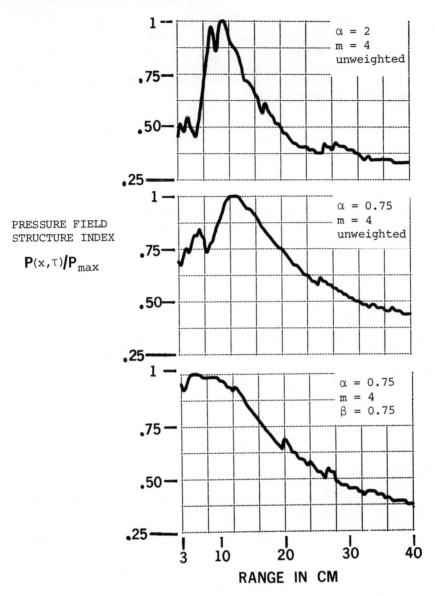

PRESSURE FIELD
STRUCTURE INDEX

$P(x,\tau)/P_{max}$

Figure 9.   The variation of pressure field structure-
index along the symmetry axis.

diffraction effects have not been considered here because of
the fixed element spacing and fixed pulse character of the
actual array. As discussed in the introduction, appropriate
variation of element spacing, element frequencies or elemental
pulse spectra would be most effective in optimizing for first
order diffraction effects.

SUMMARY

Two methods of modeling and visualizing the pressure
fields of pulsed planar arrays have been developed. One of
these methods has been applied to interactive optimization of
the zeroth order diffraction pattern for efficient cardiac
imaging in the near-field of an array.

The trends observed as a result of progressive changes
in element phasing and weighting are shown to be consistent
with the trends that one would expect from optics and classical work on the far-field of continuous-wave arrays (i.e.,
appropriate weighting will reduce side structure on the
zeroth order beam and appropriate phasing will give an extended depth of focus). In addition, the effect of pulse
length upon field structure is demonstrated. This effect
cannot be predicted from classical continuous wave approaches
to the optimization problem.

Beyond variations of weighting, phasing and pulse length,
the methods developed here are potentially useful for studying
the effects of variations in elemental spectra, and element
spacing.

Comparison is also shown between schlieren pulse photographs from an actual array and gray-scale pulse maps obtained
using optimized parameters. Both a ten centimeter point focus,
where ten centimeters corresponds to the approximate center
of the near field for the given aperture, and an optimized
case are presented. It is apparent from comparison of the
two beams in both model (Figure 1) and schlieren (Figure 2)
figures that an approximately constant lateral beam dimension
is found over at least twice the range in the optimized case
as in the case of the point focus.

ACKNOWLEDGEMENTS

This work has been supported in port by UPHS grants
HL-12715 HL-41, 131 and by the Bureau of Radiological Health.

REFERENCES

1. F.L. Thurstone, O.T. von Ramm, "A New Ultrasound Imaging
    Technique Employing Two-Dimensional Electronic Beam
    Steering," Acoustical Holography and Imaging, Vol. 5
    (1973) (in press)

2. G.D. Ludwig, "The Velocity of Sound through Tissues and
    the Acoustic Impedance of Tissues," Journal of the
    Acoustical Society of America, 22(6): 862-866 (1950)

3. D.E. Goldman and T.F. Hueter, "Tabular Data of the Velo-
    city and Absorption of High Frequency Sound in Mammalian
    Tissues," Journal of the Acoustical Society of America,
    28(1): 35-37 (1956)

4. P.M. Morse and K.U. Ingard, Theoretical Acoustics, pp.
    285-291, McGraw-Hill Book Company (1968)

5. F.G. Friedlander, "On an Improperly Posed Characteristic
    Initial Value Problem," Journal of Mathematics and Mechan-
    ics, 16(8): 907-915 (1967)

6. J.H. McLeod, "The Axicon: A New Type of Optical Element,"
    Journal of the Optical Society of America, 44(8): 592-
    597 (1954)

7. S. Fujwara, "Optical Properties of Conic Surfaces. I.
    Reflecting Cone," Journal of the Optical Society of
    America, 52(3): 287-292 (1962)

8. J.W.Y. Lit and R. Tremblay, "Focal Depth of a Transmitting
    Axicon," Journal of the Optical Society of America, 63(4):
    445-449 (1973)

ACOUSTIC IMAGING WITH THIN ANNULAR APERTURES

David Vilkomerson

RCA Laboratories

Princeton, New Jersey   08540

I.   INTRODUCTION

It was shown in a previous analysis that a mosaic of
piezoelectric acoustic detectors would be sensitive enough
to image tissue structures deep within the body and would be
able to resolve details as small as one millimeter[1]. We will
discuss here a method of achieving the effect of a full
piezoelectric aperture without having to build one. The
basic idea is that if we measure the amplitude and phase of
a wavefront on the annulus bounding an aperture, it is possible to use these measurements to produce an image of the
object producing that wavefront. Moreover, the resolution
of the image produced by an annulus around an aperture will
equal the resolution produced by the full aperture.

The theory of imaging with thin annular apertures will
be discussed, the experimental implementation described, and
the resulting images presented. We will conclude with a
discussion of the practical role of annular-aperture imaging.

The paper describes annular-aperture imaging in terms
of acoustic imaging; however, the same techniques can be applied to imaging with any waves whose amplitude and phase
can be measured, e.g. microwaves, seismic waves, and optical
waves.

II. THEORY OF IMAGING WITH THIN ANNULAR APERTURES

The sum of the complex signals around a thin annulus, produced by a point source, can be shown to be proportional to a Bessel function of first kind and lowest order (J_0) whose argument is proportional to the angular distance from the source to the axis of the annulus. This response shape is not usable for imaging. A new one, suitable for imaging, can be synthesized from higher-order Bessel functions (J_1, J_2, J_3, etc.) obtained by shifting the phase of the measured signal; adding these Bessel functions, squared and properly weighted, to the square of the unaltered response, J_0^2, produces a response which is both narrow and smoothly decreasing away from the axis. This narrow response can be thought of as a small aperture over which the intensity is averaged; if we scan this sampling aperture over the object plane and plot the response as a function of sampling position, we form the image of the object.

We will derive the response of a thin annulus to a point source as a function of the source angular distance from the axis, and show in what manner the synthesis of the improved response is carried out. We will then describe how this sampling aperture is scanned by adding phase shifts to the signal, and derive the change in effective size of the sampling function as it moves away from the physical axis of the annulus. The general case of multiple point sources in the object plane is then examined.

In this section we will concentrate on the theory and ignore practical considerations which will be discussed in the next section. Some of the mathematical details will be relegated to appendices.

A. Imaging One Point

Figure 1 shows a point source radiating at a frequency ω a wavefront of wavelength λ. The point is at a distance ρ from the origin in the object plane and at an angle α from the x-axis (which is parallel to $\Theta = 0$), or in cartesian coordinates at x_0, y_0, 0. At a point Θ around the annulus of radius r, corresponding to x', y', z in cartesian coordinates, the distance from point x_0, y_0, 0 to x', y', z is

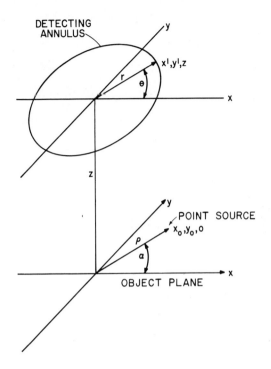

Figure 1.   Diagram showing the relation of the point source
at x_0, y_0, 0 and a point on the detecting annulus.

$$R = \sqrt{(x'-x_0)^2 + (y'-y_0)^2 + z^2} \tag{1}$$

In cylindrical coordinates,

$$x_0 = \rho \cos \alpha; \; y_0 = \rho \sin \alpha$$

$$x' = r \cos \theta; \; y' = r \sin \theta$$

Inserting into eq. 1 and combining terms gives R^2:

$$R^2 = z^2 + r^2 + \rho^2 - 2\rho r \cos(\Theta-\alpha). \qquad (2)$$

Defining D as $(z^2 + r^2)^{\frac{1}{2}}$,

$$R^2 = D^2 \left[1 + \left(\frac{\rho}{D}\right)^2 - \frac{2\rho r \cos(\Theta-\alpha)}{D^2} \right]$$

Using the expansion

$$(1 + x)^{\frac{1}{2}} = 1 + \tfrac{1}{2}x - \tfrac{1}{8}x^2 + \tfrac{3}{48}x^3 + \ldots, \qquad (3)$$

$$R = D \left\{ 1 + \tfrac{1}{2} \left[\left(\frac{\rho}{D}\right)^2 - \frac{2\rho r}{D^2} \cos(\Theta-\alpha) \right] \right.$$

$$- \tfrac{1}{8} \left[\left(\frac{\rho}{D}\right)^4 + \frac{4\rho^2 r^2}{D^4} \cos^2(\Theta-\alpha) \right.$$

$$\left. - \frac{4\rho^3 r}{D^4} \cos(\Theta-\alpha) \right]$$

$$\left. + \tfrac{3}{48} \left[\left(\frac{\rho}{D}\right)^6 + \ldots \right] + \ldots \right\} \qquad (4)$$

or

$$R = D + \frac{\rho^2}{2D} - \frac{\rho r}{D} \cos(\Theta-\alpha) + R_{ex} \qquad (5)$$

where we have lumped the higher order terms in ρ/D as R_{ex}.
If we assume that the amplitude change due to the different
distances from the point to the annulus is negligible, we
can write the amplitude and phase around the annulus due to
the point as

$$A(\Theta) = Ae^{jkR}$$

$$= A \exp \left\{ jk \left[D + \frac{\rho^2}{2D} - \frac{\rho r}{2D} \cos(\Theta-\alpha) + R_{ex} \right\} \right. \qquad (6)$$

where

$$k = 2\pi/\lambda.$$

D, the distance from the origin to the annulus, is independent of both ρ and Θ; it can be removed from eq. (6) by
multiplication by e^{-ikD} to yield

$$A'(\Theta) = A \exp \{ jk[\frac{\rho^2}{2D} - \frac{\rho r}{D} \cos(\Theta-\alpha) + R_{ex}]\} \qquad (7)$$

The response O to the point is obtained by summing the complex signal around the annulus.

$$O \triangleq \int_o^{2\pi} A'(\Theta) \, d\Theta$$

$$= A \, e^{jk \frac{\rho^2}{2D}} \int_o^{2\pi} e^{jk [\frac{\rho r}{D} \cos (\Theta-\alpha) + R_{ex}]} \, d\Theta \qquad (8)$$

In Appendix I we show that the factor $e^{jkR_{ex}}$ in eq. (8)
can be approximated as unity within certain limits of ρ and
D. We will continue the development neglecting that term,
whose effect is discussed in Appendix I.

The integral in eq. (8) is a Bessel integral; the
defining integral is[2]

$$J_n(u) = \frac{j^{-n}}{2\pi} \int_o^{2\pi} e^{ju\cos\Theta} \, e^{jn\Theta} \, d\Theta. \qquad (9)$$

Eq. (8) can be written, with $\Theta' = \Theta-\alpha$

$$O = 2\pi \cdot Ae^{jk \frac{\rho^2}{D}} \int_{-\alpha}^{2\pi-\alpha} e^{jk \frac{\rho r}{D} \cos \Theta'} \, d\Theta'$$

$$= 2\pi A \, e^{jk \frac{\rho^2}{D}} J_o (\frac{2\pi}{\lambda} \frac{\rho r}{D}) \qquad (10)$$

We note that the response is dependent on ρ/D, not α,
i.e. the response is radially symmetric, as expected from
the radial symmetry of the annular aperture.

The J_o response of the annular aperture is unsuitable
for imaging, as the sidelobes of the response decreases so

slowly that nearby points will be obscured.[3] J. P. Wild of
Australia showed how the sidelobes could be suppressed by
the addition of properly weighted Bessel functions of higher order.[4] (That these higher-order Bessel functions can
be easily generated will be shown.) This process be called
J^2-synthesis. He showed formally that a realizable point
source response $f(\rho/D)$ of an annular aperture of radius r
could be formed by

$$f(\rho/D) = \sum_{K=o}^{\infty} t_K J_K^2 (2\pi\rho r/D) \tag{11}$$

The realizability of the response requires that the spatial
frequency spectrum resulting from the Fourier transform of
$f(\rho/D)$ contains no frequencies $\geq 2r$. (The autocorrelation
function of the aperture is the Fourier transform of the
spatial frequency power spectrum of the point response
function[5]; with an aperture radius of r, the autocorrelation
must be zero outside of 2r.) The proof of eq. 11 and the
method of determining the t_K's is reproduced from Wild's
paper in Appendix II.

It may be noted that the J^2-synthesis theorem, eq. (11),
allows many possible point source responses; this freedom
to choose the response will be discussed later in the paper.
For the remainder of this section the point response will be
assumed to be:

$$f(\rho/D) = \left[\frac{2 J_1(2\pi\rho r/\lambda D)}{(2\pi\rho r/\lambda D)} \right]^2$$

$$\overset{\Delta}{=} \Lambda_1^2 (2\pi\rho r/\lambda D) \tag{12}$$

which is the point response of a diffraction-limited lens of
the same diameter.[6] By following the procedure of Appendix
II, the coefficients t_K for this response can be calculated;
they are

$$1, 1, -1/3, -3/8, -4/15, -1/6 \ldots \tag{13}$$

The higher-order Bessel functions needed for J^2-synthesis are derived from the signal by adding the proper phase shift at each point on the annulus. The definition of the Bessel function

$$J_n(z) = \frac{j^{-n}}{2\pi} \int_0^{2\pi} e^{jz\cos\Theta} e^{jn\Theta} \, d\Theta \tag{14}$$

shows that to obtain the nth-order Bessel function, we multiply the signal $A \exp[j2\pi \, \rho r\cos\Theta - \alpha/\lambda D]$, by $e^{in\Theta}$. Operationally, this can be accomplished by adding a 10° phase shift to the signal at 10° around the annulus from (arbitrary) zero, 20° phase shift at 20°, etc. to form $J_1(jk\rho r/D)$, adding a 20° phase shift to the signal at 10° around the annulus, etc., for $J_2(jk\rho r/D)$, and so on for any desired higher-order Bessel function. Summing the modified signal from each point around the annulus is equivalent to performing the integration of eq. 14, and results in the desired J_n.

The total response is calculated by squaring the absolute value of each signal-derived Bessel function, multiplying it by the appropriate t_K, and summing the result. The squaring operation removes the phase factors outside the integrals, such as those found in eq. 10. The intensity is

$$I = A^2 \sum_{K=0}^{\infty} t_K J_K^2 \left(\frac{2\pi}{\lambda} \frac{\rho r}{D}\right)$$

$$= A^2 \Lambda_1^2 \left(\frac{2\pi}{\lambda} \frac{\rho r}{D}\right) \tag{15}$$

This is the response of the J^2-synthesized aperture to a point source located in the object plane at a radial distance ρ from the axis.

The response is centered on the axis of the annulus, and is of the same form as the intensity in the focus of a lens of the same diameter and at the same distance as the annulus. If we consider the response as a sampling aperture to be scanned over the object plane, the diameter of the sampling aperture to the first zero, wherein 84% of the sampled energy falls, is

$$2\rho_0 = 1.22 \ \lambda D/r. \tag{16}$$

How this sampling aperture is scanned over the object plane is discussed next.

B. Scanning

The response of the annular aperture is highly peaked around the axis of the annulus. Scanning this sampling aperture by physically moving the annulus would produce an image, but is not practical. We can effectively move the axis of the annulus without physical movement by adding phase shifts around the annulus. Intuitively one sees that with the right distribution of added phase shift around the annulus, a point source not on the axis will produce the (net) equiphase signal distribution associated with the on-axis point. To put a particular point "on-axis", we calculate the phase distribution on the annulus that would be produced by a source located at that point; then we subtract that phase distribution from the signal so that only when a source is at the chosen aiming point will the calculated and measured phases cancel, and the equiphase distribution, equivalent to being on-axis, hold.

This argument suggests that scanning can be performed by multiplying the signal by e^{-ikL}, where L is the distance from the annulus to the aiming point. (Note that L is a function of Θ, unlike D.) Moving the aiming point changes e^{-ikL}, and the effective position of the axis of the annulus, i.e. the effective sampling aperture, will move with it.

Appendix III proves that the intuitive argument is correct. It shows that if we replace the multiplication of the signal by e^{-ikD} between eq. 6 and eq. 7 by multiplication by e^{-ikL}, the resulting signal is of the form

$$A(\Theta) = A \ \exp \left[\frac{(\Delta^2/2) + r\Delta\cos(\Theta-\alpha)}{D'} \right] \tag{17}$$

When compared to the unscanned signal, eq. 7, it appears the same except that (1) ρ, the distance from the annular axis to the point source, has been replaced by Δ, defined as the distance between the aiming point and the point source, (2) D, the distance from the annulus to the intersection of

the axis with the object plane, has been replaced by D',
defined as $D + \sigma^2/4D^2$, where σ is the vector sum of the
vector from the origin to the point source and the vector
from the origin to the aiming point and (3) α, the angle of
ρ to the $\Theta = 0$ axis, is replaced by γ, the angle of $\bar{\Delta}$ to the
$\Theta = 0$ axis. (The Δ and σ vectors are shown in Fig. II.1 of
Appendix II.)

Therefore the effective aperture can be scanned by multiplying signal by $e^{-ikL(\Theta)}$. The width of the response,
i.e. the effective size of the sampling aperture, increases
with increasing distance of the aiming point from the axis
of the annulus. The increase is proportional to

$$D' = D \left(1 + \frac{\sigma^2}{4D^2}\right) \tag{18}$$

When the aiming point is near the source point, σ is
approximately equal to 2ρ so eq. 18 can be written

$$D' \approx D \left[1 + \left(\frac{\rho}{D}\right)^2\right] \tag{19}$$

For example, the effective sampling aperture of an f/1 system when aimed r away from the physical axis of the annulus,
i.e. directly below the annulus, is 1/5 larger than at the
origin. (In an f/1.0 system, $D = \sqrt{5}\, r$.) That the resolution is higher in the center than at the edges of the object
field is typical of imaging systems.

As is discussed in Appendix II, there is also a slight
($< 1\%$) diminution of response as the aiming point moves away
from the physical axis of the annulus.

The output image of a scanned sampling-aperture imaging
system, is[7]

$$I = B * h \tag{20}$$

where I is the image, B is the object intensity, h is the
sampling aperture function, and * indicates spatial convolution. The form of h for the thin annular-aperture system
can be selected by choosing the t_K's of eq. 11; for the response previously chosen

$$h = \Lambda_1^2 \left[\frac{2\pi\Delta R}{\lambda D'} \right] \tag{21}$$

where

$$D' = (z^2+r^2)^{\frac{1}{2}} \left[1 + \rho^2/(z^2+r^2) \right] \tag{22}$$

The depth of focus of the annular aperture system is infinite along its physical axis; anywhere on the axis a point source produces an equiphase signal around the annulus. When scanning is implemented by multiplication of e^{-jkL}, however, depth of focus appears. This is because the aiming point has a definite position in z as well as ρ and Θ; a point directly above or below the aiming point will not produce the same signal on the annulus. The depth of focus produced by the scanning procedure will vary from very large near the physical axis to very small far away from it. The exact form of the depth of focus as a function of z/r and ρ has not yet been established.

C. Multi-point Images

We will assume the same general configuration as Fig. 1, but N points in the plane instead of one. The signal along the annular aperture is the sum of the signals from each of the N points; referring back to eq. (7),

$$A(\Theta) = \sum_{i=1}^{N} A_i \, e^{jkR_i + j\phi_i}$$

$$= \sum_{i=1}^{N} A_i \, e^{jk \left[\frac{\rho_i^2}{D} - \frac{\rho_i r}{D} \cos(\Theta-\alpha) \right]} e^{j\phi_i} \tag{23}$$

The ϕ's are the random phases associated with each point source. The R_{ex} term has been dropped.

If we integrate around the annulus to get the response,

$$O = \int_o^{2\pi} A(\Theta) \, d\Theta$$

$$= \sum_{i=1}^N A_i \, e^{j(k \frac{\rho_i^2}{D} + \phi_i)} \int_o^{2\pi} e^{jk\frac{\rho_i}{D} r \cos(\Theta - \alpha)} \, d\Theta$$

$$= \sum_{i=1}^N A_i \, e^{j\phi_i'} \, J_o \left(\frac{k\rho_i r}{D} \right) \tag{24}$$

In eq. (24) we have interchanged the order of integration and summation and lumped together the random phase angle ϕ_i with $k\rho_i^2/D$ to form a new random phase ϕ_i'.

If the procedure for J^2-synthesis is followed, adding proper phase shifts, etc., we obtain for the O_n^{th} term which contains the Bessel function of order n,

$$O_n = \sum_{i=1}^N A_i \, e^{j\phi_i'} \, J_n \left(\frac{k\rho_i r}{D} \right) \tag{25}$$

Now when the absolute value squared of these Bessel functions are formed, as is necessary to fulfill the J^2-synthesis procedure of eq. 11, certain extra terms are generated. For the general term

$$|O_n|^2 = \left| \sum_{i=1}^N A_i \, e^{j\phi_i} \, J_n \left(\frac{k\rho_i r}{D} \right) \right|^2$$

$$= \left. \sum_{i=1}^N A_i^2 \, J_n^2 \left(\frac{k\rho_i r}{D} \right) \right\} \mathrm{I}$$

$$+ 2 \underset{\ell \neq m}{\sum_{m=1}^N \sum_{\ell=1}^N} A_\ell A_m \cos(\phi_\ell' - \phi_m') J_n \left(\frac{k\rho_\ell r}{D} \right) J_n \left(\frac{k\rho_m r}{D} \right) \tag{26}$$

he first term on the right (in the bracket labelled I) is he sum of the responses for each point; with the proper κ's the response would be a simple sum of the responses for each single points source. This term we shall call the true image.

The terms in bracket II are terms extraneous to the true image. They represent point-point interference effects that depends upon the relative phase between the points. This sort of interference is commonly observed in imaging systems using coherent illumination. It is called speckle, because the "noise" it produces looks like speckling on top of the image.

To remove the speckle from the image, the term in bracket II must be made small compared to the true image. There are several ways to do this, all based on the cos $(\phi_\ell - \phi_m)$ term. If the argument $(\phi_\ell - \phi_m)$ varies randomly over 2π, the sum of cos $(\phi_\ell - \phi_m)$ approaches zero with a root-mean-square deviation of $\sqrt{N/2}$.

A simple way to vary the phase is to make a number of exposures with the relative phase between the sources different. If we sum M measurements of the response of the annulus with nothing changing except the relative phases of the point sources, the true image, the I bracket of eq. 26, will be M times larger, as it is independent of the relative phase of the point sources; the II bracket term, which can be called the speckle noise, will be $\sqrt{M}$ larger. Therefore the signal-to-noise ratio, which is given by the "power" of the signal divided by the square of the RMS deviation, increases by a factor of M. When M is 10 or so, it is likely that the electronic noise of the detecting system will be greater than the speckle noise, so that further averaging would not be needed.

If a diffuse source of ultrasound is used to insonify the object, by the nature of a diffuser the points in the object will have random phase differences from each other. If the diffuser, and diffuser only, is rotated, the relative phase relations will change randomly. This reduces speckle, as has been demonstrated by summing a number of optical holograms made with a diffuser rotated between exposures; the image shows markedly reduced speckle[8].

Another way of introducing a sum of cosines with random arguments into the II term of (26) is to use several wavelengths of insonification simultaneously; the different relative phase shifts at the different frequencies will have

he same randomizing effect on $(\phi_\ell-\phi_m)$ as multiple exposures.
This speckle-noise reduction method has also been proven in
optical holography[9].

III. DISCUSSION OF THE THEORY

Imaging with thin annular apertures was developed in
the early 60's by Wild and his coworkers[10,4] to image sun-
spots. The first images from his annulus of 96 radio antennas, using solar radio emissions at 80 Mhz, were published in 1968[11]. The spatial incoherence of the solar radio
sources permits time-averaging to eliminate the second part
of eq. 26; we eliminate that part of the intensity by producing artificial spatial incoherence and time-averaging by
the methods described. Wild showed images of 3000 elements
using an array of 96 antennas. This was considered a paradox ("the Culgoora paradox") until Toraldo di Francia and
coworkers showed[12] that with incoherent illumination the thin
annulus could indeed produce an image with as many resolution
elements as a full aperture would.

There are a number of engineering considerations that
affect the theory. For instance in Section II the theory of
imaging with a thin annulus was described in terms of integrals. In a realizable system, the annulus must be broken
into short segments and the amplitude and phase measured at
each segment. The integrals of Section II must be replaced
by sums.

Figure 2 shows that this replacement neglibibly affects
the system point response. The difference between the response of an ideal diffraction limited lens, Λ_1^2, and the
J^2-synthesized Λ_1^2 response, agree to within the truncation
errors. This is to be expected from the well-behaved characteristics of the signal and correction terms here being
summed.

The J^2-synthesis response shown in Fig. 2 used up to
J_{10}^2 in the synthesis; agreement between the J^2-synthesized
and the desired response is almost perfect at low values of
the argument, but when the argument reaches 10, the synthesis
fails as shown in Fig. 3. This is consistent with the property of Bessel functions of the first kind that for arguments smaller than the order, the value of the function is
zero. As the argument, which in the imaging case is the

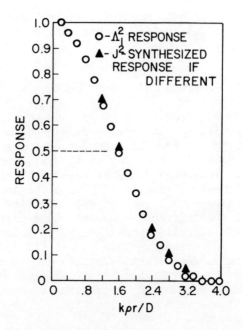

Figure 2.   The calculated J^2-synthesis compared to the calculated value of Λ_1^2; truncation in the plotting
routine emphasizes small differences.

distance from the (translated) axis of the annulus, grows
larger, higher and higher order functions are required.
Figure 4 shows the same synthesis as Fig. 3 but with Bessel
functions with orders up to 20 included in the synthesis:
for arguments less than the highest order Bessel function
included, agreement is excellent.

     The truncation error that occurs for a finite number of
Bessel functions has the effect of limiting the field of view
of the imaging system.   If the largest argument that need be
considered is less than the highest order Bessel function
used, exact synthesis can be achieved.   (Limiting the field
of view of optical systems to reduce aberrations caused by
high angle rays is commonplace.)   While more orders require
more computation, i.e. time or money, large field of view is
important, too.   We believe a new set of weighting functions
(t_K's) can be found to help the trade-off between field of

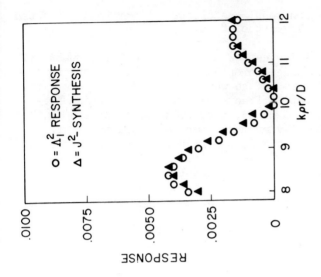
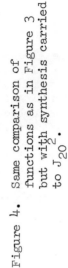

Figure 4. Same comparison of functions as in Figure 3 but with synthesis carried to J_{20}^{2}.

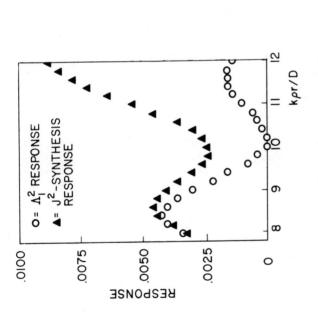

Figure 3. The J^{2}-synthesis of Λ_{1}^{2} using the expansion up to J_{10}^{2}, compared to Λ_{1}^{2}.

view and computation requirements, but as yet have not found the proper optimizing set of t_K's.

The field-of-view can be limited also by the number of detector elements placed around the annulus. If the detector elements undersample the wavefront, i.e. the wavefront contains higher spatial frequencies than the spatial frequency of detector elements, a spurious ring appears in the image. The field-of-view must then be restricted so that the ring does not appear in the image. Wild provides a good discussion of this.[4] (In acoustical imaging we can use enough detectors to avoid undersampling; for radio astronomy, where Wild used a radio antenna for each detector element, using enough detectors is more of a problem.)

The depth of focus of an annular imaging system varies from infinity along the axis of the annulus to the depth of field of a full aperture at high angles from the axis, as was discussed in the previous section. To reduce the depth of imaging, the annular system can be used only off-axis or in a pulsed, range-gated mode. For biological imaging, where there often are many strongly reflecting structures overlying the region to be imaged, a pulsed range-gated system is required to remove strong out-of-focus reflections even in a short depth-of-focus imaging system.

One restriction on the annular aperture imaging system is that it requires a spatially diffuse wavefront from the object; if the object is not spatially diffuse it will not be properly imaged, e.g. a collimated beam that passes through the middle of the annulus, which would not even record its existence. As diffuse illumination is necessary anyway to ensure that reflected acoustic wavefronts are returned to the detection aperture, the requirement for it is not objectionable.

The sensitivity of the annular aperture system is less than that of the full aperture system because it has less energy-receiving area; the annulus must be thin for the amplitude and phase of the signal not to vary over the width of the annulus, so the energy-receiving area can not be large. For example, if the total aperture is 100 wavelengths across, as it might be for ultrasonic imaging, the ratio of the full aperture area to a one wavelength thick annulus (which is suitable for wavefronts of up to 30° to the normal) is 25:1. For the diffuse illumination used, the energy distribution

is approximately uniform so that the signal energies from
the full and annular aperture will be in the same 25:1 ratio.

A further lowering in sensitivity occurs when we "remove
the speckle" by summing spatially incoherent images. With a
coherent system, increasing the signal-gathering time proportionately decreases the noise power (if the noise is assumed
white). Breaking a certain available exposure time T into
M segments for separate illuminations (as proposed in the
previous section) produces an image whose signal-noise ratio
is proportional to $T/\sqrt{M}$: the M signals of power T/M are
summed to give T and the noise is summed quadratically in the
root-mean sense to give $\sqrt{M}$ times the noise. If we had used
the whole time T for a single exposure, the signal-noise ratio
would be proportional to T. Therefore the sensitivity is reduced by a factor of $\sqrt{M}$ by the speckle-removing process.
Because the completeness of removal of the interfering terms
of equation 26, here called speckle, is proportional to M,
there will be a trade-off between sensitivity and completeness of speckle suppression.

IV. EXPERIMENTAL RESULTS

Figure 5 shows the apparatus used to measure the amplitude and phase around a ring aperture. A single detector is
stepped around a circle; at each position the signal is multiplied by quadrature references coherent with the object
insonification. (Though the diagram shows an insonifier,
the experiments reported here all use as an object small
transducers which were driven to produce quasi-point sources
of ultrasound.) The resulting two signals, corresponding to
the amplitude of the signal multiplied by the sine and cosine
of the phase angle between the master oscillator and the signal, are sampled-and-held (which provides a range-gate),
digitized, and digitally integrated over several pulses to
increase the S/N ratio. After these two signals are punched
into paper tape, the detector is stepped to the next position
around the annulus and the process is repeated. After a full
circle has been made, the relative phases of the object-
transducers is altered and another ring of measurements made.

In the present apparatus, the annulus is 81 mm in diameter. The detector is 2 mm in diameter, and 400 steps com-
prise the annulus. (The overlap between detector positions
is for convenience).

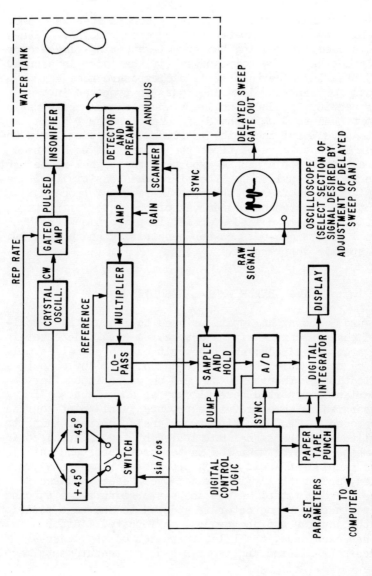

Figure 5. Electronic equipment used to measure the amplitude and phase around the thin annulus.

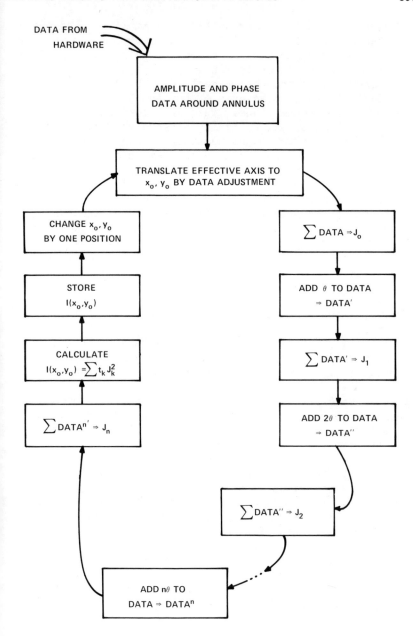

Figure 6. Processing of the annular aperture data to obtain the image.

The set of numbers representing the amplitude and phase
around the annulus is entered into the memory of a computer.
Figure 6 shows how the data is processed to produce a point-
by-point map of the intensity in a focal plane by the procedure discussed in the theory section. The data is adjusted
to translate the effective axis of the annulus to a particular point in the focal plane. This is accomplished, as was
discussed in Section II, by adding a phase shift at each detector position of such magnitude that if there were a source
at that point, the phase would be the same around the annulus.

The sum of the data around the annulus is proportional
to the J_O component of the intensity; after a phase shift
proportional to the detector position around the annulus (θ)
is added to the data, the sum is proportional to J_1; after
adding 2θ, the sum is proportional to J_2, etc., until the J_n
term is obtained by adding $n\theta$ to the data. (The magnitude
of n depends upon the field of view, as discussed in Section
III.) The individual J_K terms are squared. They are weighted
by the set of t_K's chosen for a particular point source response, and then summed. The result is proportional to the
image intensity at the particular point, x_O, y_O, z_O at which
the annulus was "aimed". That point's intensity is recorded
and the effective axis is translated to the next point of
interest. In this way a complete image is scanned out.

This image represents a subframe; to achieve the needed
incoherence several such subframes are added together. As
noted before, the insonifying wavefront must be changed between subframes, so each subframe has a unique set of data.

Clearly, there is a lot of computation to do to obtain
one frame. However, examination of the computing procedure
shown in Fig. 6 reveals that most of the steps can be done
in parallel. With enough computing elements, i.e. money,
the image can be formed rapidly.

Figures 7 and 8 show a scan of two quasi-point sources.
Both scans are derived from the same amplitude and phase information, but different sets of t_K were used in calculating
the intensity. Figure 7 results from using the set of t_K
equivalent to $\Lambda_1{}^2(k\rho r/D)$ the response of an ideal lens, and
Fig. 8 results from the set of t_K equivalent to $\Lambda_1(2k\rho r/D)$.
The half-power points of Fig. 8 are half the width of Fig. 7.
(The negative parts of the Λ_1 response are here plotted as
positive.) Superresolution [13] is possible in annular imaging

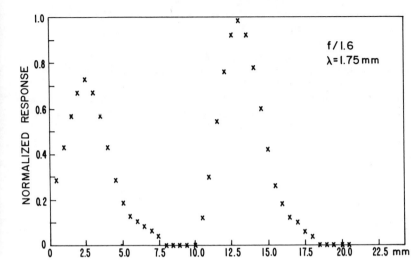

Figure 7. One line in the image of two point sources, using
a set of t_K's to give a Λ_1^2 response; note that the
half-power width of 3.2 mm corresponds to the
Rayleigh resolution limit of a perfect lens of this
f/number.

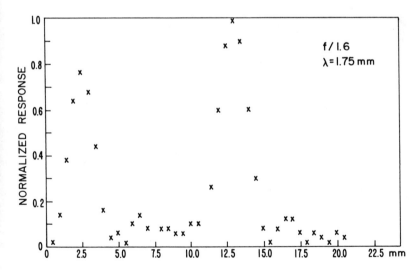

Figure 8. One line in the image of two point sources, using
the $|\Lambda_1|$ response with the same data as Figure 7.

because negative intensities are computable. In optical
imaging, negative intensities are forbidden.

The usefulness of the superresolution possible with
annular imaging remains to be proven. Lower signal-noise
ratio counterbalances the greater resolution. Images using
different sets of t_K should indicate which impulse response
is most useful. Also, the truncation error effects discussed in Section III should be evaluated for different
choices of t_K.

Figure 9 is the intensity map of a two-dimensional
image. Here the computer changed type-face when different

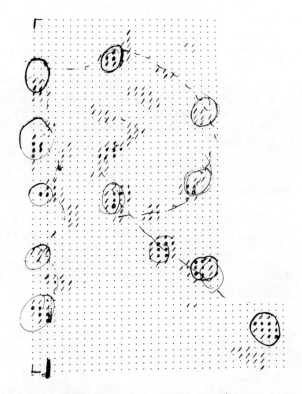

Figure 9.   Image formed with $\lambda = 1$ mm at $f/2.0$. The circles
             show the position of the point sources used; the
             object was examined at $\frac{1}{2}$ mm intervals. Changes
             in intensity are indicated here by change in type-
             face.

threshold levels were passed. The location of the transducers in the object is shown by circles. The dotted line
is the path of the ground return wire between the transducers
that (to our surprise) was imaged. There are several areas
where speckle noise is evident.

Similar two dimensional images have been made of areas
60 mm on a side using an 81 mm aperture at f/1.5. This indicates fields of view of the size of the aperture should be
realizable.

The annular aperture imaging system of Figs. 5 and 6
has high sensitivity as well as high resolution. With as
little as 5 nanowatts for 5 mS driving the transducers used
as object points, the output images had greater than 100:1
S/N ratios. Five nanowatts per element corresponds to an
acoustic intensity at the thin annulus aperture of less than
10^{-11} W/cm^2. Significantly, five nanowatts per resolvable
element of the object is the expected level of signal power
from biological objects near the center of the body[1].

V. CONCLUSIONS

We have described and built an annular aperture imaging
system that can use a coherent source of illumination. This
system has experimentally demonstrated a number of interesting imaging characteristics: selectable point responses,
including superresolution; a varying depth-of-field with
position in field; wide field-of-view; and low f-number operation. The analysis presented here, while incomplete, indicates how these characteristics come about.

While the imaging characteristics of the thin annular
aperture system are of inherent interest, the practical
utility of the system rests in its use of a small number of
detectors to synthesize an aperture that would require a
large number of such detectors. The saving in detectors must
be paid for in sensitivity and in complexity of the required
signal processing. The ultimate practicability of the system probably depends upon the relative difficulty of making
detector elements versus computation elements.

ACKNOWLEDGMENTS

The author wishes to acknowledge the valuable aid of several colleagues, in particular the theoretical suggestions and practical transducer designs of R. Mezrich, and the helpful review of the manuscript by W. Stewart.

REFERENCES

1. Vilkomerson, D. "Analysis of Various Ultrasonic Holographic Imaging Methods for Medical Diagnosis", Acoustical Holography, Vol. IV, Plenum Press, N.Y. (1972).

2. Jahnke, E. and Ende, F. "Tables of Functions" Fourth Ed., page 149, Dover Publications, N.Y. (1945).

3. McLean, D. J. "The Improvement of Images Obtained with Annular Apertures" Proc. Roy. Soc. A 263, 545-551 (1961).

4. Wild, J. P. "A New Method of Image Formation with Annular Apertures and an Application in Radio Astronomy", Proc. Roy. Soc. 286 A, 499 (1965).

5. Goodman, J. W. "Intro. to Fourier Optics", Chapter 6, McGraw-Hill, N.Y. (1968).

6. Born, M. and Wolf, E. "Principles of Optics" Third Ed., Chapter 8, Pergamon Press, N.Y. (1965).

7. Goodman, J. W. "Intro. to Fourier Optics", Chapter 1, McGraw-Hill, N.Y. (1968).

8. Martienssen, W. and Spillter, S. "Holographic Reconstruction Without Granulation", Phys. Lett. 24 A, 126 (1967).

9. George, N. and Jain, A. "Speckle Reduction Using Multiple Tones of Illumination", Appl. Optics 12, 1202 (1973).

10. Carter, A. W. L. and Wild, J. "Some Correction Processes for Annular Aerial Systems", Proc. Roy. Soc. A 282, 252 (1964).

11. Wild, J. "Eighty Mhz Photography of Eruption of a Solar Prominence", Nature 218, 536 (1968).

12. Toraldo Di Francia, G., "Degrees of Freedom of Image", J. Opt. Soc. Am. 59, 799 (1969); also Gori, E. and Guattari, G., J. Opt. Soc. Am. 61, 36 (1971).

13. Harris, J. L., "Diffraction and Resolving Power", J. Opt. Soc. Am. 54, 931 (1964).

14. Sokolnikoff, I., and Redheffer, R., "Mathematics of Physics and Modern Engineering", page 129, McGraw-Hill, N.Y. (1950).

15. Courant, R. "Differential and Integral Calculus", Vol. I, page 131, Interscience Publishers, N.Y. (1957).

APPENDIX I. THE EFFECTS OF THE R_{ex} TERMS

In equation (4) the expression for the distance was expanded as

$$R = (1+x)^{\frac{1}{2}}$$

$$\approx 1 + \frac{1}{2} x - \frac{1}{8} x^2 + \frac{3}{48} x^3 - \frac{5}{128} x^4$$

where x is

$$x(\rho,\theta) = \left(\frac{\rho}{D} \right)^2 - \frac{2\rho r}{D^2} \cos (\theta - \alpha)$$

which is less than unity assuming

$$z \geq 2r \text{ and } \rho < r$$

Then

$$R = D + \frac{D}{2} \left[\frac{\rho^2}{D^2} - \frac{2\rho r}{D^2} \cos (\theta - \alpha) \right]$$

$$- \frac{D}{8} \left[(\frac{\rho}{D})^4 + \frac{4\rho^2 D^2}{D^4} \cos^2 (\theta - \alpha) - \frac{4\rho^3 r}{D^4} \cos (\theta - \alpha \right]$$

$$+ \frac{3D}{48} \left[\left(\frac{\rho}{D}\right)^6 - \frac{8\rho^3}{D^6} r^3 \cos^3 (\Theta - \alpha) \right.$$

$$\left. + 8 \left(\frac{\rho}{D}\right)^4 \frac{2\rho r}{D} \cos (\Theta - \alpha) \right]$$

$$+ \dots \tag{I.1}$$

We wish to evaluate the terms of 2nd and higher order in x that were lumped together in the terms R_{ex} after eq. 7 of the text. The term in x^2, which will be called R_{ex2}, is

$$R_{ex2} = - \frac{1}{8} \left[\frac{\rho^4}{D^3} + \frac{4\rho^2 r^2}{D^3} \cos^2 (\Theta - \alpha) - \frac{4\rho^3 r}{D^3} \cos (\Theta - \alpha) \right]$$

$$= - \frac{1}{8} \left[\frac{\rho^4}{D^3} + \frac{2\rho^2 r^2}{D^3} (\cos 2(\Theta - \alpha) - 1) - \frac{4\rho^3 r}{D^3} \cos (\Theta - \alpha) \right]$$

$$= - \frac{1}{8} \left\{ \left(\frac{\rho^4 - 2\rho^2 r^3}{D^3} \right) - \frac{4\rho^3 r \cos(\Theta - \alpha)}{D^3} \right.$$

$$\left. + \frac{2\rho^2 r^2}{D^3} \cos [2(\Theta - \alpha)] \right. \tag{I.2}$$

For a convergent series of alternating sign, such as the expansion for $(1+x)^{\frac{1}{2}}$, the remainder after n terms has a value that is between zero and the first term in the series not taken.[14] If we find the effect of the R_{ex2} term, then, the total error in R, R_{ex}, is less than that of R_{ex2}.

The error caused by the R_{ex2} term is a change in the phase of the signal of

$$d\phi = k R_{ex2} = \frac{2\pi 4\rho^2 r}{8\lambda D^3} \left[r \cos^2(\Theta - \alpha) - \rho \cos (\Theta - \alpha) \right] \tag{I.3}$$

where the term ρ^4/D^3 independent of Θ has been taken out. The usual criterion for affecting the image is a phase error of $\pi/2$;[6] we find within what region the R_{ex2} term can be ignored by this criterion:

$$\frac{\pi}{2} \leq k \ R_{ex2}$$

or

$$\lambda \leq \frac{\rho^2 r}{D^3} \ [r \cos^2(\theta - \alpha) - \rho \cos(\theta - \alpha)] \qquad (\text{I}.4)$$

The maximum change occurs when the cosine term changes from zero to minus one; therefore

$$\lambda \leq \frac{2\rho^2 r}{D^3} \ (r + \rho) \qquad (\text{I}.5)$$

We can evaluate this condition for various systems; for example, an f/1.5 system has

$$D = \sqrt{10} \ r \qquad (\text{I}.6)$$

If we set

$$\rho = xr \qquad (\text{I}.7)$$

we have

$$\lambda \leq \frac{x^2(1+x)r}{10\sqrt{10}} \qquad (\text{I}.8)$$

If $r = 50\lambda$ (reasonable aperture size for acoustic imaging at 1 mm) we can solve for x:

$$x \leq .66r \qquad (\text{I}.9)$$

i.e. for $o < \rho < .66 \ r = 33\lambda$ we can ignore the R_{ex2} term, and all the higher order terms as well.

For higher f/number systems, the area where the effect of R_{ex} is negligible is still larger, e.g., for f/2.0 system, $\rho \leq .87r$.

The point source response of the system is quite small this far from the axis, and we would expect that adding the relatively small deviation in phase caused by R_{ex2} to the rapidly changing phase of the main term would have small effect. We have not yet solved exactly the effect of R_{ex2} on the values of the integral. We can show that the integral including the R_{ex2} term does not blow up by use of Schwartz's inequality for integrals[15]:

$$\left[\int_a^b f(x)g(x)dx \right]^2 \leq \int_a^b [f(x)]^2 dx \int_a^b [g(x)]^2 dx \qquad (I.10)$$

We use, from eq. (I.2)

$$f(x) = \exp \left[\frac{j2\pi}{\lambda} \left(\frac{\rho r - \rho^3 r}{D} \frac{}{D^3} \right) \cos (\theta - \alpha) \right]$$

$$g(x) = \exp \left[\frac{j2\pi}{\lambda} \left(\frac{\rho^2 r^2}{4D^3} \cos 2(\theta - \alpha) \right] \right. \qquad (I.11)$$

and obtain

$$\int_o^{2\pi} \exp \left[\frac{j2\pi}{\lambda} (R + R_{ex2}) d\theta \right.$$

$$\leq \left\{ J_o \left[\frac{4\pi}{\lambda} \left(\frac{\rho r}{D} - \frac{\rho^3 r}{D^3} \right) \right] J_o \left[\frac{4\pi}{\lambda} \left(\frac{\rho^2 r^2}{4D^3} \right) \right] \right\}^{\frac{1}{2}} \qquad (I.12)$$

Examination of eq. I.12 shows that the integral including R_{ex2} is larger than without R_{ex2}, i.e. $J_o(k\rho r/D)$, but that still is a small, smoothly decreasing function which should not appreciably affect the imaging properties.

Awaiting further study is rigorous calculation of the effect of the expansion on all terms of the J^2-synthesis at large distances from the axis.

APPENDIX II. PROOF OF J^2-SYNTHESIS

The proof of J^2-synthesis is given here for convenience; it is taken directly from Wild.[4]

The theorem to be proved is that any circularly symmetrical function $f(\rho/D)$ whose Fourier components are restricted to the range $0 < \sigma < 2r$ can be expressed as

$$f(\rho/D) = \sum_{K=0}^{\infty} t_K J_K^2 (2\pi\rho r/D) \qquad (II.1)$$

The symmetrical function $f(\rho/D)$ and its Fourier-Bessel transform, $F(\sigma)$, are related by

$$f(\rho/D) = \int_0^{2\pi} F(\sigma) J_0 (2\pi r\sigma) \sigma d\sigma \qquad (II.2)$$

Putting

$$\sigma = 2r \cos \alpha, \quad 0 \leq \alpha \leq \frac{\pi}{2} \qquad (II.3)$$

we define

$$G(\alpha) = (2\pi r)^2 F (2r \cos \alpha) \sin \alpha \cos \alpha \qquad (II.4)$$

so that we can write

$$f(\rho/D) = \frac{2}{\pi} \int_0^{\pi/2} G(\alpha) J_0 (4\pi r\rho\cos\alpha/D) \, d\alpha \qquad (II.5)$$

Since $G(\alpha)$ is defined in the restricted range $0 \leq \alpha \leq \pi/2$, we can expand it as a Fourier series:

$$G(\alpha) = \sum_{k=0}^{\infty} S_k \cos 2k\alpha \qquad (II.6)$$

where

$$S_k = C \cdot \frac{2}{\pi} \int_0^{\pi/2} G(\alpha) \cos 2k\alpha \, d\alpha \qquad (II.7)$$

where $C = 2$, except when $k = 0$, $C = 1$. We rewrite (II.5) as

$$f(\rho/D) = \frac{2}{\pi} \sum_{k=0}^{\infty} S_k \int_0^{\pi/2} J_0 (4\pi r\rho\cos\alpha/D)\cos 2k\alpha \, d\alpha \qquad (II.8)$$

Using Neumann's formula

$$\int_o^{\pi/2} J_{p+q}(2z\cos\Theta)[\cos(p-q)\Theta]d\Theta = \frac{\pi}{2} J_p(z)J_q(z)$$

We find

$$f(\rho/D) = \sum_{k=o}^{\infty} S_k \, J_k \, (2\pi r\rho/D) \, J_{-k} \, (2\pi r\rho/D)$$

$$= \sum_{k=o}^{\infty} S_K \, (-1)^k \, J_K^{\ 2} \, (2\pi r\rho/D)$$

$$(II.9)$$

which proves the theorem and gives the coefficients

$$t_K = (-1)^k \, S_k \tag{II.10}$$

$$= C \cdot (8_\pi r^2)(-1)^k \int_o^{\pi/2} F(2rc\cos\alpha)\cos2\kappa r\cos\alpha\sin\alpha d\alpha$$

APPENDIX III. DERIVATION OF THE SCANNING FUNCTION

We wish to prove that the effective axis of the annulus can be scanned over the object plane by mathematical manipulation of the signal data.

We propose that we can move the effective axis of the annulus by replacing, as the multiplier of the signal in eq. 7, e^{-jkD} by e^{-jkL}, where L is the distance from the annulus to the point in the object plane where we wish the axis to be centered. Multiplication by e^{-ikL} of the signal, which is of the form e^{ikR}, produces

$$A'(\Theta) = Ae^{jk(R-L)} \tag{III.1}$$

We calculate $(R-L)$:

$$R-L = \frac{R^2-L^2}{R+L} \tag{III.2}$$

Referring to Fig. 1,

$$R^2 = D^2 + x_1{}^2 + y_1{}^2 - 2x_1 r\cos\theta - 2y_1 r\sin\theta \qquad (III.3)$$

Similarly for the distance L from the aiming point at x_2, y_2 to the annulus,

$$L^2 = D^2 + x_2{}^2 + y_2{}^2 - 2x_2 r\cos\theta - 2y_2 r\sin\theta \qquad (III.4)$$

Therefore

$$R^2 - L^2 = (x_1 - x_2)^2 + (y_1 - y_2)^2 - 2(x_1 - x_2)r\cos\theta$$
$$\qquad (III.5)$$
$$- 2(y_1 - y_2)r\sin\theta$$

We calculate R from R^2 by using the expansion for $(1+x)^{\frac{1}{2}}$:

$$R = D + \frac{x_1{}^2 + y_1{}^2}{2D} - \frac{x_1 r\cos\theta}{D} - \frac{y_1 r\sin\theta}{D} + R_{ex} \qquad (III.6)$$

R_{ex} are the higher order terms of the expansion; the previous appendix discussed why these terms can be dropped.

Similarly,

$$L = D + \frac{x_2{}^2 + y_2{}^2}{2D} - \frac{x_2 r\cos\theta}{D} - \frac{y_2 r\sin\theta}{D} \qquad (III.7)$$

Therefore

$$R + L = 2D + \frac{(x_1{}^2 + x_2{}^2) + (y_1{}^2 + y_2{}^2)}{2D} - \frac{(x_1 + x_2)r\cos\theta}{D}$$
$$\qquad (III.8)$$
$$- \frac{(y_1 + y_2)\ r\sin\theta}{D}$$

We can simplify the expressions for $R^2 - L^2$ and $R + L$ by using the vectors shown in Fig. III.1: $\bar{\Delta}$, the vector

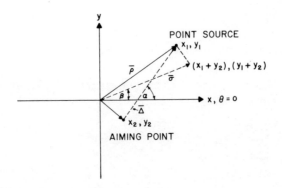

Figure III.1.   Diagram of the object plane, showing the
$\bar{\rho}$, $\bar{\Delta}$ and $\bar{\sigma}$ vectors.

between the point source at x_1, y_1, and the aiming point at
x_2, y_2, and $\bar{\sigma}$, the vector sum of the vectors from the origin
to the two points.   These vectors are defined by

$$\bar{\Delta} = (x_1 - x_2) \; \bar{i}_x + (y_1 - y_2) \; \bar{i}_y \qquad\qquad (III.9)$$

$$\bar{\sigma} = (x_1 + y_2) \; \bar{i}_x + (y_1 + y_2) \; \bar{i}_y$$

We can rewrite the equation for $R^2 - L^2$, eq. III.5 in
terms of $\bar{\Delta}$, which is at the angle γ to the $\Theta = 0$ axis:

$$R^2 - L^2 = |\bar{\Delta}|^2 - 2 \; r \; |\Delta| \; \cos\gamma\cos\Theta - 2r|\Delta|\sin\gamma\sin\Theta$$

$$\qquad\qquad\qquad (III.10)$$

$$= \Delta^2 - 2r\Delta \; \cos \; (\Theta - \gamma)$$

where $\Delta = |\bar{\Delta}|$.   We can rewrite R+L, eq. III.8, in terms of
$\bar{\sigma}$, which is at angle β to the $\Theta = 0$ axis:

$$R+L = 2D + \frac{|\bar{\sigma}|^2}{2D} - \frac{r|\bar{\sigma}|}{D} \; \cos \; \beta \; \cos \; \Theta - \frac{r|\bar{\sigma}|}{D} \; \sin\beta\sin\Theta$$

$$= 2D + \frac{\sigma^2}{2D} - \frac{\sigma r}{D} \; \cos \; (\Theta - \beta) \qquad (III.11)$$

where $\sigma = |\bar{\sigma}|$.   Then the expression for R-L is

$$R-L = \frac{R^2-L^2}{R+L}$$

$$= \frac{(\Delta^2/2) - r\,\Delta\,\cos\,(\theta-\gamma)}{D + \frac{\sigma^2}{4D} - \frac{\sigma r}{2D}\cos\,(\theta-\beta)} \qquad\qquad (III.12)$$

The denominator of III.12 is a function of θ which would complicate the calculation of the integral in $d\theta$. However when

$$\frac{\sigma r}{2D} < D + \frac{\sigma^2}{4D} \qquad\qquad (III.13)$$

we can expand the denominator III.12 as

$$\frac{1}{D^2 + \frac{\sigma^2}{4D}\left[1 - \frac{2\sigma r}{4D^2+\sigma^2}\cos(\theta-\beta)\right]} = \qquad\qquad (III.14)$$

$$\frac{1 + \left[\frac{2\sigma r\,\cos(\theta-\beta)}{4D^2 + \sigma^2}\right] + \frac{1}{2}\left[\frac{2\sigma r\,\cos\,\theta-\beta)}{4D^2 + \sigma^2}\right]^2 + \ldots}{D^2 + \frac{\sigma^2}{4D}}$$

The second term, $2\sigma r\cos(\theta-\beta)/4D^2 + \sigma^2$ will multiply the numerator term $r\Delta\,\cos(\theta-\gamma)$ of eq. III.12 to produce a $\cos(2\theta-\gamma-\beta)$ which when placed in the integral of eq. 8 will produce a second J_o term multiplying the desired one.

However, for much of the object plane the effect will be negligible. The argument of the Bessel function produced by the $\cos(\theta-\beta)$ term in the expansion of III.14 is much smaller than that produced by the main $\cos(\theta-\gamma)$ factor, and so is close to unity.

For example, in an F/1.5 imaging system, $D^2 = 10r^2$; if σ is equal to $2r$, corresponding to an image point directly below the annulus, the ratio of the arguments of the main term and of the secondary term arising from the expansion of III.14 is

$$\frac{\text{Secondary}}{\text{Main}} = \frac{2\sigma r}{4D^2 + \sigma^2} = \frac{2r^2}{22r^2} \approx .09$$

We examine the J_0 function, and find that when the main term has decreased to half of the peak value (argument = 1.33), the $J_0(1.33 \times .09)$ arising from the secondary term is .9964. The change in response caused by the secondary term is less than .4%. For higher f-number systems or areas closer to the axis, the error will be less. The general procedure above can be used to find the reduction of response as the aiming point moves away from the axis of the annulus.

The higher order terms of eq. III.14 will have much smaller arguments and the resulting Bessel functions, derived like those in Appendix I, can be regarded as equal to unity.

If we disregard the slight-effect of the $\cos(\theta-\beta)$ term in the denominator of III.12, we can write

$$R-L = \frac{(\Delta^2/2) - \Delta r \cos(\theta-\gamma)}{D'} \qquad (III.15)$$

Comparison of eq. III.15 with the argument of eq. 7

$$R = \frac{(\rho^2/2) - \rho r \cos(\theta-\alpha)}{D}$$

shows that the modified signal $e^{-ik(R-L)}$ obtained contains Δ, the distance between the aiming point and the source point, in place of ρ, the distance from the real axis of the annulus. Therefore the response of the annulus is as if the axis were shifted to the aiming point, except that the effective distance D' is greater than D. As the point source response is of the form $f(2\pi\Delta r/\lambda D')$, D' being larger broadens the response. The amount of broadening can be calculated from D':

$$D' = D + \frac{\sigma^2}{4D} \qquad (III.16)$$

Besides being broadened, the response is lowered by the term arising in the expansion of III.14. This reduction in response is calculable, however, and therefore can be adjusted for.

AN ELECTRONICALLY FOCUSED ACOUSTIC IMAGING DEVICE

J.F. Havlice, G.S. Kino, J.S. Kofol
and C.F. Quate

Microwave Laboratory
W.W. Hansen Laboratories of Physics
Stanford University
Stanford, California. 94305

INTRODUCTION

We describe in this paper a new technique for processing acoustic information from a piezoelectric array. The device we shall describe is capable of presenting dynamic, nearly real time images of acoustic objects whether they be internal body organs in medical applications, wreakage in the sea, or flaws in non-destructive testing. The present device is capable of 1 mm´ resolution at distances of 20 cm and operates without the use of an external focusing element or an intermediary hologram. In the experimental results reported here two dimensional images are obtained using electronic scanning in one dimension and mechanical scanning in the other dimension. We will describe near the end of this paper how a fully electronic two dimensional scan may be implemented. The sensitivity of the imaging apparatus is expected to be sufficient to insure low sound power levels while still obtaining high quality images.

ELECTRONIC SCANNING AND FOCUSING OF ACOUSTIC BEAMS

In our system the acoustic image information is received by an array of piezoelectric detectors. The major difficulty in constructing such an acoustic imaging device is to scan this array in an efficient and economical manner. The technique we have chosen uses an acoustic surface signal wave traveling along an acoustic surface wave delay line as the basic scanning element. A schematic-pictorial diagram of the essential elements of the device is shown in Fig. 1. We will confine our attention for the moment to a one dimensional array and consider the two-dimensional problem later. A series of equally spaced taps is placed along the delay line, each tap corresponding to an individual transducer. The taps locally sample the acoustic surface wave amplitude and phase and these local signals are mixed by simple diodes with the signals from the corresponding detectors. The tapped surface wave line performs two functions: i) it converts a time varying signal to a spatially varying signal by its delay properties and ii) it samples this spatially varying signal at various points via the taps and applies these signals simultaneously to all of the diodes. This allows us to add the electrical signals from the individual elements while performing the necessary signal processing locally.

Also across each diode is connected one of the elements of the receiving transducer array. Hence two separate electrical signals are applied to each diode. The first, typically at about 4 MHz, arises from the acoustic object which we wish to image and the second, typically at 50 MHz, arises from the surface wave delay line which scans and processes the image information. The diodes serve as local mixers to generate sum and difference frequencies, typically 54 and 46 MHz. The electrical imaging output of the device is received at one of these two frequencies.

In order to understand the scanning operation of this device, we first consider the situation when a short acoustic pulse is sent along the delay line such that only one tap at a time is excited. In this case an output signal is obtained at the sum frequency only when the pulsed signal passes by a tap and there is a signal present on the corresponding detector.[1] As the pulse passes along the delay line, it scans each detector in turn so that the output may be used to intensity modulate a cathode ray tube, and hence display a visual image corresponding to one line of the acoustic image. The acoustic pulse thus acts like the scanning electron beam in a vidicon.

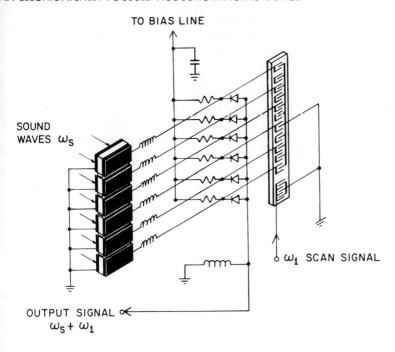

FIG. 1-- Schematic-pictorial diagram of acoustic imaging
         System

The arrays of acoustic detectors which are presently
used to record an acoustic image serve much the same function
as photographic film in recording an optical image.[2,3]
The film merely records the light intensity which falls on
a given element. An optical image of an object is formed
by using the film in combination with an optical lens.
This lens collects the optical energy falling on its entire
surface and bends the ray paths originating from a source
point in such a way that the energy is directed toward a
given point on the photographic film. At the same time the
lens alters the phase in such a way that all rays arrive
at this point with equal phase - regardless of which point
they pass through on the lens surface - and the signals
add algebraically. The individual source points form
corresponding points on the film, and the recorded image
is a faithful reproduction of the object. In just the same
way as in optics, an "acoustic lens" is required to form

an image at the acoustic detector array. In conventional
systems this is a plastic or liquid filled lens shaped
similarly to an optical glass lens.

The action of an optical lens is usually explained
by resorting to classical ray optics, and Snell's law,
which itself is derived by determining the phase delay of
a propagating wavefront.[4] Because the phase velocity of
an optical wave is less in the lens material than in air,
a ray incident on the lens suffers a phase delay proportional to the thickness of the lens at each point. Let us
consider the two ray paths shown in Fig. 2: i) a ray
passing through the axis of the lens, and ii) a ray
passing through a point distance x from the lens axis.
If the lens were not present, the phase delays for the two
rays would be

$$\Phi_1 = \frac{2\pi}{\lambda_0} [d_0 + d_i] \tag{1}$$

$$\Phi_2 = \frac{2\pi}{\lambda_0} [(d_0^2 + x^2)^{\frac{1}{2}} + (d_i^2 + x^2)^{\frac{1}{2}}] \tag{2}$$

respectively. This latter expression can be simplified for
values of x much smaller than d_0 and d_i (the so-
called paraxial approximation) to:

$$\Phi_2 = \frac{2\pi}{\lambda_0} \left[d_0 + d_i + \frac{x^2}{2} \left(\frac{1}{d_0} + \frac{1}{d_i} \right) \right] \tag{3}$$

It will be seen that ray (2) experiences an added amount
of phase delay which is proportional to x^2 .

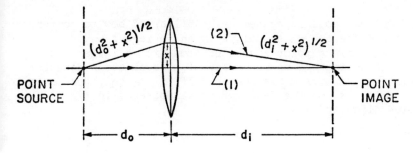

FIG. 2--Ray paths for an optical or acoustic lens

The phase delay through the lens compensates for this phase difference.[5] The thickness of a lens of the correct focal length is decreased as x increases in such a way that the phase shift, Φ_L, introduced by the lens has a value

$$\Phi_L = A - \frac{2\pi}{\lambda_0} \frac{x^2}{2} \left(\frac{1}{d_0} + \frac{1}{d_i} \right) = A - \frac{2\pi x^2}{2\lambda_0 f}$$

(4)

where f is the focal length of the lens and A is some constant. With this addition, the total phase shift from a point on the source to the corresponding image point is simply

$$\Phi \approx (2\pi/\lambda_0) (d_0 + d_i) + A$$

(5)

This phase shift is independent of the parameter of x, and all point on the lens surface produce rays which arrive in phase at the image point on the image plane of the film in a camera.

In the acoustic system we have learned how to compensate for the different phase delays in an analogous way. We consider again a point source of acoustic radiation and allow it to impinge on the array of piezoelectric detectors a distance d_0 from the source. If $x << d_0$, the phase at a point x on the detector as illustrated in Fig. 3 , varies directly as

$$\Phi_A = \omega_s t - (2\pi/\lambda_0)(d_0^2 + x^2)^{\frac{1}{2}} \approx \omega_s t - \frac{2\pi}{\lambda_0}(d_0 + \frac{x^2}{2d_0})$$

$$(6)$$

where ω_s is the radian frequency of the acoustic wave.

As before we see that the phase is a quadratic function of the parameter x . It is this term that we must compensate in order that each element of the detector will contribute components which are "in phase" to the output signal. To accomplish this we convert the acoustic signal to an electrical signal at each piezoelectric element of the detector array. We then add an electrical phase shift to each element of an amount that compensates the $(2\pi/\lambda_0)$ $(x^2/2d_0)$ term that arises from the different path lengths. In principle, this could be done with a computer after first converting to digital information. Our purpose here is to show that the acoustic surface wave delay line can be used to perform this function in a simple and direct manner.

When the piezoelectric array is illuminated by a point source, the amplitude of the imaging signals are almost constant from diode to diode, but the phase now corresponds to the sum of the phases Φ_A of the incident wave and the Φ_m of the surface wave signal. The aim is to adjust Φ_m at each tap element so as to cancel the quadratic phase term $(2\pi/\lambda_0)$ $(x^2/2d_0)$ in the incoming wave. We must therefore drive the mixing diodes with an electrical signal whose phase at any point is of the form: $\Phi_m = A - (2\pi/\lambda_0)$ $(x^2/2d_0)$. The phase of the mixing signal

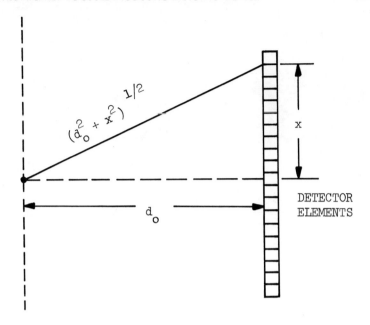

FIG. 3--Ray paths for a point source illuminating the
acoustic imaging system.

is determined by the phase of the traveling wave sent along
the delay line. As we have already seen, we need for lens
action a quadratic variation of phase ϕ_m with x and,
therefore, we must use a mixing signal which is modulated
in a particular manner. If the surface wave frequency is
constant the phase shift at each successive tap is a
linear function of x . But if we inject a signal with a
frequency which varies linearly with time, the so called
"chirp" signal as $\omega = \omega_1 + \mu t$, the phase will vary
quadratically with time. With this input the phase of the
traveling wave becomes

$$\phi_m = \left[\omega_1 + \frac{\mu}{2} \left(t - \frac{x}{v} \right) \right] (t - x/v) \qquad (7)$$

or

$$\omega_m = \omega_1 (t - \frac{x}{v}) + \frac{\mu}{2} (t^2 - \frac{2xt}{v} + \frac{x^2}{v^2}) \qquad (8)$$

The value of μ (the frequency sweep rate) is now adjusted so that the x^2 term in ω_m just cancels the x^2 term in ω_A. This is equivalent to adjusting the focal length of a lens. Rather than exciting one tap at a time as before, we use a long scan pulse, coded with the proper "chirp", so that all the taps are excited simultaneously. In this way we form the electronic equivalent of the optical lens and construct a system for summing in phase all the contributions from each array element.

An additional point to consider is that this quadratic phase signal is encoded onto a traveling surface wave. This means that the spatial position of the compensating waveform changes with time and in fact moves from one end of the array to the other at the velocity of the acoustic wave on the delay line. Hence the device interrogates a line located a distance d_0 from the array with d_0 determined by the frequency sweep rate and the lateral source position determined by the spatial position of the surface wave. The traveling surface wave acts like the scanning beam in a cathode ray tube.

A theoretical analysis of this imaging scheme has been carried out and we summarize below the technically important results. The resolution of the device is determined by the total aperture of the electronic lens, just as with any other type of lens.[6] The resolution is

$$\text{Resolution} = \frac{\lambda}{2 \sin (\theta/2)} \qquad (9)$$

or

$$\text{Resolution} \cong \frac{\lambda \, d_0}{D} \quad (\sin \theta/2 << 1 \text{ or } D << d_0)$$

$$(10)$$

where λ is the wavelength of the acoustic image signal, d_0 the distance of the object from the array, and D the total aperture of the array. θ is the angular aperture of the lens, i.e. the angle subtended by the two extreme rays.

There is one significant difference however, between the optical analogy and the acoustic device. Due to the discrete nature of the receiving array, the number of resolvable spots is limited in the acoustic case to the number of elements in the receiving array. A typical case can be illustrated by a device operating at 4 MHz with an array 10 cm wide consisting of 100 elements. The device is capable of 1 mm resolution for objects 25 cm away. The unambiguous field of view is limited to 100 resolvable elements, or 10 cm. With the same array for an object distance of 15 cm, the device is capable of 0.6 mm resolution with an unambiguous field of view of 6 mm.

Much of the theoretical work can be summarized in contour plots. These curves are three dimensional in content with a vertical dimension proportional to output imaging power and two horizontal dimensions: lateral position along the array and range (or distance from the array). A typical example is shown in Fig. 4 with the calculations based on the experimental device described later. It consisted of a 30 element receiving array with center to center spacings of 2 mm. The array is assumed to be focussed on a point 25 cm away and the power output for an acoustic source exactly at this focal point is shaded for reference. Other sources at intervals of 2.5 cm on either side of this point contribute successively less to the image. This corresponds, of course, to out of focus, blurred images of these points. The width of the signal in the lateral dimension indicates the resolution (about 1 mm for the case illustrated). The depth of the field can

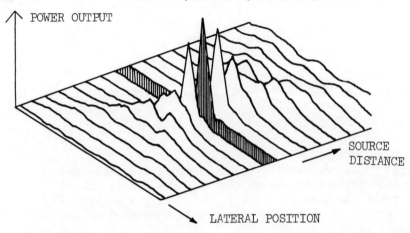

FIG. 4--Chirp rate set for 25 cm, (each line stepped by
1.25 cm).

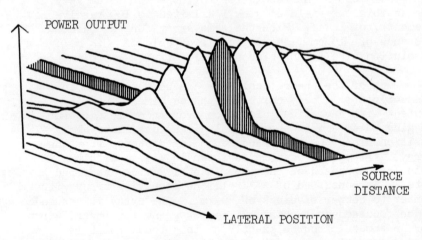

FIG. 5--Chirp rate set for 50 cm, (each line stepped by
2.5 cm).

easily be estimated as well as side-lobe structure.
Another example of a contour plot is shown in Fig. 5
where the same array is assumed focussed at 50 cm .
The resolution (pulse width) has worsened somewhat and
the depth of field has increased as predicted above.

EXPERIMENTAL RESULTS

In order to test these ideas a thirty element linear
array was constructed. It consisted of thirty 1.2 mm
wide receiving transducers made from PZT-5, a line of
parallel connected, forward biased 1N914 diodes, and
thirty surface wave taps 3.2 mm apart on a bismuth
germanium oxide substrate. The piezoelectric array was
mounted onto a water tank and exposed to acoustic
radiation from a 4 MHz source located in the water some
distance in front of the receiving array. In the first
experiments an acoustic point source was constructed by
placing a diverging lens in front of a plane wave transducer.
In Fig. 6 is shown the amplitude distribution of sound
across the array. Each pulse corresponds to one of the
array elements and as can be seen, approximately twenty
of the thirty elements were illuminated covering a time
interval of about 40 μsec. When a surface wave with the
proper "chirp" signal was applied, the result of Fig. 7
was obtained. In this case the output at the difference
frequency consisted of a single large pulse with a half-
width of 2 μsec and a peak power level 25 dB above that
obtained using the short scan pulse technique. This agrees
well with the theoretical predictions. As the point source
was translated parallel to the array, the image signal moved
to an earlier and then a later time as required for scanning.
Different "chirp" rates are required for different distances
of the point source from the array. Shown in Fig. 8 is a
plot of the chirp rate required for focussing versus source
distance. The solid line represents the theoretical calculation and the brackets the measured values. Very close
agreement is again obtained.

In order to demonstrate the potential of this device
we have simulated a two dimensional imaging system in
which one of the dimensions was scanned mechanically.
Eventually, of course, both dimensions will be scanned
electronically, but in order to demonstrate quickly the
type of result possible, the mechanical scan was used.

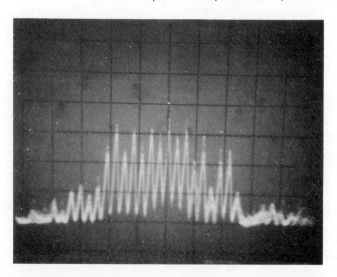

FIG. 6--Received signals from acoustic imaging device. Each pulse corresponds to one illuminated element in the array.   Total time duration is 40 μsec.

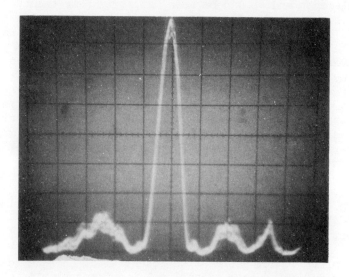

FIG. 7--Received pulse using chirp scan signal.   Power output increased by 25 db and time duration of 4 μsec.

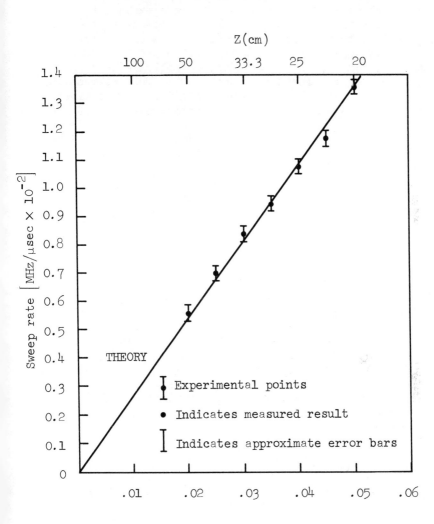

FIG. 8--Theoretical and experimental sweep rates for
4 MHz acoustic illumination.

The transducer array was illuminated by a plane wave
transducer so that all the elements were exposed to
sound. Twenty five (25) centimeters in front of the array
between the transmitter and receiver, a small crescent
wrench was immersed in the water and physically moved up
and down. The vertical position controlled a battery
potentiometer combination so that a dc voltage proportional to the vertical height of the object was obtained.
The dc voltage controlled the vertical position of a
cathode ray tube trace, thus providing a vertical scan.
The horizontal scan was obtained electronically as before.
The result is shown in Figs. 9a and 9b. The images are
a reasonably good representation of the object. The
handle of the wrench was approximately 9 mm wide and
the open position at the top approximately 10.5 mm .
The resolution of the system was measured to have been
better than 1 mm as compared to a theoretical prediction
of 0.75 mm.

In Fig. 9a the image signals were displayed directly
by intensity modulating a cathode ray tube. The bright
background corresponds to the plane wave illumination of
the array and the dark region that portion of the plane
wave obstructed by the wrench. In Fig. 9b the image
signals were reversed in contrast electronically. Hence
the background appears dark and the wrench bright. The
dynamic range is greater in this latter image and internal
structure to the wrench appears as a central dark region.
This dark region corresponds to an opening in the wrench
to provide motion for the carriage that opens and closes
the open end.

TWO DIMENSIONAL ELECTRONIC SCANNING AND FOCUSING

A two dimensional device may be constructed by
replacing the linear array with a square array as shown
in Fig. 10. In the proposed scheme, on one side is
deposited a series of parallel metallic strips and on the
opposite side a two dimensional matrix of metallized regions.
A separate diode is connected to each element of the matrix
so that the appropriate signal processing can occur locally
as before. One set of strips is connected to the taps on
one of the surface wave delay lines. The focusing and
scanning occurs in one dimension just as before except
that many lines are processed simultaneously. The outputs
are taken from a single line of the matrix to another

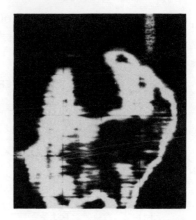

FIGS. 9a and 9b--Acoustic images of a small crescent wrench located 25 cm from the array. The photograph in Fig. 9b shows the internal structure of the wrench.

independent series of diodes and the taps of a second surface wave delay line. This second surface wave is also a "chirp" signal which scans and focuses the individual outputs from each line in the previously unfocussed dimension. If the acoustic image signal is at ω_s, the first surface wave at ω_1, and the second at ω_2, the output signal appears at $\omega_s + \omega_1 + \omega_2$. The output signal at a time t is generated from a point on the object with co-ordinate $x_0 = v\,t$ and $y_0 = v(t - \tau)$. Thus a line at 45° to the x-axis is scanned. This line may be displaced by changing the delay time τ between the two surface wave signals and a complete raster is traced out.

CONCLUSIONS

A system has been described and demonstrated for linear electronic scanning and focusing of acoustic beams. Acoustic images of an object 25 cm in front of the array have been presented with a resolution of 1 mm. A two dimensional version has been described which will allow a fully electronic two dimensional system. It is expected that TV frame rates will be easily obtained in this system.

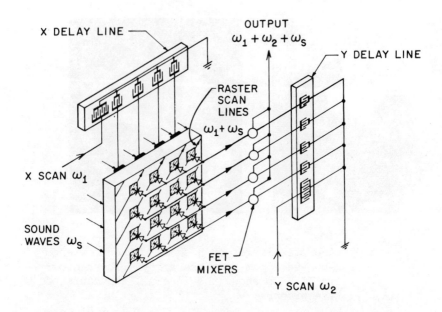

FIG. 10--Proposed two dimensional imaging scheme.

ACKNOWLEDGEMENTS

The authors gratefully acknowledge the fine technical
assistance of D. J. Walsh and L. C. Goddard. We would also
like to thank Harper Whitehouse of the Naval Undersea
Laboratory for many stimulating conversations on the
concepts in this paper.

This work was supported by the Naval Undersea Center,
Contract N000123-72-C-0866 and the Office of Naval Research
Contract N00014-67-A-0112-0039. We wish, also, to acknowledge
the support of the U. S. Army Research Office, Durham, for
support of the basic work on acoustic scanning that formed
a foundation for this work.

REFERENCES

1.   J.F. Havlice and T.M. Reeder, "Scanning of Acoustic
     Arrays", in 1972 Proceedings of Ultrasonics Symposium
     Oct. 4 - 7, 1972, p.463.

2.   W.H. Wells, "Acoustic Imaging with Linear Transducer
     Arrays", in Acoustical Holography, Vol. 2, Plenum Press,
     (1970).

3.   N. Takagi, et al., "Solid State Acoustic Image Sensor",
     in Acoustical Holography, Vol. 4, Plenum Press,   (1972).

4.   R.W. Wood, Physical Optics, Dover Publications, (1934).

5.   J.W. Goodman, Introduction to Fourier Optics, McGraw-
     Hill, (1968).

6.   op.cit.

RIGOROUS ANALYSIS OF THE LIQUID-SURFACE ACOUSTICAL HOLOGRAPHY SYSTEM

P. Pille

Box 758

Waterdown, Ontario, Canada

B. P. Hildebrand

Battelle-Northwest

Richland, Washington 99352

The liquid-surface focussed-image acoustical holography system has attained a certain level of maturity with the availability of commercial units.[1] The system was developed experimentally, with analysis following later. The first analysis was performed by B. B. Brenden simply to provide a better understanding of the concept.[2] Later, T. J. Bander extended the analysis to include pulsed operation which Brenden has experimentally shown to be much superior to continuous wave operation.[3] This analysis showed that the liquid surface acted like a classical low-pass filter whose bandwidth increased inversely with pulse length. In this analysis we solve the linearized Navier-Stokes equation by the finite difference method of Harlow and Welch to obtain the response of the liquid surface to time and space variations in radiation pressure.[4] We also include the effect of the isolation or mini-tank invented by Brenden as a means of isolating the detecting surface from environmental disturbances.

ANALYSIS

The basic imaging system is shown in Figure 1.

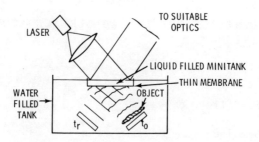

Figure 1. The Basic Imaging System

The acoustic field distribution in the fluid will be represented by the velocity potential $\phi(x,y,z,t)$. Since ϕ is a scalar and satisfies the scalar wave equation, the scalar laws of diffraction apply directly to ϕ. Also, ϕ will be considered to be a complex quantity, and since the ultrasonic field will be monochromatic, there will be an implied time factor of the form $e^{j\omega t}$ (where $j = \sqrt{-1}$). However, this time dependence will not be explicitly shown, and the time dependence of $\phi(x,y,z,t)$ will refer to time changes in amplitude, for example a step change

$$\phi(x,y,z,t) = \phi(x,y,z)u(t), \tag{1}$$

$$\text{where } u(t) = 1 \, , \ t > 0$$

$$= 0 \, , \ t < 0 \ .$$

The system can be described in terms of the block diagram shown in Figure 2. Some of the blocks are linear, and as such can be treated with transfer function concepts. The block labelled minitank turns out to be linear in the space variables but not in time. The radiation pressure-to-surface deformation block is a non-linear system but can be made piece-wise linear. Some of the blocks are definitely non-linear and must be treated as such. We will concentrate on the three block unit enclosed by the dotted lines.

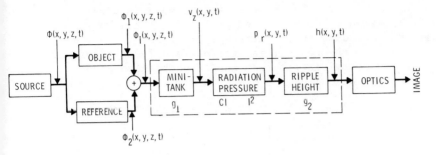

Figure 2. Block Diagram of the Imaging System

Object and Reference Structures

The object and reference structures describe the object to be imaged and any acoustical optics used in either the object or reference beams. We will, however assume that $\phi(x,y,z,t)$ is a plane wave, the object is imaged into the detection surface, and the reference is unaltered except for direction of propagation. Thus, we have at the bottom of the minitank ($z = 0$)

$$\phi_2(x,y,t) = R_2(t)\exp[j(u_2 x + v_2 y)] \quad , \qquad (2)$$

where R_2 = amplitude

$$u_2 = k_s \cos \theta_{x2}$$

$$v_2 = k_s \sin \theta_{y2}$$

$$k_s = 2\pi/\lambda_s$$

$$\cos^2 \theta_{x2} + \cos^2 \theta_{y2} + \cos^2 \theta_{z2} = 1$$

λ_s = wavelength of the sound, and $\cos \theta_2$ are the direction cosines of the plane wave.

The effect of the object on the plane wave is simply a
multiplication. That is, the amplitude of the wave is
altered according to the attenuation or reflectance distribution of the object, and the phase delayed according
to the velocity distribution and/or pathlength changes
induced by the object. A plane wave

$$\phi(x,y,z,t) = R(t) \exp [j(ux + vy)] \ ,$$

will emerge as the wave

$$\phi_0(x,y,z,t) = R(t) \ 0(x,y,z) \ \exp[j(ux + vy + \Psi(x,y,z))] \ ,$$

where $0(x,y,z)$ = object amplitude effect,

and $\Psi(x,y,z)$ = object phase effect.

If a system of lenses is used to image ϕ_0 into the detecting plane

$$\phi_1(x,y,t) = m \ \phi_0(mx, \ my, \ t) \overset{s}{*} q(mx, \ my) \ , \tag{3}$$

where m = magnification ,

q = spread function of the imaging system

and $\overset{s}{*}$ denotes space convolution.

Since we are dealing with coherent imaging the trans-
fer function $\overline{q}(u,v,t)$ is simply a perfect low-pass function
with cutoff at spatial frequency

$$r_0 = \sqrt{u_0^2 + v_0^2} = \frac{L}{2d_i \lambda_s} \ , \tag{4}$$

where d_i = distance from exit pupil
to image,

and L = diameter of exit pupil.

Typical numbers, L = 20 cm, di = 20 cm and λ_s = 0.15 mm
yield a cutoff of 33.3 ℓ/cm. This example is for 1 MHz
sound in water. Equation 4 holds for a diffraction-
limited imaging system; any aberrations or focus error
will decrease the cutoff substantially.[5]

The effect of the reference beam is simply to shift the low-pass object information onto a carrier of spatial frequency

$$\frac{2 \sin \Theta \sin \Phi/2}{\lambda_s} , \qquad (5)$$

where Φ = angle between the propagation directions of the reference beam and object beam and

Θ = angle between the bisector of Φ and the plane of the hologram.

For example, the bandwidth of 33.3 ℓ/cm cited in the last paragraph will be shifted to a bandpass function on a spatial carrier of 34.9 ℓ/cm when Θ = 90° and Φ = 30°. Thus, the detecting surface must respond to spatial frequencies of 34.9 - 68.2 ℓ/cm.

Minitank

In the formation of acoustical holograms on a liquid surface a minitank is used as shown in Figure 3. A thin membrane separates the liquid in the minitank from the water in the main tank to prevent the surface from being disrupted by any distrubances in the water that are not associated with the sound field.

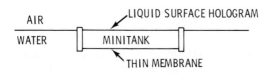

Figure 3. Minitank

Pulses of ultrasonic waves that form the hologram pass from the water through the membrane and are multiply reflected by the boundaries. In this section we shall determine the effect of this as well as determining the effective velocity potential at the surface. It will be shown that a time dependency is introduced into the system by the minitank. We shall assume that the presence of membrane has negligible effect on the passage of sound

waves, requiring that it be much thinner than the wavelength in the medium.

The linearized equations describing sound fields in
ideal fluids are[6]

$$\frac{\partial^2 \phi}{\partial t} - c^2 \nabla^2 \phi = 0 \, , \tag{6}$$

$$\vec{v} = \nabla \phi \, , \tag{7}$$

$$p' = -\rho \frac{\partial \phi}{\partial t} \, , \tag{8}$$

$$c^2 = \frac{\partial p}{\partial \rho_o} \, , \tag{9}$$

$$p' = (\frac{\partial p}{\partial \rho_o}) \rho' \, . \tag{10}$$

The symbols $\vec{v}$, ϕ, p and ρ denote the fluid velocity,
velocity potential, pressure and density respectively.
The velocity of propagation of a sound wave is given by c.
The primes denote variations about the static value, where
the static value is indicated by the subscript o.

i.e., $p = p' + p_o$,

$\rho = \rho' + \rho_o$.

The linearized equations are valid if

$v \ll c$, or $\rho' \ll \rho_o$.

When a travelling plane wave ϕ_i strikes a plane
boundary (z=0) between two different fluids we obtain the
reflected and transmitted waves ϕ_r and ϕ_t as shown in
Figure 4.

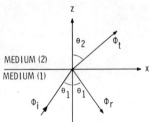

Figure 4. Waves at a Boundary

If the incident, reflected and transmitted waves are

$$\phi_i (x,z) = A_i \, e^{j(u_1 x + w_1 z)} \, ,$$

$$\phi_r (x,z) = A_r \, e^{j(u_1 x - w_1 z)} \, , \tag{11}$$

$$\phi_t (x,z) = A_t \, e^{j(u_2 x + w_2 z)} \, ,$$

then we have the following relationships between the three waves:[6]

$$\frac{A_r}{A_i} = \frac{Z_2 \cos\theta_1 - Z_1 \cos\theta_2}{Z_2 \cos\theta_1 + Z_1 \cos\theta_2} \, , \tag{12}$$

$$\equiv R_{12} \, ,$$

$$\frac{A_t}{A_i} = \frac{\rho_1}{\rho_2} \frac{2 Z_2 \cos\theta_1}{Z_1 \cos\theta_2 + Z_2 \cos\theta_1} \, , \tag{13}$$

$$\equiv T_{12} \, ,$$

and

$$\frac{\lambda_1}{\lambda_2} = \frac{k_2}{k_1} = \frac{c_1}{c_2} = \frac{\sin\theta_1}{\sin\theta_2} \, , \tag{14}$$

where

$$Z = \rho c \, , \tag{15}$$

$$\left.\begin{array}{l} u_j = k_j \sin\theta_j \\[2mm] w_j = k_j \cos\theta_j \\[2mm] k_j = 2\pi/\lambda_j \end{array}\right\} \quad j = 1, 2 \, , \tag{16}$$

and R_{12} is the reflection coefficient of a plane wave travelling in medium (1) and reflected from a plane boundary with medium (2) as defined in Equation 12. Similarly T_{12} is the transmission coefficient of a plane wave travelling from medium (1) into medium (2) as defined by Equation 13.

As a plane wave travels through a boundary the spatial frequency u_i along the x direction does not change from one medium to another. We shall write.

$$u_1 = k_1 \sin\theta_1 \tag{17}$$

$$= \eta .$$

Hence, $\cos\theta_1 = \sqrt{(1 - \eta^2/k_1^2)} .$ (18)

Then we can write R_{12}, T_{12}, w_1 in terms of spatial frequency η.

If medium (2) is a gas and medium (1) is a liquid then $\rho_1 \gg \rho_2$ and $R_{12} \simeq -1$. The velocity potential ϕ in the fluid (1) becomes the sum of the incident and reflected waves as

$$\phi(x,z) = \phi_i(x,z) + \phi_r(x,z)$$

$$= j\, 2\, A_i \sin w_1 z\, e^{ju_1 x} . \tag{19}$$

Then from Equation 7 the velocities v_x and v_z at the surface z=0 are

$$v_x \Big|_{z=0} = 0$$

$$v_z \Big|_{z=0} = j\, 2\, w_1\, A_1\, e^{ju_1 x}$$

$$= j\, 2\, w_1\, \phi_i \Big|_{z=0} . \tag{20}$$

Thus the velocity at the surface is related to the incident velocity potential by Equation 20, and the velocity at the surface due to a single plane wave of spatial frequency η becomes

$$v_z \Big|_{z=0} = j\, 2\, k_1 \sqrt{1 - \eta^2/k_1^2}\ \phi_i \Big|_{z=0} . \tag{21}$$

For an arbitrary velocity potential distribution in 2 dimensions we can then associate a transfer function relating the velocity and the incident velocity potential at the surface as follows.

$$\bar{v}_z (u,v) = j \, 2 \, k_1 \, \bar{g}_1 (u,v) \, \bar{\phi}_i (u,v) \, , \quad (22)$$

or

$$v_z (x,y) = j \, 2 \, k_1 \, g_1 (x,y) \overset{s}{*} \phi_i (x,y) \, ,$$

where

$$g_1 (x,y) \overset{s}{\leftrightarrow} \bar{g}_1 (u,v) = \sqrt{1 - \eta^2/k_1^2} \, , \quad (23)$$

$$\eta^2 = u^2 + v^2 \, ,$$

and $\overset{s}{\leftrightarrow}$ denotes a spatial Fourier transform pair. Note that $g_1 (x,y)$ is not a function of time, also, v here is a spatial frequency and not velocity. If we now introduce the mini-tank we get a further effect due to multiple reflections in the fluid layer. Consider the situation in Figure 5 where a plane wave ϕ_i in medium (1) passes into medium (2) (the minitank) and strikes the surface at time t = 0 and is reflected at the surface and interface a number of times. In medium (1) we have

$$\phi_i (x,z,t) = e^{j[u_1 x + w_1 (z+d)]} u(t) \, , \quad (24)$$

where

$$u(t) = \begin{cases} 0 \text{ for } t < 0 \\ 1 \text{ for } t > 0 \, . \end{cases}$$

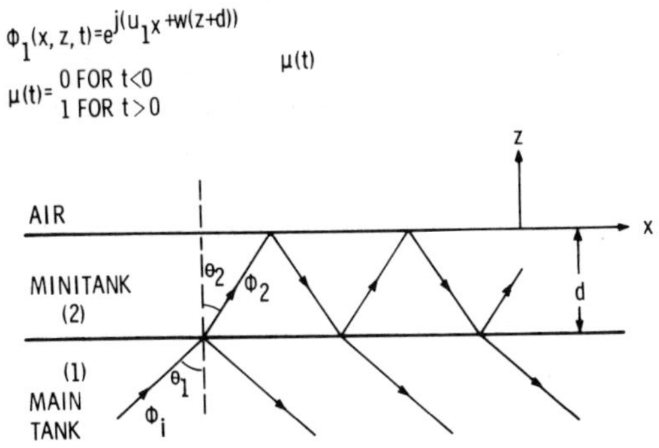

Figure 5. Reflections in the Minitank

The velocity potential incident on the surface in medium (2) is

$$\phi_2 \Big|_{z=0} = T_{12} \; e^{j(u_2 x + w_2 d)} \; u(t)$$

$$= T_{12} \; \phi_i \Big|_{z=0} \; u(t)$$

(25)

and the velocity at the surface is then, taking into account the multiple reflections

$$v_z \Big|_{z=0} = j2k_2 \; \sqrt{(1 - \eta^2/k_2^2)} \; T_{12} \; \phi_i$$

$$\cdot \sum_{n=0}^{\infty} (-1)^n R_{21}^n \; e^{j2nw_2 d} \; u(t - t_n)$$

(26)

The factor $j \, 2 \, k_2 \sqrt{(1 - \eta^2/k_2^2)} \; \phi_i$ is similar to the result in Equation 21 when no minitank was present while T_{12} is the transmission coefficient in passing from medium (1) into medium (2). Each time the wave reflects off the surface, then the interface, and reaches the surface again, it changes in phase by $(-1) R_{21} \exp(j \, 2 \, w_2 d)$ along z, where (-1) and R_{21} are the reflection coefficients at the surface and interface respectively, and d is the thickness of the fluid layer (2). The number of reflections off the surface is given by n + 1. Also,

$$t_n = n \; 2 \; d/c \; \sqrt{(1 - \eta^2/k_2^2)} \; ,$$

(27)

where c is the velocity of propagation and $2d/c \sqrt{(1 - \eta^2/k_2^2)}$ is the time between reflections off the surface.

Similarly, if the wave ϕ_i is a short pulse or impulse

$$\phi_i(x,z,t) = e^{j[u_1 x + w_1 (z+d)]} \; \delta(t) \; ,$$

(28)

then the velocity at the surface will be a sum of impulses, delayed in time, of different phase and diminishing amplitude,

$$v_z\Big|_{z=0} = j\ 2\ k_2\ \sqrt{(1-\eta^2/k_2^2)}\ T_{12}\ \phi_i$$

$$\cdot\sum_{n=-\infty}^{\infty} (-1)^n\ R_{21}^n\ e^{j2nw_2 d}\ \delta(t-t_n)\ .$$

(29)

We can relate the velocity at the surface to the incident velocity potential in terms of the three dimensional Fourier transform $\overline{\overline{g}}_1\ (u,v,\omega)$ as

$$\overline{\overline{v}}_z(u,v,\omega) = j\ 2\ k_2\ \overline{\overline{g}}_1\ (u,v,\omega)\ \overline{\overline{\phi}}_i(u,v,\omega)\ ,$$ (30)

or

$$v_z(x,y,t) = j\ 2\ k_2\ g_1(x,y,t)\ *\ \phi_i(x,y,t),$$ (31)

where $g_1(x,y,t)$ is the effective value of the velocity potential when the incident velocity potential is an impulse in space and time. Then with

$$g_1(x,y,t) \leftrightarrow \overline{\overline{g}}_1(u,v,\omega) = \overline{\overline{g}}_1\ (\eta,\omega)$$

where $\leftrightarrow$ denotes a Fourier transform pair in space and time, we have

$$\overline{\overline{g}}_1\ (\eta,\omega) = \sqrt{(1 -\eta^2/k_2^2)}\ T_{12}(\eta)$$

$$\cdot\sum_{n=0}^{\infty} (-1)^n\ R_{21}^n\ (\eta)\ e^{j2ndw_2(\eta)}\ e^{j\omega t_n(\eta)}$$ (32)

where $e^{j\omega t_n}$ is the Fourier transform in time of $\delta(t-t_n)$.

One final item that should be mentioned is that Equations 12 and 13 are valid for

$$0 < \eta < (\text{smaller of } k_1, k_2).$$

We therefore desire the properties of the fluid media (1) and (2) to be such that

$$k_2 > k_1\ (\text{or } c_2 < c_1)\ ,$$

otherwise no spatial frequencies from the object higher than k_2 will pass through the interface, from the water, into the fluid in the minitank. In other words, if $k_2 < k_1$, all plane waves striking the interface (1) - (2) with angles θ greater than the critical angle θ_c, where

$$\theta_c = \text{arc sin } (k_2/k_1) ,$$

will be totally reflected at the interface.

We now consider $\bar{\bar{g}}_1(u,v,\omega)$ in more detail when $\omega = 0$, i.e., when there is no time variation of the incident ultrasound field. This is equivalent to looking at the time step response after it has reached a steady state ($n = \infty$).

The simplest case occurs when liquids (1) and (2) are the same. Then $T_{12} = 1$, $R_{21} = 0$ and

$$\bar{\bar{g}}_1 (\eta,o) = \sqrt{1 - \eta^2/k_2{}^2} , \tag{33}$$

as there would be no internal reflections in the minitank, but the high spatial frequencies would be attenuated. If the two liquids are different, then $T_{12} \neq 1$, $R_{21} \neq 0$ and we must sum the effects of the internal reflections as given by Equation 32.

Some values of $|\bar{\bar{g}}_1(\eta,0)|$ are plotted in Figures 6 to 9 for various depths of fluid d. In Figures 6 and 8 the curve for $\sqrt{(1 - \eta^2/k_2{}^2)}$ is the value of $|\bar{\bar{g}}_1(\eta,0)|$ if the fluids (1) and (2) are the same. It is apparent in Figure 6 that the minitank has the effect of flattening the curves for $|\bar{\bar{g}}_1(\eta,0)|$ for some depths, but not for others, and that the curves are fairly sensitive to depth. For the greater depths d in Figure 8 the curves become oscillatory for the higher spatial frequencies.

The phases of $\bar{\bar{g}}_1(\eta,0)$ are plotted in Figures 7 and 9 corresponding to the absolute values in Figures 6 and 8. In Figure 7 the phase distortion does not appear significant except for the high spatial frequencies. However, these high spatial frequencies might not be present at the surface because of the finite apertures used in the imaging system. For greater depths in Figure 9 the phase distortion becomes

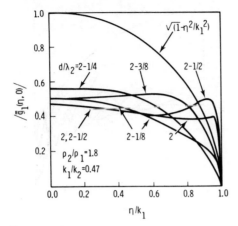

Figure 6. $\left|\bar{\bar{g}}_1(\eta,0)\right|$ Versus η/k_1 for Values of d/λ_2
from 2 to 2 1/2

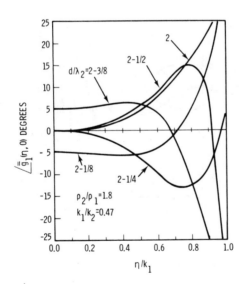

Figure 7. $\angle\bar{\bar{g}}_1(\eta,0)$ Versus η/k_1 for Values of d/λ_2
from 2 to 2 1/2

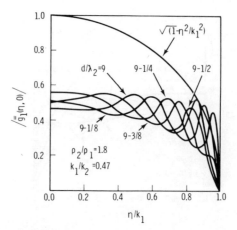

Figure 8. $|\bar{\bar{g}}_1(\eta,0)|$ Versus η/k_1 for Values of d/λ_2
             from 9 to 9 1/2

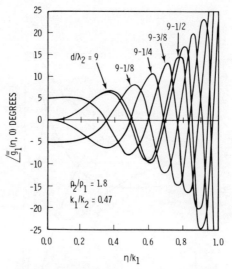

Figure 9. $\angle\,\bar{\bar{g}}_1(\eta,0)$ Versus η/k_1 for Values of d/λ_2
             from 9 to 9 1/2

oscillatory for the higher spatial frequencies. Again, the
curves in Figures 7 and 9 are fairly sensitive to the
depth d.

     Although the curves in Figures 6 to 9 were for n = ∞,
the values of $\bar{\bar{g}}_1(\eta,0)$ for n ≃ 2 or 3 are almost the same,
except for the very high spatial frequencies. That is, it

takes only a few reflections of the sound waves in the minitank to reach steady state for the densities and wavelengths
indicated. The smaller the reflection coefficient $R_{21}(\eta)$,
the more quickly $\bar{\bar{g}}_1(\eta,0)$ reaches a steady state. The
reason why the higher spatial frequncies take more reflections, n, to reach a steady state is because R_{21} increases
with spatial frequency.

Thus we see that the effect of the fluid layer on the
imaging process is to cause amplitude and phase distortions
in the image obtained, especially at high spatial frequencies. Although there may be some depths of the minitank
for which the amplitude versus spatial frequency is quite
uniform, there is a phase distortion at high spatial frequencies. Again, however, the high frequencies may be cut
off by the finite aperture of the imaging system.

Also we can see that for the lower spatial frequencies
the resonance effect caused by constructive interference of
the internal reflections increases the amplitude of $\bar{\bar{g}}_1$.
However, this positive aspect does not appear significant,
as the more important consideration is the uniformity of $\bar{\bar{g}}_1$
with spatial frequency.

It does not appear obvious that there is any optimum
depth of the fluid in the minitank. However, we can at
least say that the curves in Figures 6 and 7 are fairly uniform over a wider range of spatial frequencies than in Figures 8 and 9. Of the curves shown; the best may be for
$d/\lambda_2 = 2 \ 3/8$.

It should also be mentioned that in deriving $\bar{\bar{g}}_1$ the
liquid layer was assumed to be of infinite lateral extent.
That is, we assumed that there were no reflections off the
sides of the finite minitank. It may be desirable to have
a sound absorbing material along the inner sides of the minitank to absorb the ultrasonic waves.

Radiation Pressure

We are now in a position to determine the radiation
pressure applied to the surface of the minitank.

In a sound field, the radiation pressure p_r, exerted
on the fixed surface of an interface between different fluids
or on the surface of an obstacle in the fluid is given by the
time averaged momentum flux per unit area of the surface.
For an ideal fluid this becomes[7]

$$p_r = \overline{p'} \, \vec{n} + \rho \, \overline{\vec{v} \, (\vec{v} \cdot \vec{n})} \quad . \tag{34}$$

The bar denotes a time average, and $\vec{n}$ is a unit vector normal to the surface, pointing out of the fluid.

If the sound field consists of waves that vary sinusoidally in time then $\overline{p'}$ from linear theory is zero. However, this is only a first order approximation, and to obtain a more accurate value for $\overline{p'}$ we must use the equations of non-linear acoustics.

Then for an ideal fluid we have[6,7]

$$\overline{p'} = \frac{c^2}{2\rho_o} \overline{\rho'^2} - \frac{\rho_o}{2} \overline{v^2} \tag{35}$$

which includes terms up to second order. The values of ρ' and v can be obtained from the linearized equations.

If the velocities at the surface in question are normal to the surface ($z=0$) then the radiation pressure p_r is also along the normal, and from Equations 34 and 35 is

$$p_r = \frac{c^2}{2\rho_o} \overline{\rho'^2} + \frac{\rho_o}{2} \overline{v_z^2} \quad . \tag{36}$$

We can now determine the radiation pressure on the liquid surface due to an arbitrary sound field. From Equation 9 $\phi|_{z=0} = 0$ and thus from Equations 8 and 10 $\rho'|_{z=0} = 0$. Then from Equations 36 and 22, with v_z a complex function, the radiation pressure on the surface is

$$p_r(x,y) = \rho_o k_1^2 \, |g_1(x,y) * \phi_i(x,y)|^2 \quad . \tag{37}$$

Ripple Height

In the steady state there is a balance between the liquid surface height, the radiation pressure, gravity and surface tension as follows

$$p_r(x,y) - \rho g \, h(x,y) + \gamma \, \nabla^2 h(x,y) = 0 \quad , \tag{38}$$

where g is the acceleration due to gravity, h the height of the surface above the quiescent level, and γ the coefficient of surface tension.[6]

If we take the Fourier transform of the terms in Equation 38 and solve for $\bar{h}$ we obtain

$$\bar{h}\,(u,v) = \bar{p}_r(u,v)\,\bar{g}_2\,(u,v) \quad , \tag{39}$$

where

$$\bar{g}_2(u,v) = \bar{g}_2(\eta)$$
$$= 1/(\rho g + \gamma \eta^2) \quad . \tag{40}$$

Thus, the height, h, of the surface deformation is related to the radiation pressure p_r by the low-pass transfer function $\bar{g}_2(\eta)$. The amplitude of the function $\bar{g}_2(\eta)$ is down by a factor of 1/2 when the spatial frequency η is given by

$$\eta = \sqrt{(\rho g/\gamma)} \tag{41}$$

For γ = 16 dynes/cm, ρ = 1.8 gm/cm^3 the half amplitude frequency is 1.7 ℓ/cm. It appears then that the resolution we can expect to obtain in an image will be limited by the low pass response of the liquid surface.

Also, we can see that there is a zero frequency component to the surface deformation. This represents a uniform levitation or bulge of the surface, and may actually be a source of distortion in the image. If the ultrasonic beams from transducers t_r and t_o of Figure 1 cover only part of the surface of the minitank then there will be a discontinuity of radiation pressure at the edges of the area covered. Thus, rather than being uniform, the bulge may be curved, especially near the edges, and distort the surface pattern containing the information of the hologram or violate the condition that the space varying component of h should not be large compared to a wavelength of light.

It should be possible to avoid the bulge if the ultrasonic beams cover the entire surface, since then the zero frequency component of h cannot raise up the surface. In this case, the average pressure in the fluid will simply decrease.[9]   However, there may still be a distortion or curvature of the surface at the sides of the minitank due to surface tension. It remains a matter of experiment to see if better images can be obtained by irradiating the entire surface.

An analysis of the transient motion of the liquid surface response to the radiation pressure is considerably more involved than for the steady state case. In the rest of this section we will solve this problem by observing the response of the surface to a pulse of radiation pressure of sinusoidal space variation. The final solution will be obtained numerically, by a finite difference scheme, and the results will be shown to fit an analytic expression.

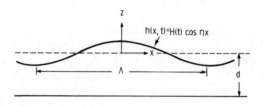

Figure 10.   Liquid Surface Deformation

As shown in Figure 10 we have a liquid layer of depth d which is assumed to be infinite in extent in the x direction. The surface is deformed into a sinusoidal pattern, $h(x,t)$, with spatial frequency η, by the sinusoidal radiation pressure $p_r(x,t)$. It is assumed that the air above the liquid has negligible effect on the liquid motion and thus may be replaced by a vacuum.

The problem is described by Equations 42 through 49. Throughout the fluid

$$\frac{\partial \vec{v}}{\partial t} = \nu \, \nabla^2 \, \vec{v} - \frac{1}{\rho} \, \nabla p + \vec{g} \quad , \tag{42}$$

$$\nabla \vec{v} = 0 \quad . \tag{43}$$

The boundary conditions are

$$p = -p_r \, -\gamma \, \frac{\partial^2 h}{\partial x^2} + 2 \, \rho \nu \, \frac{\partial v_z}{\partial z} \quad , \tag{44}$$

$$\frac{\partial v_x}{\partial z} + \frac{\partial v_z}{\partial x} = 0 \quad , \tag{45}$$

at the surface (z=h)

and $$\vec{v}\big|_{z=-d} = 0 \tag{46}$$

at the bottom (z=-d).

Initial conditions (t = 0) are

$$h(x)\big|_{t=0} = 0 \qquad\qquad (47)$$

and

$$\vec{v}(x,z)\big|_{t=0} = 0 \quad . \qquad\qquad (48)$$

The radiation pressure on the liquid surface (z=h) will take the form

$$p_r(x,t) = P_r \cos \eta x \, [u(t) - u(t-\Delta t)] \quad . \qquad\qquad (49)$$

P_r is the constant radiation pressure amplitude, η is the spatial frequency, $2\pi/\Lambda$, Δt is the pulse duration and

$$u(t) = \begin{cases} 1 \text{ for } t > 0 \\ 0 \text{ otherwise} \end{cases} .$$

Also, we have the condition that h << Λ. The only new symbol is ν, the kinematic viscosity.

Equation 42 is the linearized Navier-Stokes equation for the motion of an incompressible viscous fluid, and must be satisfied by all fluid particles.[10]   The linearization is valid if h << Λ and h << d. Equation 43 is the incompressibility condition.

The boundary conditions are given by Equations 44 to 46. Equation 44 is obtained from considerations of the balance of the radiation pressure, the viscous forces, and the surface tension at the liquid-vacuum interface when h << Λ.[11]   Equation 45 states that the shear stress at the liquid surface is zero. Equation 46 is the condition that a viscous fluid has zero velocity at a fixed wall. The initial conditions are given by Equations 47 and 48.

We will solve Equation 42 numerically using a simplified form of the marker and cell finite difference method of Harlow and Welch for incompressible fluids.[4]   Effects due to acoustic streaming in the ultrasound field will not be accounted for. Only the velocities will be determined numerically, while the pressure can be determined as follows.

If we take the divergence of Equation 42 and use Equation 43 we find that

$$\nabla^2 p = 0 \qquad\qquad (50)$$

where ∇^2 is the Laplacian operator. From Equations 42 and 46 we obtain the boundary condition

$$\left(\frac{\partial p}{\partial z} - \rho \nu \frac{\partial^2 v_z}{\partial z^2}\right)\Bigg|_{z=-d} = -\rho g \tag{51}$$

Now we shall assume that since the driving function p_r varies sinusoidally along x, the pressure p and the velocity v_z will also vary sinusoidally along x. Since $h \ll \Lambda$ and $h \ll d$ we assume that we can set $h = 0$ in determining the pressure distribution $p(x,z,t)$. Thus, at some time t we set

$$p(x,z,t) = P(z,t) \cos \eta x - \rho g z \quad , \tag{52}$$

$$v_z(x,z,t) = V_z(z,t) \cos \eta x \quad , \tag{53}$$

where $P(z,t)$ and $V_z(z,t)$ are the amplitudes of the pressure and z component of velocity respectively at some time t.

Equations 50 to 53 describe a boundary value problem for the pressure p in the fluid layer $z = -d$ to $z = 0$. The solution at the time t is

$$p(x,z,t) = -\rho g z + \frac{\cos \eta x}{\cosh \eta d}\left\{P(0,t) \cosh \eta(z+d)\right.$$
$$\left. + \frac{\rho \nu}{\eta} \frac{\partial^2 V_z(-d,t)}{\partial z^2} \sinh \eta z\right\} \tag{54}$$

This equation represents the fluid pressure p in terms of the pressure amplitude $P(0,t)$ at the fluid surface, and the second derivative of the z component of the velocity at the bottom.

Now we can solve for the velocity distribution in the fluid since we know the initial conditions and the form of the pressure distribution. Writing Equation 42 in finite difference form, and using Equation 54, the velocities can be solved for a sequence of times using a digital computer as described in Appendix A.

A general solution of the problem may be quite complicated. However, for the case of interest where the dissipation of energy in the fluid due to viscosity is small, we

can fit an analytic expression to the results that were
obtained using the numerical method. We shall comment on
these numerical results by making use of analytical results
that are available for the propagation of gravity capillary
waves on the surface of a fluid.

It is known that when a liquid surface is subject to
radiation pressure of some spatial frequency η, the surface
will be deformed and may oscillate and decay in time.[12]
Figures 11 to 14 are examples of the variation in time of
the amplitude $H(t)$ of the sinusoidal surface deformation
$\eta(x,t) = H(t) \cos \eta x$ when subject to radiation pressure of
the form $P_r \cos \eta x$ for some duration Δt. A number of computer runs were made as indicated by Table A-1, for various
values of the liquid density ρ, surface tension γ, kinematic
viscosity ν, spatial frequency $\eta = 2\pi/\Lambda$ and fluid depth given
by the ratio of depth to wavelength, d/Λ.

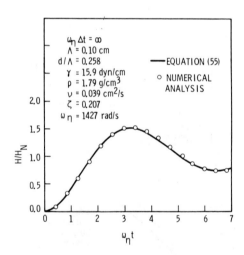

Figure 11.   Response of Liquid Surface to Step
Function Radiation Pressure

The results obtained by the numerical method are similar to the familiar transient response of a second order
system as described in any book on linear systems. These
curves are characterized by the two parameters, ζ and ω_n.
The damping ratio ζ determines the overshoot of the response
$H(t)$ above the steady state value, and ω_n, the natural

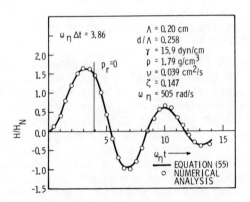

Figure 12. Response of Liquid Surface to a Pulse of Radiation Pressure

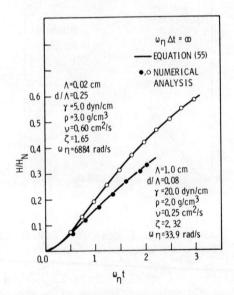

Figure 13. Overdamped Liquid Surface Response to Step Function Radiation Pressure

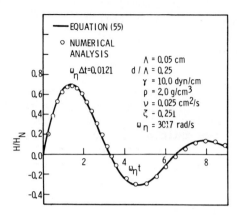

Figure 14.   Response of Liquid Surface to an Impulse
of Radiation Pressure

frequency of oscillation, determines the time scale. For a
radiation pressure of duration Δt the response $H(t)$ can be
written as

$$H(t) = \frac{P_r}{pg+\gamma n^2} \left\{ \left[1 - \frac{1}{\beta} e^{-\zeta\omega_n t} \sin(\omega_d t + \theta)\right] u(t) \right.$$

$$\left. - \left[1 - \frac{1}{\beta} e^{-\zeta\omega_n (t-\Delta t)} \sin (\omega_d(t-\Delta t) + \theta)\right] u(t-\Delta t)\right\} \quad ,(55)$$

where

$$\beta = \sqrt{(1 - \zeta^2)} \quad ,$$

$$\omega_d = \beta\omega_n \quad ,$$

$$\theta = \tan^{-1} \beta/\zeta \quad .$$

Equation 55 can be written in other forms if β is imaginary
or if $t > \Delta t$. These are given in Appendix B.

It was found that close agreement with the numerical
results was obtained if

$$\zeta = 2 n\rho\nu \sqrt{\left\{ n/ [\rho(\rho g + \gamma n^2)]\right\}} \ [\tanh (nd)]^E \qquad (56)$$

where $\quad \omega_n^2 = (\rho g + \gamma \eta^2) \frac{\eta}{\rho} \tanh (\eta d) \quad$,

and $\qquad E = \begin{cases} -2.3 \text{ if } d/\Lambda < 1/4 \\ 0 \text{ if } d/\Lambda > 1/4 \end{cases}$. $\hfill$ (57)

The solid curves in Figures 11 to 14 are obtained from Equations 55, 56 and 57, while the numerical results are indicated by the circles. The differences between the numerical results and Equations 55 to 57 are shown in Table A-1 for the peak values of the curves H(t) and the times at which the peak occurs. The differences are quite small for the values of the parameters used.

Figure 11 shows the response H(t) to a step function of radiation pressure. Figure 12 is similar except that the radiation pressure is of duration $\omega_n \Delta t = 3.86$. Two cases of the response H(t) when ζ is greater than one are shown in Figure 13. The response H(t) to an impulse of radiation pressure (actually a short pulse, $\omega_n \Delta t = 0.0121$) is shown in Figure 14.

The response of the liquid surface to radiation pressure is similar to the propagation of gravity capillary waves on a liquid surface. The similarity comes from the fact that in both situations there is a sinusoidal deformation of the surface, the amplitude of which decays in time. In Equation 57, the natural frequency of oscillation ω_n is the same as the result that can be obtained analytically for the frequency of oscillation of capillary gravity waves propagating on the surface of an ideal fluid of depth d.[13]

Also, except for the dependence on depth, the damping coefficient as given by Equations 56 and 57 is the same as that for the damping of gravity waves propagating on the surface of a viscous fluid of infinite depth. It is shown in Reference 13 that for gravity waves this value of damping coefficient is valid only if it is small, so that the fluid motion is approximately that of an ideal fluid. For fluids of infinite depth or of depth greater than about $\Lambda/4$ this condition is

$$\nu \eta^2 \ll \sqrt{g\eta} \quad . \hfill (58)$$

If we include the effect of surface tension by replacing g by $g + \eta^2 \gamma / \rho$ Equation 58 becomes

$$\nu\eta^2 \ll \sqrt{(g\eta + \eta^3 \gamma/\rho)} \quad . \tag{59}$$

We would then expect that Equations 55 to 57 apply if Equation 59 is satisfied. For freon E-5 or water, Equation 59 is easily satisfied for spatial frequencies η from zero up to values of interest, for example 1000 rad/cm. At 10 MHz in water the wavelength if 0.015 cm, for which the wave number k = 420 rad/cm. Actually, for the curves in Figure 13, Equation 59 is not satisfied, as both sides of the inequality are approximately the same. This seems to indicate that this condition is too restrictive when Equations 55 to 57 are used.

For fluids of depth less than about $\Lambda/4$ a factor $(\tanh \eta d)^{-2.3}$ applies as indicated in Equation 56. This was found to fit the numerical results well for values of d/Λ at least as low as 0.02, but no smaller depths were used in the numerical analysis. If the depth d is held constant with the spatial frequency η varying, the dependence on depth becomes more significant at the lower spatial frequencies and the validity of Equations 55 to 57 may become doubtful for very low values of η. However no studies were made to determine at what point this may be reached.

The damping coefficient for fluids where the dissipation of energy is very large may be quite different from that given by Equations 56 and 57. This is shown by the results for the decay of sinusoidal deformation on a fluid surface.[6,13] However, Equations 55, 56, and 57 seem to be sufficient for most fluids and spatial frequencies we may use in real-time liquid-surface holography. The numerical method in Appendix A can be used to determine the surface motion for any particular fluid and spatial frequency that is of interest.

In the numerical analysis described in Appendix A the acceleration due to gravity, g, was set to zero. This was done because the numerical method used was not suitable for accounting for the small variations in pressure that occur at the surface due to gravity. However, the effect of gravity is negligible for all but very low spatial frequencies, and in Equations 56 and 57 g can be set to zero for most spatial frequencies of interest.

Finally, for a sinusoidal deformation of spatial frequency η we can write the response of the surface amplitude $H(t)$ to be an impulsed of radiation pressure of amplitude $P_r \delta(t)$ as

$$H(t) = \frac{P_r}{\rho g + \gamma \eta^2} \frac{\omega_n}{\beta} e^{-\zeta \omega_n t} \sin(\beta \omega_n t) \quad . \tag{60}$$

Writing the three-dimensional equivalent of Equation 39 we have

$$\bar{\bar{h}}(u,v,\omega) = \bar{\bar{p}}_r(u,v,\omega) \, \bar{\bar{g}}_2(u,v,\omega) \quad , \tag{61}$$

where $g_2(x,y,t)$ is the response of the surface to an impulse $\delta(x,y,t)$ of radiation pressure. Taking the Fourier transform in time of Equation 60 we see that

$$\bar{\bar{g}}_2(\eta,\omega) = \frac{1}{\rho g + \gamma \eta^2} \frac{\omega_n^2}{\omega_n^2 + 2j \zeta \omega_n \omega - \omega^2} \quad . \tag{62}$$

Recalling that $\eta^2 = u^2 + v^2$, it is evident that

$$\bar{\bar{g}}_2(u,v,\omega) = \bar{\bar{g}}_2^{\,*}(-u,-v,-\omega)$$

If $p_r(x,y,t)$ is given by

$$p_r(x,y,t) = P_r \, \delta(x,y) \, [u(t) - u(t-\Delta t)] \quad , \tag{63}$$

$$p_r(x,y,t) \overset{S}{\leftrightarrow} P_r[u(t) - u(t-\Delta t)] \quad .$$

Then from Equation 55 the Fourier transform of the response of the surface $h(x,y,t)$ is

$$\bar{h}(u,v,t) = \bar{p}_r(u,v,t) \overset{t}{*} \bar{g}_2(u,v,t) \quad ,$$

$$= H(\eta,t) \tag{64}$$

Figures 15 and 16 show the response $H(t)$ when $\Delta t = \infty$ (step response) for freon E-5 and water, respectively, at various times t over a frequency range from 1 to 2000 rad/cm. These curves are for an infinite fluid depth d, or at least greater than 1/4 of the largest wavelength $\Lambda = 2\pi$ cm, so $d > 1.5$ cm approximately. In all figures H is normalized by $H_N = P_r/\rho g$.

We can see in Figures 15 and 16 that the high spatial frequencies reach a steady state much sooner than the low frequencies. The spatial frequency response rises at the rate of 10 db per decade of spatial frequency (db - 10 log H/H_N) and 20 db per time decade until it is close to the

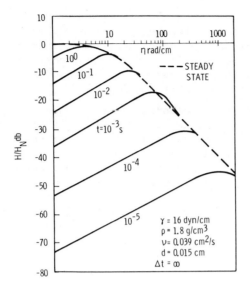

Figure 15. Liquid Surface (Freon E-5) Spatial Frequency
Response to Step Function Radiation Pressure

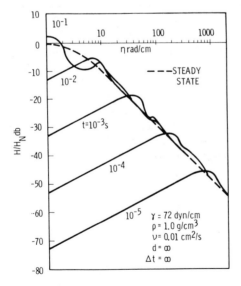

Figure 16. Liquid Surface (Water) Spatial Frequency
Response to Step Function Radiation
Pressure

steady state value. Then the steady state drops at the rate
of 20 db per decade of frequency at the higher frequencies.
The oscillations of the curves near the steady state, especially at the lower spatial frequencies, correspond to the
oscillations in the response once the steady state value is
passed (overshoot) as in Figure 11.

The steady state response is seen to be very poor, with
the higher spatial frequencies dropping off at 20 db per
decade. However, the transient response at the higher frequencies is relatively flat over a restricted frequency range
where the curves peak near the steady state curve. By
adjusting the angle between reference and object beams,
advantage can be taken of the transient motion by shifting
the object information into a region where the spatial frequency response is momentarily flat and sampling the surface
with the laser at this moment. Evidently, the greater this
angle, the earlier must be the sample time.

Consider that we have an imaging system in which the
spatial frequencies at the surface from the object range
from zero to a maximum of η_m. The maximum amount by which
the object information can be shifted in spatial frequency
is k_w, where $k_w = 2\pi/\lambda_w$ and λ_w is the wavelength of the
ultrasound in water. Thus the sampling time is determined
by any of the curves that has its flat region between η_m
and $\eta_m + k_w$.

For example, for ultrasound at 10 MHz in water
$k_w = 420$ rad/cm, but say that the apertures of the system
limit η_m to 250 rad/cm. In Figure 15 with freon E-5 in the
minitank, we see that the curve for $t = 0.5 \times 10^{-4}$s has a
flat region from about 275 to 525 rad/cm. Thus, if we shift
zero frequency to $\eta_r = 275$ rad/cm then the sample time
would be 0.5×10^{-4}s. The angle ψ, between reference beam
and object beam when the bisector θ, is normal to the
surface is then

$$2 \sin^{-1} \eta_r/2 \ k_w = 36° \quad . \tag{65}$$

Another possible sample time is 0.1×10^{-3}s with $\eta_r = 150$ rad
per cm and $\psi = 21.6°$.

With the transducers t_r and t_o producing pulses of
ultrasound, the pulse length can be set so that the pulse
terminates after the suface is sampled. Then after sufficient time to allow the surface deformation to decay,
another image can obtained with another pulse of

ultrasound. Figure 17 shows the response of the
surface to a pulse of radiation pressure of duration
$\Delta t = 10^{-4}$ s. The curves for the various interrogation times
are terminated at those values of η where the response
$H(\eta,t)$ becomes negative, as the response decays and oscil-
lates as shown in Figure 12. The high frequencies begin to
decay rapidly, but the lower frequencies to the left of the
peak region for $t = 10^{-4}$ s continue to rise, and then eventually decay.

It is apparent from Figure 17 that we can sample the
surface for times $t > \Delta t$. For example, at 5 MHz in water
$\kappa_w = 210$ rad/cm. If $\eta_m = 105$ rad/cm, $\Delta t = 10^{-4}$ s,
$t = 0.2 \times 10^{-3}$ s, and $\eta_r = 105$ rad/cm, the angle $\psi = 28.8°$.

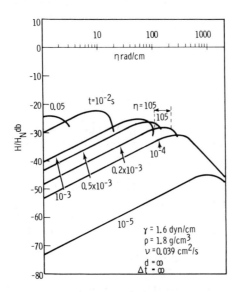

Figure 17.   Liquid Surface (Freon E-5) Spatial Frequency
Response to a Pulse of Radiation Pressure

The time between the pulses of ultrasound is determined
by the rate at which the surface decays after the pulse ends.
If $t = 0$ at the beginning of the pulse, then a sufficient
condition for the time t_1 at which the surface at a given
spatial frequency has decayed to within a fraction f, of its
height at the sample time t_s where $t_s \geqslant \Delta t$, is from Equation B-1,

$$e^{-\zeta\omega_n t}1 \leqslant f\, e^{-\zeta\omega_n t}s\, \sin\,(\omega_d t_s + \theta + \delta) \quad . \qquad (66)$$

Then the time at which the next pulse can begin can be taken as $t_1 - t_s$. Using the values of the previous example and applying Equation 66 to the lowest spatial frequency within the bandwidth of interest, i.e., $n_r = 105$ rad/cm, we obtain for $f = 0.05$, $t_1 \geqslant 4.5 \times 10^{-3}$s, or the period of the pulses $t_1 - t_s \geqslant 5.3 \times 10^{-3}$s.

In Figures 15 to 17 the depth d of the minitank was assumed to be $d > 1.5$ cm so that there was no effect of the depth on the surface motion for the spatial frequencies shown. Figures 18 and 19 are similar to Figures 15 and 17 respectively except that $d = 0.015$ cm. This means that for those spatial frequencies $\eta < 2\pi/4d = 100$ rad/cm the surface motion is more heavily damped. The fact that a thin fluid layer damps the motion of low spatial frequencies is a desirable feature since this contributes to the stability of the fluid surface.

Liquid-air Interface

The effect of a reflecting surface containing a ripple pattern has been adequately analyzed by a number of people.[14]   The essential feature is that for $\lambda_\ell/\Lambda \ll 1$ and $h/\lambda_\ell \ll 1$ we have, for the first order image a flat transfer function with a cutoff far beyond any possible spatial wavelengths present on the surface. Hence, we may safely neglect the optical train as a contributor to imaging problems although it can be useful for spatial filtering.

CONCLUSION

Figures 15 to 19 show that the pulsed method of imaging is necessary for any imaging system that uses the liquid surface since its spatial frequency response to radiation pressure in the steady state is very poor. It is the liquid surface response given by the function g_2 (which acts as a bandpass filter) that is the most important determinant in the method of operation of the imaging system. Graphs of the liquid surface response such as Figures 15 to 19 and Equation 55 can be used to determine the angle of incidence of the acoustic reference, the pulse length of the acoustic beams, the sampling time of the light beam and the time between the pulses of the acoustic beams.

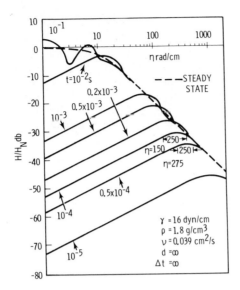

Figure 18. Liquid Surface (Freon E-5) Spatial Frequency
Response to a Step Function Radiation Pressure
with a Finite Fluid Depth

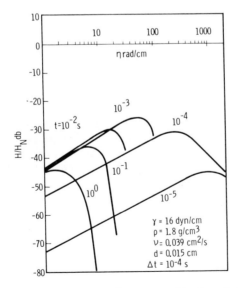

Figure 19. Liquid Surface (Freon E-5) Spatial Frequency
Response to a Pulse of Radiation Pressure
with a Finite Depth of Fluid

The function g_1, which accounts for the angle of incidence of the acoustic waves to the liquid surface and the
presence of the minitank is of less importance than the
function g_2. However, g_1 shows that at large angles of
incidence of the ultrasound a phase distortion is introduced
into the acoustic field at the surface if the fluid in the
minitank is different from the fluid in the main tank. For
thinner fluids this effect is less pronounced.

Also, a thin fluid layer in the minitank is desirable
in order to damp the motion of surface ripples of low spatial
frequency. Furthermore, the velocity of propagation of the
ultrasound in the minitank should be less than or not much
greater than the velocity in the main tank.

APPENDIX A

NUMERICAL SOLUTION OF LIQUID SURFACE MOTION

The numerical method was used to determine the liquid
surface motion when subject to radiation pressure is based
on a method by Harlow and Welch.[4]   With this method the
Navier-Stokes equation is written in finite difference form,
and the velocities of the fluid particles throughout the
fluid are determined numerically for a sequence of time
steps and the subsequent motion of the surface is followed
by updating the position or marker particles at the fluid
surface after each time step.

The liquid is divided up into rectangular cells as in
Figure A.1 with the fluid pressure defined inside the center
of each cell and the horizontal velocities u and the vertical velocities v defined on the cell boundaries as in Figure A.2. The marker particles are initially set at the
center of the second from the top row of cells to mark the
position of the surface. The surface motion is contrained
to stay within the second from the top row of cells.

The deformation of the surface h is assumed to take
the form $H(t) \cos \eta x$ when the radiation pressure takes the
form $P_r \cos \eta x [u(t) - u(t-\Delta t)]$. Since the surface deformation h is assumed to be much less than the wavelength Λ or
the depth of the fluid d, and the acceleration due to gravity
is set to zero, then from symmetry considerations it is only
necessary to apply the numerical method to a length of fluid
(along x) of $\Lambda/4$ (Figure A.3) of depth d.

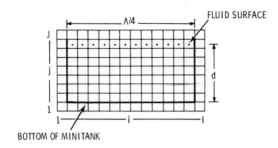

Figure A.1.   Fluid Divided into a Rectangular
Mesh of Cells

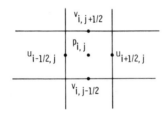

Figure A.2.   Position of Pressure and Velocities
for Cell i, j

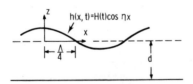

Figure A.3.   Deformation of the Liquid Surface

The linearized Navier-Stokes equation can be used, so
that the nonlinear terms in the finite difference equation
can be dropped. Also, since we have an analytic expression
(Equation 54) for the pressure, the iteration required to
determine the fluid pressure distribution can be replaced
by using Equation 54. With the pressure at the surface, the
surface deformation and the velocity distribution varying

along x as cos ηx, (Equation 44) can be used to determine
the pressure at the surface for each time step of the
numerical method.

With the heavy line bounding the fluid on the left in
Figure A.1 considered to be a free-slip wall, and the heavy
horizontal line a no-slip wall, the velocities required by
the numerical method outside the wall are given in Reference 4. At the right 'wall' the horizontal velocities on
either side of the wall are set equal, while the vertical
velocities and pressures on either side of the right wall
are set to be the negative of each other. Note that the
right 'wall' is neither a free-slip wall nor a no-slip wall,
since fluid is allowed to pass through it. The horizontal
velocities on the right wall are determined by the numerical
method. The velocities at the surface are set as in Reference 4, with the added condition that the shear stress at
the surface be zero, which determines the horizontal velocities outside the surface in the top row of cells.

Thus, with the modifications that have been discussed,
the motion of the liquid surface can be determined by using
a simplified form of Harlow and Welch's method. Since the
shape of the surface is known to have the form cos ηx,
surface tension forces can be accounted for, and since
$h \ll \Lambda$, shear and normal stresses at the surface can be
accounted for since the surface can for this purpose be
assumed to be flat.

A number of computer runs were made for various parameter values as listed in Table A-1. The results were
typically those of the response of a second order system to
a pulsed or step forcing function, as shown by Figures 11 to
14. The results were seen to fit closely Equations 55, 56
and 57 and the percentage overshoot (P.O.) and the damped
frequency of oscillation (ω_d) taken from the numerical
results are also listed in Table A-1.

The peak value of the response H(t) as determined by
Equation 56 was compared with the numerical results, and the
difference is shown in Table A-1 under Column A as a percentage of the numerical results. A positive percentage indicates that Equation 56 gave a higher value. Also, the
percentage difference in the damped frequency ω_d, obtained
is listed under Column B, where the damped frequency is
determined by the time at which H(t) reaches its peak value.
The damping ratio ζ listed was determined from Equation 56.

Table A-1.

	λ cm	d/λ	ρ g/cm^3	γ dyn/cm	$\nu \times 10^2$ cm^2/s	P.O.	ω_d rad/s	A %	B %	ζ
1	0.20	0.258	1.79	15.9	3.90	65.3	491	-1.5	+1.6	0.147
2	0.10	0.258	1.79	15.9	3.90	53.3	1372	-1.2	+1.8	0.207
3	0.05	0.258	1.79	15.9	3.90	39.4	3880	-0.9	-0.5	0.293
4	0.40	0.258	1.79	15.9	3.90	76.0	178	-2.3	+1.7	0.104
5	0.05	0.258	1.79	15.9	1.95	64.5	3910	-1	+2	0.147
6	0.10	0.258	7.16	15.9	3.90	75.2	697	-1.8	+1.9	0.104
7	0.20	0.258	1.79	3.98	3.90	43	236	-3.5	+2.1	0.293
8	0.05	0.50	1.79	15.9	1.95	73	4110	-6.2	+0.9	0.147
9	0.05	0.129	1.79	15.9	1.95	30.9	3090	-1.6	+3.4	0.369
10	0.20	0.50	1.79	15.9	3.90	73.5	511	-6.2	+1.4	0.147
11	0.05	0.04	1.79	15.9	0.0	103.5	2065	-1.7	+0.9	0.0
12	0.05	0.129	1.70	15.9	0.0	105.6	3270	-2.7	+4.9	0.0
13	0.20	0.04	1.79	15.9	0.390	30.4	224	-5.4	+8.0	0.369
14	0.20	0.02	1.79	15.9	0.9075	19.6	151	+1.7	+10.6	0.438
15	1.0	0.07	1.00	72.0	1.00	85.9	82.6	+0.5	+4.0	0.045
16	0.05	0.25	2.00	10.0	10.0	---	2870	---	+1.7	0.25
17	0.02	0.25	3.00	5.00	6.00	0.0	---	---	---	1.65
18	1.0	0.08	2.00	20.0	25.0	0.0	---	---	---	2.32
19	0.05	1.00	1.79	15.9	1.95	73.4	4110	-6.1	+0.9	0.147

Computer runs to determine the liquid surface response $H(t)$ to a step function radiation pressure were made with the parameters indicated in the first five columns. The resulting percentage overshoot (P.O.) of $H(t)$ and damped frequency of oscillation ω are shown; also, the values of P.O. and ω_d were obtained from Equations 56 and 57 are given in columns A and B respectively as a percentage of the values in the columns P.O. and ω_d. Numbers 1, 2 and 16 are shown in Figures 11, 12 and 14 respectively, and number 17 and 18 in Figure 13.

APPENDIX B

LIQUID SURFACE RESPONSE

The equation for the motion of a liquid surface when subject to radiation pressure (Equation 55) can also be written as

$$H(t) = \frac{P_r}{(\rho g + \gamma n^2)} \sqrt{(A^2 + B^2)} \frac{e^{-\zeta \omega_n t}}{\beta} \sin(\beta \omega_n t + \theta + \delta) \quad (B-1)$$

for $t \geq \Delta t$, $\zeta < 1$, where

$$\beta = (1 - \zeta^2),$$

$$A = -e^{\zeta \omega_n \Delta t} \sin(\beta \omega_n \Delta t),$$

$$B = e^{\zeta \omega_n \Delta} \cos(\beta \omega_n \Delta t) - 1,$$

$$\tan \theta = \beta/\zeta,$$

$$\tan \delta = A/B.$$

Equation (B-1) still holds if β is imaginary ($\zeta > 1$) as we can use the following relations

$$\cos jx = \cosh x,$$

$$\sin jx = j \sinh x,$$

$$\tan jx = j \tanh x.$$

REFERENCES

1. Holosonics Corporation, 2950 George Washington Way, Richland, Washington, 99352.

2. B. B. Brenden, "A Comparison of Acoustical Holography Methods," Acoustical Holography, Vol. 1, Plenum Press, New York, N.Y., Ch. 4, pp. 57-71 (1969).

3. T. J. Bander and B. P. Hildebrand, "Analysis of the Liquid Surface Motion in Liquid Surface Acoustical Holography," J. Acoust. Soc. Am., Vol. 47, pp. 81 (1970).

4. F. H. Harlow and J. E. Welch, "Numerical Calculation of Time-Dependent Viscous Incompressible Flow of Fluid with Free Surface," The Physics of Fluids, Vol. 8, No. 12, p. 2182 (1965).

5. J. W. Goodman, Introduction to Fourier Optics, McGraw Hill, pp. 60, 80 (1968).

6. L. D. Landau and E. M. Lifshitz, Fluid Mechanics Course of Theoretical Physics, Vol. 6, Pergamon Press, London, Chapter 8 (1959).

7. Z. A. Gol'dberg, "Acoustic Radiation Pressure," in High-Intensity Ultrasonic Fields, edited by L. D. Rozenberg, Plenum Press, New York, (1971).

8. P. S. Green, "Acoustical Holography with the Liquid Surface Relief Conversion Method," Lockheed Internal Report No. 6-77-67-42 (1972).

9. R. F. Beyer, "Radiation Pressure in a Sound Wave," Am. Jour. Phys., Vol. 18, No. 1 (1950).

10. L. D. Landau and E. M. Lifshitz, Fluid Mechanics Course of Theoretical Physics, Vol. 6, Pergamon Press, London, Sections 12, 15 (1959).

11. L. D. Landau and E. M. Lifshitz, Fluid Mechanics Course of Theoretical Physics, Vol. 6, Pergamon Press, London, Sections 15, 60 (1959).

12. B. P. Hildebrand and B. B. Brenden, An Introduction of Acoustical Holography, Plenum Press, New York, N.Y., p. 157 (1972).

13. L. D. Landau and E. M. Lifshitz, Fluid Mechanics Course of Theoretical Physics, Vol. 6, Pergamon Press, London, Sections 12, 25, 61 (1959).

14. B. P. Hildebrand and B. B. Brenden, An Introduction to Acoustical Holography, Plenum Press, New York, N.Y., p. 145 (1972).

ELIMINATION OF SPURIOUS DETAIL IN ACOUSTIC IMAGES

Adrianus Korpel, Robert L. Whitman
and Mahfuz Ahmed

Zenith Radio Corporation
Chicago, Illinois 60639

1. INTRODUCTION

Many methods of acoustic imaging employ monophonic (single frequency) and strongly directional (spatially coherent) insonification of the object to be examined. Depending on the nature of the object, such methods frequently suffer from the presence of highly objectionable spurious image detail. Typically this detail consists of complex patterns of high contrast which are unrelated to the actual physical structure of the object in the plane under observation. The spurious patterns are due partly to near field detail in the sound emanating from the transducer and partly to interference effects caused by out-of-focus structures in the object (Talbot images)[1,2] as well as multiple reflections from interfaces.

In the familiar visual world, the illumination is normally highly diffuse and polychromatic; consequently out-of-focus objects appear blurred with soft outlines. Spurious information of this kind can easily be disregarded by the brain in favor of the crisp in-focus patterns with sharp outlines. In contrast to this, illumination (or insonification) with space-and-time coherent radiation causes crisp interference patterns which break up and overlay the outlines of real physical detail in the plane of observation. As a result the final picture is extremely difficult to interpret.

In acoustic imaging the situation is even more serious because the acoustic images of interest are often <u>unfamiliar</u>. Especially in medical applications this is frequently the case: the structure that is displayed may have no immediate resemblance to the one shown in anatomical textbooks or visible on x-ray plates.

In view of these considerations it becomes extremely important to eliminate spurious detail in acoustic imaging. From what was said before, it appears that, in order to accomplish this, uncontrollable interference effects should be eliminated. In principle there are two ways of doing this: either interference is avoided altogether through the use of polyphonic radiation or else an average is constructed of many pictures, each one insonified differently and exhibiting a different interference pattern.

It should be noted at this point that it is not enough to insonify the object with monophonic but <u>diffuse</u> sound. This merely causes a random kind of interference pattern called speckle. Such speckle is familiar to everyone who has ever looked at diffuse objects illuminated by laser light; holograms of statuettes are a prime example. In optical pictures such as these the speckle is not too objectionable; speckle detail size is of the order of the resolution element and this is typically much smaller than the structures of interest (optical microscopy, however, would be an exception). In acoustic imaging, on the other hand, there usually is a high premium on resolution, i.e. devices most often operate at their resolution limit. Consequently, speckle is very objectionable here as it obscures vital detail.

It would appear then that simply diffusing the insonification is not sufficient but that some temporal variation and averaging of the insonification is necessary for the elimination of spurious detail. There are many ways of doing this: one may use noise to drive the transducer, the frequency source may be swept in some predetermined fashion, a scatter plate may be moved through the insonifying

soundfield, etc. In general these methods are not completely equivalent. Sweeping the frequency, for instance,
does not exclude beam collimation whereas a scatterplate,
by definition, destroys strong directionality. There is one
thing, however, that these techniques do have in common:
they make holography, if not outright impossible, extremely difficult to apply.

This paper will discuss the first two methods of spurious detail elimination; i. e. , the use of noise or a swept
frequency as a source of insonification. It can be shown,
that, in general, spurious details due to reflections from
multiple interfaces are eliminated more readily than Talbot
images. Therefore, this paper will only discuss the theory
of Talbot images which require the greatest frequency
change for elimination. A convenient model is that of a
sinusoidal amplitude transmission pattern, $\hat{f}$, shown in
Fig. 1,

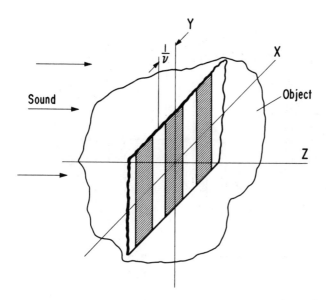

Fig. 1. A sinusoidal amplitude transmission function
insonified by a plane wave.

$$\Gamma = (1-a) + a \cos (2\pi \nu x) \qquad (1)$$

insonified by a plane wave propagating perpendicularly to the X axis (i.e. in the Z direction). The pattern given by eq. (1) may be thought of as a particular Fourier component of the transmission function $\Gamma(x,y)$ of a thin slice through the object located at $z = 0$. For simplicity re-scattering will be neglected, i.e. it will be assumed that $a \ll 1$.

If a pattern such as (1) is insonified by a plane wave then, as is well known, the phenomenon of periodic focusing occurs. That is to say, at certain characteristic positions $z = nz_T$ (n is an integer) the resulting field is (but for a possible contrast reversal) <u>identical</u> to that at $z = 0$. This is illustrated in Fig. 2. These recurring field patterns are called "Talbot" images,[1] after Fox Talbot who described their existence in 1836.[2] It is clear that they, in fact, are spurious images in the sense defined before.

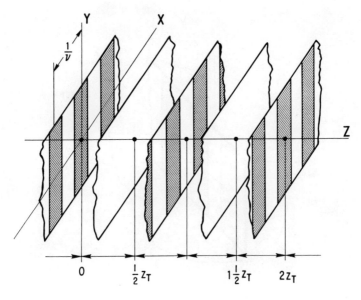

Fig. 2. Talbot images of a sinusoidal amplitude transmission function.

The "Talbot distance" z_T is given by

$$z_T = \frac{1}{\Lambda \nu^2} \tag{2}$$

where Λ is the wavelength of the incident radiation.

A further point of interest is that the Talbot images at n = odd have reversed contrast, i.e. instead of (1) one finds $(1-a) - a \cos (2\pi \nu x)$. For points on the Z axis in between two Talbot image planes the patterns exhibit reduced contrast: exactly halfway in between, the contrast vanishes entirely. As the Talbot distance depends on Λ and hence on the frequency of the radiation, it will be clear that a change in frequency will change the spurious image contrast in a given plane. It seems plausible that the contrast will be washed out in an averaging process if the frequency is swept by an amount sufficient to vary the contrast through a complete cycle (e.g. positive to positive image) from the beginning (f_1, Λ_1) to the end (f_2, Λ_2) of the frequency sweep. To obtain a quantitative expression for this required frequency change, consider an arbitrary plane at a distance z from the object [described by (1)] responsible for the spurious image. For $\Lambda_2 > \Lambda_1$, the conditions

$$z = n \, z_T \, (\Lambda_1) \quad \text{and} \quad z = (n + 2) \, z_T \, (\Lambda_2)$$

must hold. It then follows that

$$\frac{z}{(1/\Lambda_2 \nu^2)} - \frac{z}{(1/\Lambda_1 \nu^2)} = 2 \tag{3}$$

$$\text{or} \quad (\Lambda_2 - \Lambda_1) = \frac{2}{z \nu^2} \tag{4}$$

Conversely, it is plausible to assume that for a wavelength change $\Lambda_2 - \Lambda_1$ all spurious images in planes further from the object Γ than

$$z = \frac{2}{(\Lambda_2 - \Lambda_1)\nu^2} \qquad (5)$$

will have substantially reduced contrast. Consequently the distance defined by (5) may be called the "spurious image penetration distance $\ell(\nu)$ for detail of size $1/\nu$ given a wavelength sweep $\Lambda_2 - \Lambda_1$ ":

$$\ell(\nu) = \frac{2}{\Delta\Lambda \nu^2} = \frac{2}{\Lambda_o' \nu^2} \cdot \frac{f_o'}{\Delta f} \qquad (6)$$

where $\Delta\Lambda = \Lambda_2 - \Lambda_1$, $f_o' = [f_1 f_2]^{\frac{1}{2}}$, $\Delta f = f_1 - f_2$

and Λ_o' is the wavelength corresponding to f_o'. Note that, if $\Delta f \stackrel{<}{=} 2f_o'$ the spurious image penetration distance has decreased to the point where it equals the Talbot distance for Λ_o'. In this case, then, virtually all spurious images are eliminated. Such a large frequency sweep is not always practical and it is frequently of interest to consider what may be achieved with a more modest approach. Knowing the thickness of the object and the resolution of the imaging system (i.e. the maximum relevant ν) the value of $\ell(\nu)$ gives a rough indication of the effectiveness of the frequency sweep used. Figure 3 shows a plot of $\ell(\nu)/\Lambda_o'$ vs. $(1/\Lambda_o' \nu)$ for various values of $f_o'/\Delta f$.

As an example of application, consider an acoustic imaging system in which a thick object immersed in water is insonified with an acoustic frequency that is swept over 10% from 2.85 MHz to 3.15 MHz. At the nominal frequency f_o' of 3 MHz the wavelength $\Lambda_o' = 0.5$ mm. For fine detail of size $2\Lambda_o'$ (i.e. $\nu = 1$ cycle/mm) the Talbot planes are spaced by approximately 2 mm. According to (6) the frequency sweep $(f_o'/\Delta f \approx 10)$ will substantially reduce the spurious fine-detail contrast contribution from parts of the object that lie farther away than 4 cm from the plane under observation. For coarser detail this distance is larger. The particular frequency sweep used in the example is thus

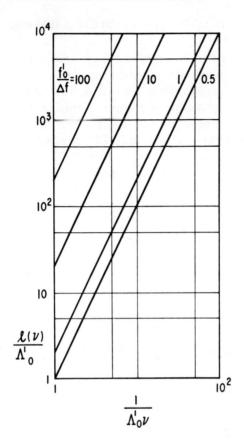

Fig. 3.   Plot of $\ell(\nu)$ showing the effectiveness of the frequency sweep.

relatively ineffective for objects thinner than ≈ 8 cm. Even so, some improvement in the image is clearly visible as will be shown later.

The next section will justify the heuristic reasoning used above, provide more detailed quantitative information and analyze the effect of the precise temporal nature of the insonification.

2. ORIGIN OF THE SPURIOUS IMAGE

Figure 4 shows the XZ plane cross section of Fig. 1.
A plane wave of sound with strain amplitude S_o is incident
from the left upon an amplitude grating as characterized
by (1). For small modulation index a, the grating generates two additional plane waves (orders -1 and +1) in addition to the incident one (order 0). The amplitudes of
these three orders are $\frac{1}{2}aS_o$, S_o and $\frac{1}{2}aS_o$ respectively.
The -1 and +1 orders propagate at angles $-\varphi$ and $+\varphi$ relative
to the zeroth order and

$$\sin \varphi = \Lambda \nu \tag{7}$$

Thus, to the right of the grating, the total sound field is
given by

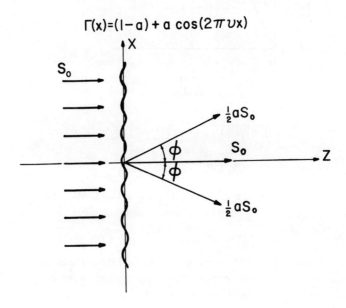

Fig. 4. Diffraction of a plane wave of sound by an amplitude
        grating.

$$S(x, z) = \frac{1}{2} a S_o \exp (j K x \sin \varphi - jKz \cos \varphi) +$$

$$+ S_o \exp (-jKz)$$

$$+ \frac{1}{2} a S_o \exp (-jKx \sin \varphi - jKz \cos \varphi). \tag{8}$$

where $K = 2\pi / \Lambda$.

The sound intensity distribution is then given by

$$I(x, z) = S(x, z) S^* (x, z) \tag{9}$$

$$\cong S_o^2 \{1 + 2a \cos (Kx \sin \varphi) \cos [K(1-\cos \varphi)z]\}$$

if it is assumed that $a \ll 1$.

With (7) this may be written as

$$I(x, z) = S_o^2 [1 + 2a' (z) \cos 2\pi \nu x] \tag{10}$$

where $\dfrac{a' (z)}{a} = \cos [K(1-\cos \varphi) z] \tag{11}$

The quantity $\dfrac{a' (z)}{a}$ describes the degree of spatial modulation at frequency ν in the image at any distance z relative to that pertaining at $z = 0$. It is somewhat analogous to the modulation transfer function used in optics to describe the image forming properties of lenses. [3] In the case under consideration here it describes the image forming properties of "free space". From (11) it is readily shown that

$$\frac{a' (z)}{a} = 1 \text{ when } z = 2n z_T \tag{12}$$

$$\frac{a' (z)}{a} = 0 \text{ when } z = (2n + \tfrac{1}{2}) z_T \tag{13}$$

$$\frac{a' (z)}{a} = -1 \text{ when } z = (2n+1) z_T \tag{14}$$

where $n = 0, 1, 2, 3 ---$ and

$$z_T = \frac{\frac{1}{2}\Lambda}{1 - \cos\varphi} \tag{15}$$

In the paraxial approximation $1 - \cos\varphi \approx \frac{1}{2}\sin^2\varphi$ and from (7) and (15) it then follows that

$$z_T = \frac{1}{\Lambda \nu^2} \tag{16}$$

Thus we recognize in z_T the Talbot distance first defined in eq. (2).

3. RANDOM NOISE INSONIFICATION

In this section we will assume that the insonification consists of a continuous distribution of plane waves, all propagating in the Z direction but at different wavelengths and with random phases. Let the intensity distribution as a function of wavelengths be characterized by $S_o^2(\Lambda)$. It is then readily seen with (10) that the resulting spatial intensity distribution is given by

$$I(x, z) = \int S_o^2(\Lambda) [1 + 2a'(z, \Lambda)\cos 2\pi\nu x]d\Lambda \tag{17}$$

where, in the paraxial approximation,

$$a'(z, \Lambda) = a \cos(\pi \Lambda \nu^2 z) \tag{18}$$

As a tractable example we shall choose for $S_o^2(\Lambda)$ a uniform band limited distribution:

$$S_o^2(\Lambda) = \frac{A^2}{\Lambda_2 - \Lambda_1} \quad \text{for } \Lambda_1 < \Lambda < \Lambda_2 \tag{19}$$

$$= 0 \quad \text{otherwise}$$

Substitution of (18) and (19) into (17) leads to the following result:

$$I(x,z) = A^2[1 + 2a'(z,\Lambda_o)q(z,\Delta\Lambda)\cos 2\pi \nu x] \qquad (20)$$

where

$$\Lambda_o = \tfrac{1}{2}(\Lambda_1 + \Lambda_2) \qquad (21)$$

$$\Delta\Lambda = (\Lambda_2 - \Lambda_1) \qquad (22)$$

$$a'(z,\Lambda_o) = a\cos(\pi \Lambda_o \nu^2 z) \qquad (23)$$

$$q(z,\Delta\Lambda) = \frac{\sin(\pi \Delta\Lambda \nu^2 z/2)}{(\pi \Delta\Lambda \nu^2 z/2)} \qquad (24)$$

Inspection of (20) shows us that the effect of periodic imaging, as expressed in the term $a'(z,\Lambda_o)$, is now reduced by the factor $q(z,\Delta\Lambda)$. The latter is plotted in Fig. 5. Note that the first zero of $q(z,\Delta\Lambda)$ occurs at the value $z = \dfrac{2}{\nu^2\Delta\Lambda}$. Upon inspection of (6) we see that this is exactly the value of what we have previously called the spurious image penetration distance $\ell(\nu)$, thus justifying the heuristic reasoning used in the Introduction.

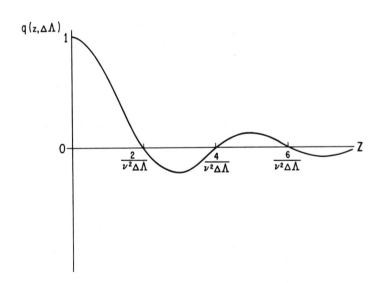

Fig. 5. Plot of $q(z,\Delta\Lambda)$ vs. z.

4. FREQUENCY SWEPT INSONIFICATION

In this section we shall assume that the wavelength of the insonifying plane wave is a slowly varying function of time. The final image is obtained by integration of the time-varying brightness patterns over the effective time duration of the total wavelength variation. Upon reflection it will be clear that the mathematical formulation of this process is quite similar to that treated in Section 3 if we replace $S_o^2(\Lambda)\,d\Lambda$ by

$$S_o^2(\Lambda)dt = S_o^2(\Lambda)\ \frac{d\Lambda}{dt}^{-1}d\Lambda \qquad (25)$$

where now $S_o^2(\Lambda)$ represents the intensity of the frequency swept signal and (d_Λ/dt) the rate of sweep. Thus (17) should be recast in the form:

$$I(x,z) = \int S_o^2(\Lambda)[1+2a''(z,\Lambda)\cos 2\pi\ \nu x](\frac{d\Lambda}{dt})^{-1}d\Lambda \qquad (26)$$

It is immediately clear that, if the sweep rate (d_Λ/dt) is constant, eq.(26) is identical to (17). Thus, with the wavelength varying uniformly with time from Λ_1 to Λ_2, and with constant amplitude $S_o(\Lambda)$, the results (20) $\to$ (24) of the previous section apply. In some cases of practical interest the frequency, rather than the wavelength, varies uniformly with time. Then $(d\Lambda/dt)^{-1} \propto 1/\Lambda^2$ and the integral (26) leads to functions that are not readily interpretable physically. We shall therefore not treat this case. A more tractable example of practical interest is that of sinusoidal frequency modulation with constant amplitude A. For the sake of simplicity we shall represent this by

$$\Lambda = \Lambda_o + \tfrac{1}{2}\ \Delta\Lambda\ \sin\omega_m t \qquad (27)$$

where ω_m is the frequency of modulation. This is equivalent to sinusoidal frequency modulation for small $\frac{\Delta\Lambda}{\Lambda_o}$. The integral (26) will be evaluated over one complete cycle of the modulation, i.e. from $t = 0$ to $t = 2\pi/\omega_m$:

$$I(x,z) = A^2 \int_0^{2\pi/\omega_m} \{1+2a\cos[\pi v^2 z(\Lambda_0 + \tfrac{1}{2}\Delta\Lambda\sin\omega_m t)]\cos 2\pi vx\}dt \tag{28}$$

Upon evaluating this integral we find:

$$I(x,z) \propto A^2[1+2a'(z,\Lambda_0)\, r(z,\Delta\Lambda)\cos 2\pi vx] \tag{29}$$

where $r(z,\Delta\Lambda)$ again represents a reduction factor, now given by

$$r(z,\Delta\Lambda) = J_0(\pi\,\Delta\Lambda\, v^2\, z/2) \tag{30}$$

with J_0 the zeroth order Bessel function. A plot of $r(z,\Delta\Lambda)$ is shown in Fig. 6. Note that the first zero is again directly related to the spurious image penetration distance of (6):

$$z \approx \frac{1.5}{v^2\Delta\Lambda} \tag{31}$$

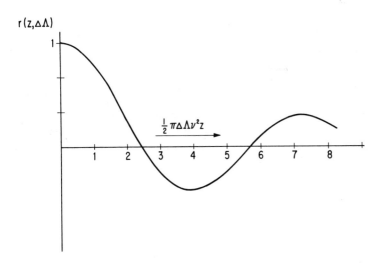

Fig. 6.   Plot of $r(z,\Delta\Lambda)$ vs. z.

5. DISCUSSION

The appearance and quenching of periodic images has been treated by considering an amplitude grating as a model for a real physical structure. It would have been possible to choose a phase grating instead. The results would not have been greatly different: the first spurious amplitude image appears at $\frac{1}{2}z_T$ instead of z_T with further images at $1\frac{1}{2}z_T$, etc. Mathematically the cosine term in (18) has to be replaced by a sine term of the same argument. Performing the necessary algebra with this modification shows that both in (20) and (29) the cosine term in $a'(z, \Lambda_o)$ must be replaced by a sine term while the important expressions for $q(z, \Delta\Lambda)$ and $r(z, \Delta\Lambda)$ remain unchanged.

Although other forms of noise and frequency sweep could have been considered, the two cases derived show that the degree of spurious image suppression is not very critically dependent upon the exact details. Thus the heuristic theory given in the Introduction gives a good indication of what can be achieved with a given frequency band.

6. EXPERIMENTAL RESULTS

The results presented in this section are not intended to provide quantitative and detailed confirmation of the theory outlined before. Rather, they show the effect of a modest frequency sweep on the appearance of some typical biological and metallic structures such as might be encountered in diagnostic applications and in nondestructive testing. The spurious images may be due to multiple reflections as well as Talbot images. All images were obtained with the laser scanning acoustic camera described elsewhere.[4,5] The camera was used at 2.25 MHz in the non-pulsed 1:1 imaging mode. The objects were imaged in transmission with slightly convergent insonification from a 7.5 cm circular quartz transducer radiating through a lucite field lens of 40 cm focal length. At the operating frequency and with water as a sound medium the minimum resolvable detail seen with this camera is of the order of 1.5 mm with a corresponding depth of field of approximately 3 mm. In operation the acoustic frequency was varied sinusoidally over a range of about 10-15%. The rate of modulation was of the order of 500 Hertz and carefully chosen so as to avoid synchronization with the television monitor on which the image was displayed. A picture was taken of the monitor's screen, with an exposure time of 4 sec thus ensuring integration over 240 television fields.

The first structure of interest to be examined was a sheep's kidney. This is of a flat oblong shape, approximately 12 cm long, 7 cm high and 3 cm thick. In order to have the maximum amount of out-of-focus detail the kidney was positioned with the long dimension almost parallel to the sound propagation direction.

(a)

(b)

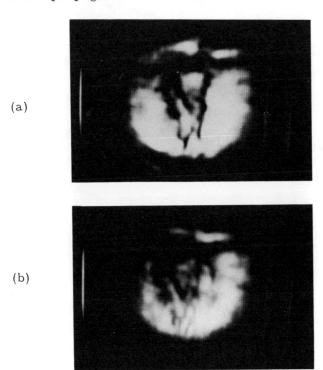

Fig. 7.   Acoustic image of a sheep's kidney (a) monophonic insonification (b) polyphonic insonification.

Figure 7a shows the result of monophonic insonification. The region of the hylus where there exists a concentration of veins and arteries is shown. It would be a mis-
take to attempt to label each black line as corresponding
to a definite physical structure. Some of these lines are
evidently spurious as will be clear from inspecting Fig. 7b
which shows the same object but this time insonified with
polyphonic sound. (The fine vertical lines in each picture
are residual hologram fringes and should be ignored.) It
is clear that even a small change in frequency results in a
dramatic change of appearance in this case.

The next object of interest was part of a femoral artery
severely affected by arteriosclerosis. Figure 8a shows
this object held in a plastic bag with saline solution. The

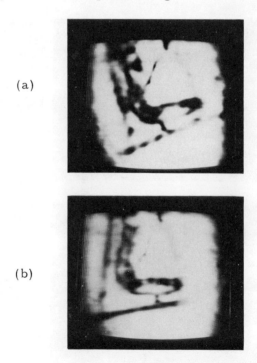

(a)

(b)

Fig. 8.   Acoustic image of a section of arteriosclerotic
          femoral artery (a) monophonic insonification
          (b) polyphonic insonification.

edges of the bag show up as diagonal lines to the left and
bottom of the (curved) artery. Although the sclerotic de-
posits within the artery are evident they appear to be somewhat distorted by interference effects. This is confirmed
by Fig. 8b which shows the same artery under polyphonic
insonification.

(a)

(b)

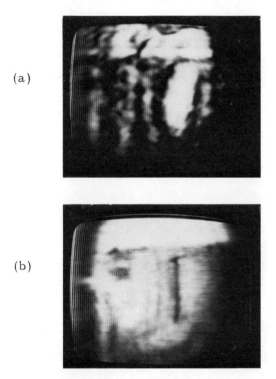

Fig. 9. Acoustic image of a hole in an aluminum plate.
(a) monophonic insonification (b) polyphonic
insonification.

The object shown in Fig. 9 is a 1/2 inch thick aluminum
plate with a 1/8 inch diameter hole drilled into it. When
insonified monophonically the multiple reflections from the
two surfaces of the plate generates spurious images which
mask the actual location of the hole, as shown in Fig. 9a.

A 10% frequency sweep is, however, sufficient to eliminate these spurious images and the hole is clearly seen in
Fig. 9b.

ACKNOWLEDGMENTS

We would like to acknowledge the cooperation of
Prof. F. Barnes of the University of Colorado and of
Dr. P. Steel of the V.A. Hospital in Denver and thank
J. Van Roon for his assistance with the experiments.

REFERENCES

1. A. W. Lohmann and D. E. Selva "A Talbot Interfero-
    meter with Circular Gratings", Optics Communications
    Vol. 4, No. 5, January 1972.

2. F. Talbot, Phil. Mag. 9 (1836) 401.

3. J. W. Goodman "Introduction to Fourier Optics"
    McGraw-Hill, New York, N.Y. 1968, Ch. 6.

4. R. L. Whitman, M. Ahmed and A. Korpel "A Progress
    Report on the Laser Scanned Acoustic Camera" in
    Acoustic Holography, Vol. 4, ed. G. Wade,
    pp. 11-33, Plenum Press, New York, N.Y., 1973.

5. R. L. Whitman, A. Korpel and M. Ahmed "Novel Tech-
    nique for Real Time Depth-Gated Acoustic Image
    Holography", Appl. Phys. Lett., Vol. 20, No. 9, p. 370,
    May 1972.

METHODS FOR INCREASING THE LATERAL RESOLUTION OF B-SCAN

C.B. Burckhardt, P.-A. Grandchamp, H. Hoffmann

F. Hoffmann-La Roche & Co., AG

Basel, Switzerland

ABSTRACT

B-scan displays a picture of a cross-section through the body and is widely used in ultrasonic diagnosis. The main deficiency of B-scan is the low lateral resolution, which has its origin in the large diameter of the ultra-sound beam.

The first system to be described is a synthetic aperture sonar system working at 2 MHz (λ = 0.75 mm in water). The important features of the system and the optical processing are given. A second system uses an annular transducer for focussing the ultrasound beam over a depth of 20 cm. The side lobes are reduced by subtracting an echo signal from a second pulse which contains echoes from reflectors in the region of the side lobes only. All systems have a lateral resolution of 2-3 mm and are thus almost one order of magnitude better than conventional B-scan.

B-SCAN

B-scan is a technique which is widely used in medical ultrasonic diagnosis and which is summarized in Fig. 1 [1, p. 151]. The method gives a picture of a cross-section through the body. The advantage of B-scan is that a linear scan of a simple transducer gives a two-dimensional image.

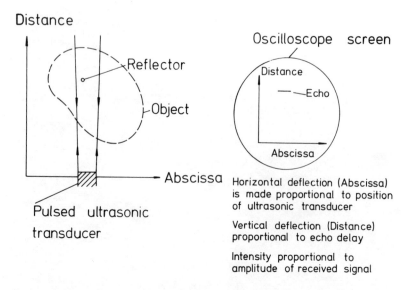

Fig. 1.   Principles of B-scan

Its main disadvantage is the low lateral resolution. As can
be seen from Fig. 1 the coordinate labelled abscissa is made
proportional to the lateral position of the ultrasonic transducer and the lateral resolution is limited by the width
of the ultrasonic beam. At a frequency of 2 MHz which is
often used for scanning the abdomen or the female breast
the ultrasonic beam typically has a diameter of 1 to 2
centimeters. This is one order of magnitude worse than the
longitudinal resolution which typically is 1 to 2 millimeters.
The wave-length at 2 MHz is 0.75 millimeters in the body.
A second disadvantage of B-scan is the poor detection of
specular reflectors. Many interfaces in the body reflect
sound specularly. If such an interface is not perpendicular
or almost perpendicular to the beam, the sound will be
reflected at such an angle that it will not be detected by
the transducer.

    Good lateral resolution and improved detection of spec-
ular reflections can be obtained by focussing the ultrasound beam over a large aperture [2, 3] . The concept is

shown in Fig. 2. The difficulty with this scheme is the
small depth of focus over which good lateral resolution is
obtained.

In this paper we will describe two methods of focussing
which do not have the depth of focus problem. Good lateral
resolution can therefore be obtained over a large depth.

SYNTHETIC APERTURE SONAR

Theory

The synthetic aperture sonar is the ultrasound analog
of the synthetic aperture radar systems [4, 5, 6]. It has
been proposed before [7] and some experimental results were
reported [8] while our work was in progress. The depth of
focus problem is solved in synthetic aperture sonar by separating the operations of sound recording and focussing.
Synthetic aperture sonar is a hybrid between acoustical
holography and traditional B-scan and combines the advan-

HIGH RESOLUTION
REQUIRES A LARGE
APERTURE AND FOCUSSING
OF THE BEAM.
→ PROBLEM OF DEPTH
   OF FOCUS.

Fig. 2.   Focussing the beam with a large aperture

vantages of both systems. In the distance coordinate, where
B-scan works well, it works like B-scan. In the abscissa
coordinate, where there is a problem, it works like a holographic system. B-scan is transformed into a synthetic
aperture sonar system by making the following three modifications:

1. Instead of envelope detection, where phase information is lost, one uses a phase-sensitive detection method.
The echoes are multiplied with a continuous wave reference
signal which is coherent with the emitted pulse. Phase information is thus preserved and focussing at a later time
is possible.

2. A beam of wide angular divergence is used. This
means that the echoes of a given object are recorded over
a large aperture. As observed earlier this is necessary to
obtain good lateral resolution.

3. The record thus obtained is processed to give the
desired image. This is analogous to the reconstruction step
in holography.

The recording of two point objects is shown in Fig. 3.
The objects are scanned with an ultrasound beam of wide
angular divergence as shown in Fig. 3a. In the direction
perpendicular to the paper the beam is made as "thin" as
possible in order to obtain a well-defined cross-section.
The recordings are shown in Fig. 3b. They are well con-
fined and separated from each other in the distance coordinate. The recordings are alternately light and dark
because sometimes the echo is in phase and sometimes out
of phase with the reference signal. The phase variation is
slow when the object is in the center of the beam and becomes more rapid when the object is at the edge of the beam.
An analysis shows that these records are one-dimensional
Fresnel zone plates. Their focal lengths are different and
are proportional to the distance coordinate. (In holography
a point object is recorded as a two-dimensional Fresnel zone
plate). The processing step consists in focussing these
zone plates in the abscissa coordinate, whereas no processing is needed in the distance coordinate. The processed
image is shown in Fig. 3c.

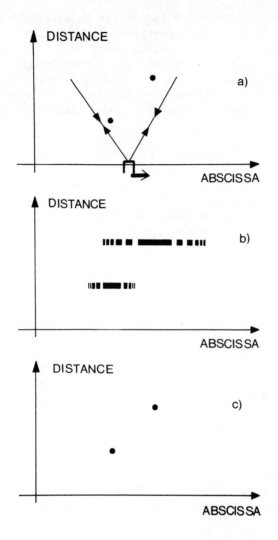

Fig. 3.   Synthetic aperture sonar
          a) Recording arrangement and objects
          b) Record
          c) Processed image

There is one remaining problem. The zone plates shown
in Fig. 3b have two focal points, one real and one virtual.
This causes an undesirable background in the processed

image. The real and virtual images can be separated during
processing if the off-axis portions of the zone plates are
recorded instead of their central portions. (This is analogous to the off-axis reference beam method in optical
holography, see e.g. Ref. [9, pp. 52-54]). The off-axis
portions of the zone plates are recorded if the phase of
the reference signal is changed between successive pulses.

Experimental System

The important data of the experimental system are
summarized in table I. Fig. 4 shows a block diagram of the
system which we will now describe. The control unit controls
the different functions and timing of the whole system. It
also generates crystal stabilized signals of 2 MHz and
100 Hz. The 2 MHz signal is used for driving the power
amplifier and together with the 100 Hz signal serves as
input to the reference generator.

Frequency of operation	2 MHz
Wavelength in water	0.75 mm
Minimum range	20 cm
Maximum range	40 cm
Lateral scan width	30 cm
Resolution	1.5 mm
Pulse length	2 μsec
Acoustical beam aperture angle	15°
Pulse repetition frequency	1000 pulses/sec
Scan velocity	10 cm/sec
Spatial carrier frequency	1 cycle/mm
Temporal carrier frequency	100 Hz

Table I. System Data

The switchable power amplifier drives the ultrasonic transducer during emission. The reference generator produces the
2.0001 MHz reference signal which is multiplied with the
received echo signal. The 100 Hz offset from the emission
frequency leads to a phase change of the reference signal
between successive pulses and the off-axis portions of the

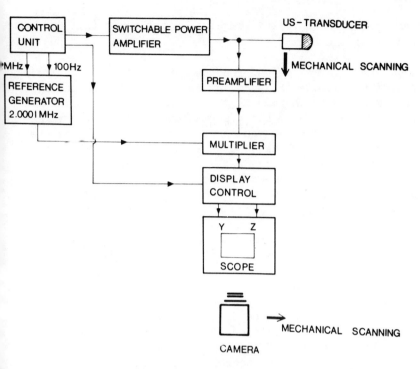

Fig. 4.   Block diagram of the experimental synthetic
          aperture sonar system

zone plates are recorded. This unit is a voltage controlled
crystal oscillator (VCO). Its output signal is mixed with
the 2 MHz signal. Any deviation from the nominal frequency
of 100 Hz is sensed and the frequency of the VCO is adjusted accordingly.

The received echoes are amplified by the preamplifier.
The preamplifier is protected from the high voltage applied
to the transducer during emission by a capacitor connected
in series with two antiparallel diodes. The amplified echo
signal is then multiplied with the 2.0001 MHz reference
signal. Note that the phase of the signal is preserved in
this demodulation process. The display control unit adds
a bias voltage to the output of the multiplier which is
then displayed on the oscilloscope. This unit also pro-

duces the sawtooth voltage required for the vertical deflection of the beam.

The information was recorded on 35 mm film. In the
distance coordinate the beam was deflected electronically
as was mentioned, whereas in the abscissa coordinate the
recording camera was translated mechanically. This arrangement was chosen because the resolution of the oscilloscope
in the abscissa coordinate was not sufficient to display
the whole record. Because of the spatial carrier frequency
the resolution requirement is considerably higher in the
abscissa coordinate than in the distance coordinate. Kodak
High Contrast Copy Film was used for recording.

We will now describe the design of the optical
processor. The processor has to transform the record
Fig. 3b into the image Fig. 3c. In the distance coordinate
a straight imaging operation is required whereas in the
abscissa coordinate one reconstructs a hologram. A further
complication arises because the focal lengths of the zone
plates vary with the distance coordinate at which they
occur. The principles of the processor are described in
Refs. [4, 5] and our processor was built along those lines.
Fig. 5 shows two views of the processor. Collimated laser
light is incident on the synthetic aperture sonar record in
plane P_1, where there is also a cylindrical lens whose
focal length varies with height y. It has been shown[4, 5]
that a pie-shaped section of a lens known as axicon can be
used for this purpose. The focal points of the axicon are
made to coincide with the focal points of the zone plates.
To the right of the axicon we therefore have a wave with
no curvature in the x-direction. The wave then passes
through a conventional cylindrical lens L_2 whose focal point
is in plane P_1 of the record. To the right of L_2 we now
have a plane wave. This plane wave is focussed by spherical
lens L_3 into plane P_3. Thus our goal of processing the one-
dimensional zone plate into a point is achieved. As suggested in Ref. [5] only a narrow vertical strip of the image
is processed at any one time. In order to obtain the full
image the synthetic aperture sonar record as well as a slit
aperture in the image plane are translated. The processed
image is recorded on Polaroid film. This procedure has the
advantages that only the central region of the pie-shaped

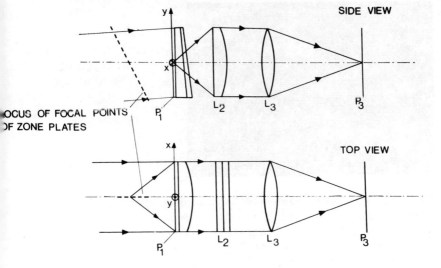

Fig. 5. Optical Processor

axicon section is used and that the distortion of the image
pointed out in Ref. [5] is avoided. One final remark should
be made: In contrast to synthetic aperture radar our resolution requirements are modest and simple and cheap lenses
can be used in the optical processor.

Experimental Results

        The experiments were performed in a water tank lined
with natural rubber. The whole recording set-up including
the water tank was put on antivibration mounts. This
greatly reduced the extreme sensitivity of the signal to
building vibrations. Care was taken to assure that the
whole system up to the recording film operated in a linear
way.

        Most tests were done with two wire grids mounted
at a distance of 7 cm from each other. Each grid had a
number of groups of three copper wires strung at different
distances. Fig. 6 shows the processed image of this object
(A cross-section through the object!). Going from left to
right in the top grid and from right to left in the bottom

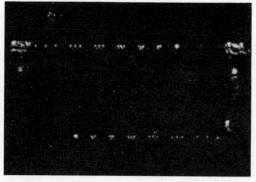

Fig. 6.   Processed image
          for an object
          consisting of
          two wire grids

grid the wires are strung at a distance of 1 cm, 5 mm, 4 mm,
3 mm, 2 mm, 1.5 mm and 1 mm. It is seen that in the top grid
wires spaced 2 mm apart can be resolved whereas in the bottom grid only a spacing of 3 mm can be resolved. These experimental values for resolution are close to the values
expected theoretically, see table I, and are almost one
order of magnitude better than those of a conventional
B-scan system.

    The main disadvantages of the synthetic aperture
sonar system are that it does not operate in real time and
that the resolution is not improved in the direction perpendicular to the drawing of Fig. 3a.

            FOCUSSING WITH AN ANNULAR TRANSDUCER

    The work that we will now describe was triggered by
McLeod's axicon [10] . The basic idea is to obscure the
central portion of the focussing lens of Fig. 2 until one
only has an annular aperture of negligible width. The lens
then becomes unnecessary as it only contributes a constant
phase shift to the wave passing through the ring and one
ends up with the focussing device shown in Fig. 7. The
incident wave is diffracted by the annular aperture. McLeod
[10] presents some heuristic arguments showing that on the
axis the wave has a large amplitude. There is no depth of
focus problem, because there is no preferred point on the
axis.

    The amplitude distribution in a plane parallel to the

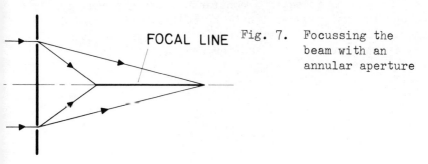

FOCAL LINE

Fig. 7.   Focussing the beam with an annular aperture

plane of the ring is found by the following considerations. It is well known (see e.g. Ref. [9, p. 117]) that the amplitude distribution in the focal plane of a lens is equal to the Fourier transform of the amplitude distribution in the plane of the lens times a spherical phase factor. Since adding a lens in the plane of the ring does not change the configuration, any plane parallel to the plane of the ring can be considered to be the focal plane. The Fourier transform of a ring is found in Ref. [11, p. 249] and we obtain for the amplitude distribution

$$a_0(r) = c_1 J_0\left(\frac{2\pi Rr}{\lambda d}\right) \exp\left(-\frac{i\pi}{\lambda d} r^2\right). \tag{1}$$

$a_0(r)$ is the amplitude as a function of distance r from the axis, J_0 is the Bessel function of the first kind and zero order, R is the radius of the ring, λ is the wavelength, d is the distance between plane of the ring and observation plane and c_1 is a constant of proportionality.

We now consider a ring shaped ultrasound transducer that acts both as an emitter and a receiver. The amplitude $b_0(r)$ received from a point reflector is then proportional to the square of the expression of Eq(1),

$$b_0(r) = c_2 J_0^2\left(\frac{2\pi Rr}{\lambda d}\right). \tag{2}$$

The phase factor has been omitted in Eq.(2), because it is unimportant. r is the distance of the point reflector from the axis and c_2 is again a constant of proportionality. In analogy to optics we will call $b_0(r)$ the "point spread function". $b_0(r)$ is plotted in Fig. 8. It is seen that it does indeed have a large, narrow central maximum.

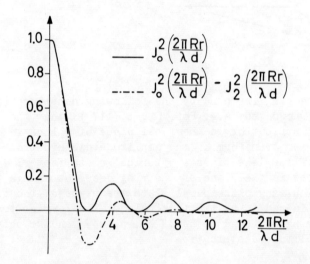

Fig. 8. The point spread functions $J_0^2\left(\dfrac{2\pi Rr}{\lambda d}\right)$,

$$J_0^2\left(\frac{2\pi Rr}{\lambda d}\right) - J_2^2\left(\frac{2\pi Rr}{\lambda d}\right).$$

An experimental annular 2 MHz ultrasound transducer of 10 cm diameter was constructed and is shown in Fig. 9. The orientation and width of the piezoelectric elements were chosen such that the ultrasound beam covers a depth between 20 cm and 40 cm, i.e. the square of the emitted amplitude on the axis falls to half value at these distances. The ring was built out of twelve segments of lead zirconate titanate Pl-60 (Quartz et Silice, Paris). Each element has a separate electrical input for reasons that will be described in the next section. The transducer is damped by a backing consisting of the usual araldite tungsten mixture (75 % tungsten by weight). Care was taken to assure good accuracy and deviations remain within $\pm \frac{1}{20}$ mm, i.e. $\pm \frac{1}{15}$ th of the wavelength.

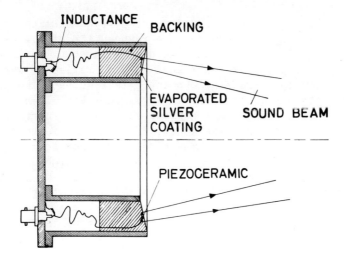

Fig. 9.   Cross section of an annular ultrasound
          transducer

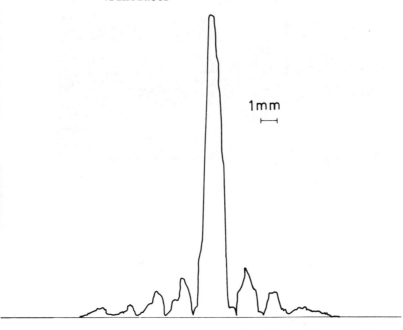

Fig. 10.   Measured point spread function of an annular
           transducer

Fig. 10 shows the measured point spread function, using a pulse length of 2 μsec. The rounded tip of a glass fiber of about 0.3 mm diameter was used as a point reflector and the distance between transducer and reflector was 24 cm. Out to the third sidelobe there is reasonably good agreement with the theoretical curve Fig. 8. Fig. 11 shows the scan of two grids of nylon wires spaced 7 cm apart. The distance to the top grid was 20 cm. The spacing of the wires within the groups of three wires was going from right to left on the top grid and going from left to right on the bottom grid 1 cm, 5 mm, 4 mm, 3 mm, 2 mm, 1.5 mm and 1 mm. The wires spaced 2 mm apart can be resolved in both grids. The gain of the system was adjusted such that the sidelobes are barely visible on the photograph. As a comparison

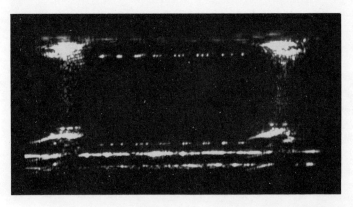

Fig. 11.   Scan of two wire grids with annular transducer

Fig. 12 shows the scan of the same grid taken with a commercial B-scanner. Not even the wires spaced 5 mm apart in the leftmost group can be resolved.

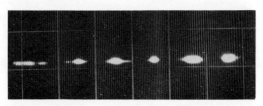

Fig. 12.   Scan of the same grid with a commercial
           B-scanner. The wires in the leftmost group are
           spaced 5 mm apart but cannot be resolved

REDUCTION OF THE SIDELOBES

Principle

The point spread function $J_0^2\left(\frac{2\pi R r}{\lambda d}\right)$ in Fig. 8 shows
rather large sidelobes. It was feared that one would have
difficulties with objects with many reflectors and echo
amplitudes of widely different magnitude, a situation
typical for biological objects. Especially disturbing is
the following thought. From Fig. 8 it can be estimated
that the height of the sidelobes decreases proportional
to $1/r$. Since the circumference increases proportional
to r the volume under each sidelobe is the same! This is
also true for large values of the argument. For $z \rightarrow \infty$
we have [12]

$$J_0(z) = \sqrt{2/(\pi z)} \; \cos (z - \pi/4), \; z \longrightarrow \infty \qquad (3)$$

and therefore

$$J_0^2(z) = (2/\pi z) \cos^2 (z - \pi/4), \; z \longrightarrow \infty \quad . \qquad (4)$$

Thus the volume under each side lobe is indeed the same.

Therefore we looked for ways to reduce the side
lobes. The method we chose is an adaption of an idea due
to Wild [13]. Wild has shown how one can obtain a point
spread function with low side lobes with an annular radio
telescope. The basic idea is the following: For a first
emitted pulse the system works as described in the last
section and we obtain echoes from reflectors on axis as
well as echoes from reflectors within the region of the

side lobes. We now emit a second pulse where we obtain
echoes from reflectors within the region of the side lobes
only. We then subtract these echoes from the stored echoes
of the first pulse and are left with the desired echoes from
the central region only.

How does one obtain echoes from reflectors in the
region of the side lobes only? This is accomplished by
emitting ultrasound with a phase shift proportional to
$k\varphi$ where φ is the angle on the ring and k is an integer.
I.e. the emitted amplitude is multiplied with the phase
factor $\exp(ik\varphi)$. The amplitude distribution in a plane
parallel to the ring then is [13]

$$a_k(r) = c_1 \exp\left[ik\left(\frac{\pi}{2} + \varphi\right)\right] J_k\left(\frac{2\pi Rr}{\lambda d}\right). \tag{5}$$

If we multiply the received echoes with a phase factor
$\exp(-ik\varphi)$ we will have the amplitude $b(r)$ received from
a point reflector

$$b_k(r) = c_2 J_k^2\left[\frac{2\pi Rr}{\lambda d}\right]. \tag{6}$$

All Bessel functions of order greater than zero have a
value of zero for $r = 0$. Wild [13] has shown that one can
synthesize very good point spread functions by adding very
many functions of the type of Eq. (6) with suitable weight
coefficients. Since we wanted to keep the equipment simple
and the time needed to make one scan as short as possible
we tried to synthesize a reasonably good spread function
with the least number of terms. The best match to the side
lobes of $J_o^2(z)$ is obtained with $J_2^2(z)$, as can be seen
from a graph of Bessel functions. For large values of z
the two functions become equal, because [12]

$$J_2^2(z) = (2/\pi z) \cos^2(z - \pi - \pi/4)$$

$$\approx (2/\pi z) \cos^2(z - \pi/4), \quad z \longrightarrow \infty. \tag{7}$$

and this is equal to Eq. (4).
The point spread function

$$b_I(r) = J_o^2\left(\frac{2\pi Rr}{\lambda d}\right) - J_2^2\left(\frac{2\pi Rr}{\lambda d}\right) \tag{8}$$

is shown in Fig. 8. It is seen that there is one large negative side lobe but that the outer side lobe decrease very
much faster than for $J_o^2\left(\frac{2\pi Rr}{\lambda d}\right)$ alone.

Experimental System

The experimental system was built such that the Bessel
functions of order zero, one, two, three, four and six
could be used for the synthesis of the spread function. For
simplicity only the system which synthesizes the spread function of Eq. (8) is shown in the block diagram Fig. 13
which we will now describe. The control unit delivers all
the necessary timing, gating and command signals. A pulse
modulator delivers pulses of the 2 MHz signal to the tapped
delay where the pulse undergoes the appropriate phase shifts.
The multiplexer then connects pulses with the appropriate
phase shifts to the power amplifiers driving the segments
of the ring. The same segments are used for reception.
Each segment has its own preamplifier which is connected
to the multiplexer. The multiplexer connects the received
signals to the appropriate taps on the delay line. The
following swept gain amplifier has a gain which increases
with time elapsed after pulse emission and thus compensates
for the absorption in biological tissue (see e.g. [1] p.
85 ff.). The gain of this amplifier is controlled by a
ramp voltage generated by the swept gain control. The
ramp is triggered by the first echo occurring within an
observation window given by the control unit. The signal
is then demodulated in the multiplier. It is important to
realize that we have a coherent system. The received echo
carries amplitude and phase information and the side lobe
reduction described in the last section has to be carried
out on the phasor of the received signal and not on its
absolute value only. The subtraction is therefore performed on the two components obtained by multiplying the
echo signal with two coherent signals in phase quadrature.
In order to save equipment the two quadrature components
are obtained in time sequence. The signal is then converted into a binary form in the following 8 bit A/D-con-
verter. The binary signal is stored in a shift register
and again converted to analog form. The undelayed signal
from the subsequent pulse is then subtracted from the

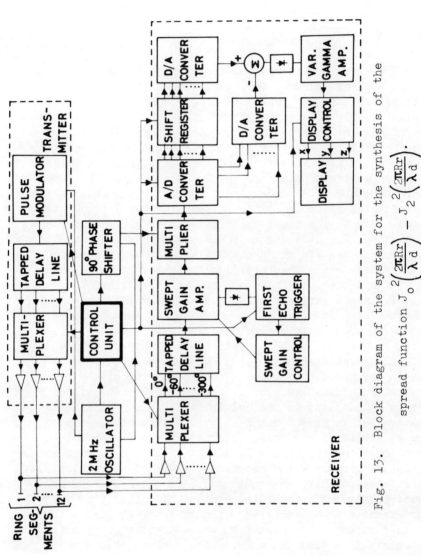

Fig. 13. Block diagram of the system for the synthesis of the spread function $J_0^2\left(\dfrac{2\pi Rr}{\lambda d}\right) - J_2^2\left(\dfrac{2\pi Rr}{\lambda d}\right)$.

delayed signal, see Eq. (8). A detector computes the absolute value and the signal then passes through a variable gamma amplifier. The output voltage u_{out} of this amplifier is given by

$$u_{out} = k \, u_{in}^{\gamma} \, . \tag{9}$$

u_{in} is the input voltage and γ an exponent which can be chosen between $1/4$ and 1. This amplifier compresses the dynamic range for better display. The display control generates the sawtooth voltages necessary for displaying the signal. The two quadrature components of the signal are added as brightness values on the oscilloscope screen. There is an error involved. If u_1 and u_2 are the two quadrature components of the signal one forms $+ \, u_1^{\gamma} + u_1^{\gamma}$ instead of $+ \, (u_1^2 + u_2^2)^{\gamma/2}$. This error could be eliminated by processing the two quadrature components simultaneously.

Experimental Results

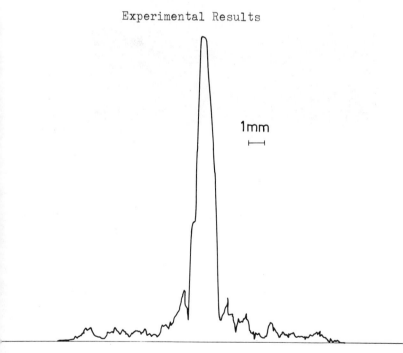

Fig. 14.   Measured point spread function of the system
            shown in Fig. 13

Fig. 14 shows the measured point spread function of
the system shown in Fig. 13. The point reflector was at
a distance of 24 cm. Comparison with Fig. 10 shows that
the side lobes beyond the first one are markedly reduced.
When comparing Fig. 14 to the theoretical point spread
function in Fig. 8 it should be borne in mind that the
signal passes through a rectifier and is therefore positive
everywhere.

Fig. 15 shows a scan of the same two grids as Fig. 11.
The two figures look quite similar but when making Fig.
15 the gain of the system was much easier to adjust to
obtain a satisfactory picture.

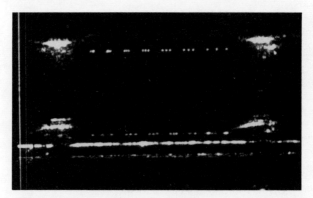

Fig. 15.   Scan of the two wire grids made with the system
          of Fig. 13. Wires spaced 2 mm apart can be re-
          solved in the top grid and wires spaced 3 mm
          apart in the bottom grid

Finally we want to show some pictures of biological
objects. Fig. 16 shows a transversal scan of a cat in the
region of the kidneys. One can recognize the spine, the
dorsal muscles, the kidneys and the air containing bowel.
Fig. 17 shows a longitudinal scan of a five months old
stillborn fetus lying on its front. The head and the spine
can clearly be recognized. Note that the vertebrae are
individually resolved. As a comparison Fig. 18 shows a
scan of a somewhat larger fetus done with a commercial
B-scanner. The vertebrae cannot be resolved.

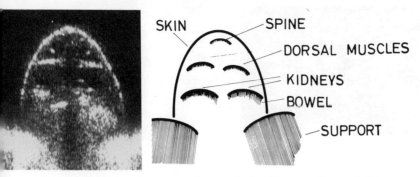

Fig. 16. Transversal scan of a cat in the region of the kidneys

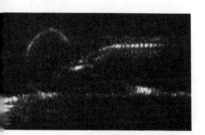

Fig. 17. Longitudinal scan of a five months old fetus lying on its front

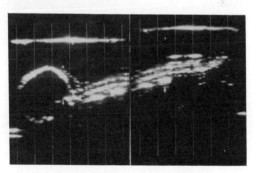

Fig. 18. Scan of a somewhat larger fetus than Fig. 17, done with a commercial B-scanner

Fig. 19 shows a longitudinal scan of a five months old fetus lying on its side. The head with the midline of the brain, the spine and the pelvic bones can be recognized.

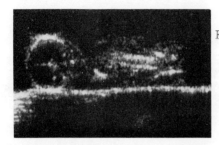

Fig. 19. Longitudinal scan of
a five months old
fetus lying on its
side

CONCLUSIONS

Methods for improving the lateral resolution of B-scan
have been demonstrated namely a synthetic aperture sonar and
focussing with an annular transducer without and with side
lobe reduction. All systems have a lateral resolution of
2-3 mm which is almost an order of magnitude better than
conventional B-scan. Focussing with an annular transducer
with side lobe reduction seems to be the most promising
system at the moment as it operates in real time and
achieves good resolution.

ACKNOWLEDGMENT

The scan of the cat was made in cooperation with
Dr. H. Weiser and the scans of the fetuses in cooperation
with Dr. M.J. Hinselmann. The authors are indebted to
A. Schirmann and H.R. Wolf for their experimental assistance.

REFERENCES

1 P.N.T. Wells, "Physical Principles of Ultrasonic Diagnosis"
(Academic Press, London, New York 1969).

2 M. von Ardenne, R. Millner, "The U.S. Focoscan Method",
IRE Trans. Bio-Med. Electron., vol. ME-9, 145-149 (1962).

3 F.L. Thurstone, W. McKinney, "Focused Transducer Arrays
in an Ultrasonic Scanning System for Biologic Tissue",
in "Diagnostic Ultrasound", C.C. Grossmann et al. Eds.
(Plenum Press, New York 1966), pp. 191-194.

4   L.J. Cutrona, E.N. Leith, L.J. Porcello, W.E. Vivian,
    "On the Application of Coherent Optical Processing
    Techniques to Synthetic Aperture Radar", Proc. IEEE,
    vol. 54, 1026-1032 (1966).

5   E.N. Leith, A.L. Ingalls, "Synthetic Antenna Data
    Processing by Wavefront Reconstruction", Appl.Opt.,
    vol. 7, 539-544 (1968).

6   R.M. Brown, "An Introduction to Synthetic-Aperture
    Radar", IEEE Spectrum, vol. 6, 52-62 (September 1969).

7   F.R. Castella, "Application of One-Dimensional Holo-
    graphic Techniques to a Mapping Sonar System", in
    "Acoustical Holography", vol. 3, A.F. Metherell Ed.
    (Plenum Press, New York, London 1971), pp. 247-271.

8   D.W. Prine, "Synthetic Aperture Ultrasonic Imaging",
    Proc. of the Symp. on Engineering Applications of
    Holography, Feb. 16-17, 1972, Los Angeles, Calif.,
    pp. 287-294.

9   R.J. Collier, C.B. Burckhardt, L.H. Lin, "Optical
    Holography", (Academic Press, New York, London 1971).

10 J.H. McLeod, "The Axicon: A New Type of Optical Element",
    J.Opt.Soc.Amer., vol. 44, 592-597 (1954).

11 R. Bracewell, "The Fourier Transform and Its Applications",
    (McGraw Hill, New York 1965).

12 M. Abramowitz, I.A. Stegun, "Handbook of Mathematical
    Functions", (Dover Publications Inc., New York 1968),
    p. 364.

13 J.P. Wild, "A New Method of Image Formation with Annular
    Apertures and an Application in Radio Astronomy", Proc.
    Roy.Soc. A, vol. 286, 499-509 (1965).

MEDICAL USES OF ACOUSTICAL HOLOGRAPHY

D.R. Holbrooke, M.D., E.E. McCurry, BSEE,
V. Richards, M.D.

Medical Engineering Research Laboratory
Children's Hospital, San Francisco, CA.

Early investigation of the potential of acoustical holography as a new modality for soft tissue diagnosis has suggested a number of possible medical applications. At the same time this work has revealed a lack of basic understanding of ultrasonics and soft tissue interactive effects using through transmission imaging.

Although the investigation of through transmission ultrasonic imaging has been retarded by the lack of suitable equipment, the last several years have seen intensive efforts to develop such equipment. These efforts have been directed toward a variety of image conversion methods capable of presenting an ultrasonic image in a manner similar to X-ray fluoroscopy. Of the image conversion methods under development, the only one presently capable of any type of realistic evaluation of sound tissue interactive effects with this modality is acoustical holography, using the surface levitation method of image conversion. This method was developed by Byron Brenden and others at Northwest Battelle in the mid-1960's and since then has been upgraded by Holosonics, Inc. of Richland, Washington.

During the past two years, our group at Children's Hospital has been working with several variants of the Holosonics liquid surface acoustical holograph, in an effort to better define sound tissue interactive effects using the through transmission modality.

415

Until the present, none of the ultrasonic equipment developed for medical usage has been able to use the phenomenon of ultrasonic acoustic absorption. Theoretical considerations, together with the results of early experimentation using crude equipment, led to the widespread belief that absorption rather than interface effects would be the predominant effect seen with through transmission imaging. However, our results have shown that within certain parameters acoustic absorption, either absolute or differential, appears to play a relatively insignificant role in the formation of through transmission ultrasonic images. This paper will attempt to confirm this. As more sophisticated equipment becomes available, investigators of through transmission ultrasonic imaging should bear this possibility foremost in their minds, because, as will be demonstrated, failure to do so can lead to gross misinterpretations of the images

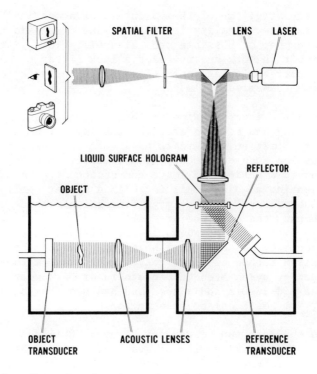

Figure 1: Functional schematic of immersion-type surface levitation holographic imaging system.

ormed and to major errors in understanding the basic
henomena under study.

METHOD

The work described in this paper was performed primar-
ly on a Holosonics GP-3 or Model 100 Acoustical Holograph,
which utilizes the liquid surface image conversion method.
A schematic of this device is shown in Fig. 1. This device
resents a real-time through transmission image which can be
displayed on a ground glass screen or recorded on video tape
or later review and analysis. The device uses a system of
liquid lenses, and the configuration used in our laboratory
as imaging frequencies of 3, 5, and 7 MHz. Since the 12 cm
diameter quartz illuminating crystal has a fundamental reso-
ance frequency of 1 MHz, the use of 3, 5, and 7 harmonics
esulted in considerable power output falloff at the higher
armonics. The measured maximum power output available on
his system at 3 MHz was approximately 20 mW/cm^2, giving a
heoretical peak power in the range of 3-4 W/cm^2. The
device is equipped with a variable pulse width, ranging
rom 30 to 120 μs. Further details of the mechanics and
heoretical operation of the system may be found in several
ublications listed in the bibliography.[8,9,10]

igure 2: Holosonics Model 100 immersion acoustical holograph imaging system in foreground and nonimmersion Model 300 unit in background.

In our experimental setup, shown in Fig. 2, freshly excised surgical specimens were immersed in a 300-gallon tank of water, and both dynamic and static images were recorded of the specimens in a variety of positions at all available frequencies. Freshly excised specimens were used to eliminate possible misleading effects of tissue degeneration and fixation. However, in experiments using fixed specimens, no significant change was noted in the sound tissue interactive effects, except when tissue was imaged in a frozen state. The immersion water bath was maintained at a temperature of 22-24° C. Again, in experiments where the temperature more closely approximated that of living tissue--37° C.--no significant changes were noted in the images obtained.

An important feature of this imaging system is its real-time or dynamic imaging capability, which allows the interpreter to process out of the image a majority of system and biological artifacts, as well as to visualize biological motion. This fact should be borne in mind in evaluating the results presented in this paper, since the static images obtained with this system are of much poorer quality than the real-time images. In previous investigations of through transmission imaging, the unavailability of this real-time imaging capability has been a major impediment to a clearer understanding of absorption vs. other interactive effects. We will attempt to demonstrate this and other problems in the discussion that follows.

DISCUSSION

The initial impression of through transmission ultrasonic imaging is very encouraging. High resolution images can be readily produced, in a format that is easily understood and with a large amount of apparent soft tissue detail.

An example is Fig. 3b; this transmission image of a 16-week fetus was produced using a linear scanner developed by Philip Green of the Stanford Research Institute in Palo Alto. The image, while requiring several hours of scanning time, shows considerable tissue detail, especially of the non-ossified regions of the skeleton.

Comparison with a positive and a negative X-ray of the same fetus (Figs. 3a and 3c) shows a remarkable correlation

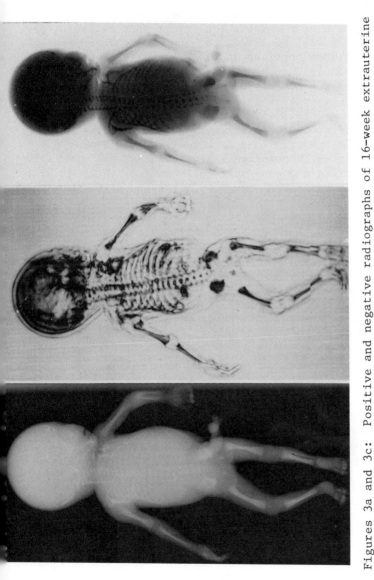

Figures 3a and 3c: Positive and negative radiographs of 16-week extrauterine in vitro fetus. Figure 3b: 3.5 MHz ultrasonic transmission image of same fetus by Stanford Research Institute piezoelectric linear scanner. Note clear delineation of ossified and non-ossified portions of skeleton and outline of other soft tissues.

between X-ray and ultrasonic images, with much better contrast obtained with the ultrasonic image. The various ossi-
fication centers of the vertebral column, the pelvis, and
the appendages are clearly delineated. In the ultrasonic
image, the head of the fetus shows a greater amount of detail which, as will be shown, is both real and artifactual.
The cartilaginous portions of the skeletal system are also
visualized, including the elbow, shoulder, knee joints, and
pelvis, which are not visible in the X-ray images. The
disappointing aspect of this ultrasonic image is the lack
of soft tissue detail within the abdominal cavity and between soft tissues of differing types in other regions of
the body. In summary, this is a remarkable ultrasonic
image of a fetus, showing excellent contrast and visualization of ossified and non-ossified regions of the skeletal
system, but a relative lack of soft tissue detail throughout the rest of the body.

Examination of images such as those in Fig. 3 requires
analysis to determine exactly why some structures are visible but others are not, and whether conditions exist under
which even better images may be produced. As with all ultrasonic tissue interaction, a variety of effects may be expected, including diffraction, refraction, reflection, and
absorption, both differential and absolute. In a new imaging
modality such as through transmission, the exact profile or
relative effect of each of these interactive effects must
be determined.

Focal Plane Imaging

In any image analysis, a number of factors must be considered; foremost of these is the effect of using a sharp or
a shallow focal plane in the formation of the ultrasonic
image.

Investigation of the focal plane problem was initially
undertaken using simple, well-defined materials. In Fig. 4
a surgical clamp was held in and then slightly out of the
focal plane of the imaging system. When the clamp was held
directly in the imaging plane (Fig. 4a), it appeared opaque,
will defined, and a one-to-one analog of the object. When
held in a slightly out-of-focus position (Fig. 4b), the
clamp appeared tubular, due to interface effects, primarily

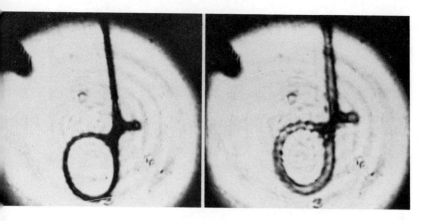

<u>igure 4a</u>: 3 MHz transmission image of surgical instru-
ment held within the focal plane.

<u>igure 4b</u>: Defocused image of the same instrument.

iffraction. It is obvious that in a complicated structure
uch as the arm, containing large numbers of tubular and
olid cylindrical structures, misinterpretation will inevi-
ably result if this diffraction effect is not taken into
ccount. Furthermore, when Fig. 4b is examined closely,
he circular system artifact appears to be visible through
he clamp, reinforcing the initial erroneous impression
hat the structure is tubular rather than solid, until the
bserver becomes cognizant of the diffractive effect of the
ut-of-focus image.

This diffractive effect is further illustrated in
ig. 5, where the clamp handle is held against a surgically
emoved specimen of large bowel. In Fig. 5a the clamp and
owel are held in the focal plane, and the imaging system
as been tuned to minimize system artifact. Both the bowel
tructures and the clamp are in focus and clearly delinea-
ed. Note the relative acoustic transparency of the bowel
tself, except where gas is present within the intestinal

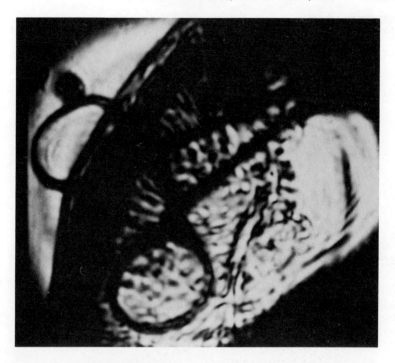

Figure 5a: 3 MHz image of resected bowel segment overly-
    ing surgical clamp. Opaque region to left of clamp is
    gas-filled bowel lumen.

lumen. The same situation exists in Fig. 5b, except that
the clamp and resected bowel are held partially out of
focus and the imaging system is poorly tuned. Although
the clamp is still visible through the bowel tissues, there
is a large amount of diffraction artifact, which markedly
degrades the image quality. It can be seen that coherent
focal plane imaging systems will have major problems with
diffraction artifact, both from system misalignment and
from biological structures just out of the focal plane.
Even in these simple experiments, the number of parameters
involved in image formation is considerable; these include
the effect of a shallow focal plane, the imaging system
itself, as well as the interactive effects of sound and
tissue.

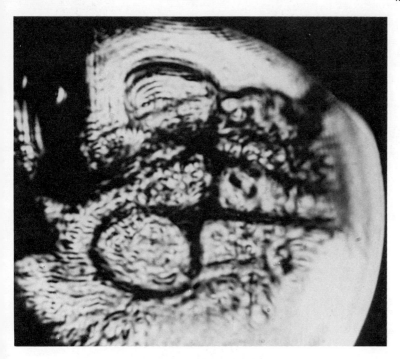

Figure 5b: Similar 3 MHz image of clamp and bowel segment
             taken on a poorly aligned imaging system.

System Artifact

System artifact makes it difficult to separate out the
biological information. An example is seen in Fig. 6, where
a sectioned 14-week pregnant uterus (fetus removed) is
transilluminated at 5 MHz in an effort to differentiate the
placenta and segment of umbilical cord from the uterine
wall. Although the acoustic transparency of the uterine
wall and attached placenta is remarkable, the umbilical
cord is easily visible (primarily by interactive or edge
effects), as are some of the edges of the placenta and
amniotic membrane. Nevertheless, both of these biological
images are heavily degraded by the clearly visible concentric ring pattern. This pattern or system artifact is due
to the fact that the focal plane lies within the "near zone"
of the illuminating transducer and is almost impossible to

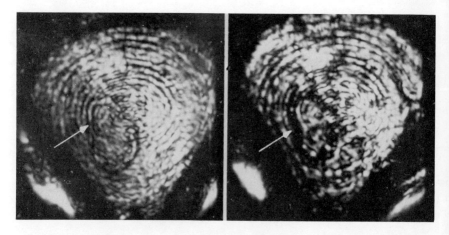

Figure 6: 5 MHz transmission images of sectioned uterus
(fundus to top) with intact placenta and segment of
umbilical cord (arrows). Figs. 6a and 6b are taken
at slightly different focal planes; other conditions
are equal.

eliminate in a coherent imaging system. These system artifacts severely hinder an accurate estimation of the true
potential of transmission imaging for medical usage, especially in the examination of· complicated tissues. Systems
must be designed with minimum mechanical artifact.

Interactive Effects

A major interpretive difficulty in transmission imaging is differentiating absorptive effects from interface
effects. Since the images are formed in a transmission
manner, it often becomes difficult to determine whether
opacity in an image is attributable to absorption of acoustical energy or to its deflection from the imaging system
aperture, either through refraction or reflection. If this
differentiation is not undertaken and the viewer is unaware
that all opacity is not attributable to absorption, he/she
is likely to make gross interpretive errors.

The problem or phenomenon of critical angle (that is,
.mpingement of acoustic energy on the tissue interface at
such an angle that the energy is deflected or reflected out
»f the recording aperture) is demonstrated by the images in
'ig. 7. These images were formed using a 1 cm thick slice
»f beef liver, sliced at a 45° angle to the liver edge.
'hrough the beveled edge can be seen the concentric ring
»attern noted in previous figures. The specimen is held
.n the focal plane in both cases. In Fig. 7a the tissue
»lice and beveled edge are held in the image plane at such
.n angle that the impinging energy is less than the critical
.ngle. As indicated by the arrow, most of the energy passes
:hrough the liver slice, revealing only the edges and some
»f the internal structures. In Fig. 7b the liver slice was
'otated very slightly but to a sufficient degree to present
:he beveled edge at such an angle that the impinging energy
ʲas reflected from the surface rather than passing through
.he tissue. This deflected ultrasound is not captured by

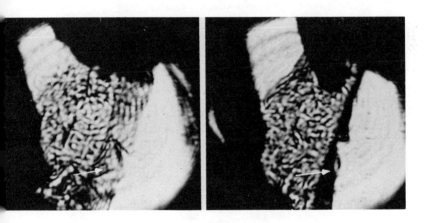

'igures 7a and 7b: 3 MHz images of 1 cm thick slice of
    beef liver with beveled edge along the right-hand mar-
    gin (arrows). In Fig. 7b the specimen has been rotated
    slightly so that the impinging ultrasonic energy is
    deflected due to critical angle.

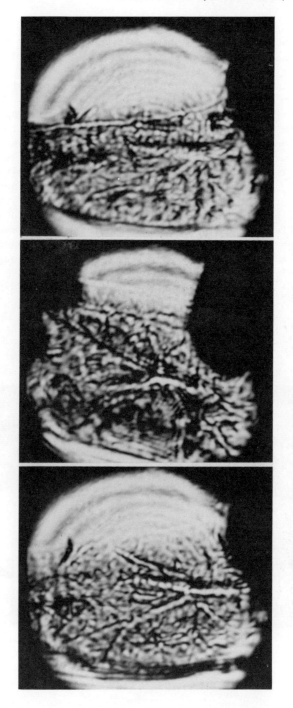

Figures 8a, 8b and 8c: 3 MHz images of beef liver slice showing variable image obtained with only slight adjustments in position and focal plane.

the detector and the image appears opaque (arrows).   If
Fig. 7a were not available, the interpretation could easily
be made that this opacity was due to absorption rather than
the reflection of the impinging ultrasonic energy.   When
specimens are imaged in real time, this problem is easily
overcome, because the movement of the tissue within the
energy field through several degrees of freedom rapidly
differentiates absorptive from interface effects.

Other examples of the interface vs. absorption phenomena are seen in Figs. 8 and 9.   In Fig. 8 the same

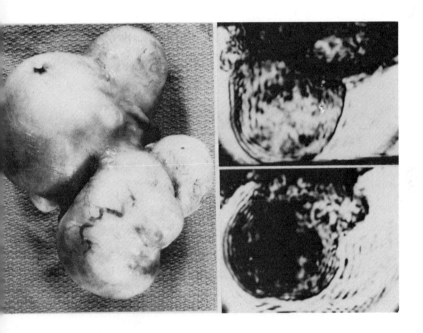

Figure 9a: Photograph of uterus with large pedunculated
    fibroids. Figures 9b and 9c: 3 MHz ultrasonic images
    of one of these fibroids, demonstrating the effect of
    critical angle and other interface effects when the
    specimen is rotated 180°. Note the relative transpar-
    ency of the upper image compared with the dark opacity
    of the lower. This is due to critical angle and thus
    rules out absorption as the major element in this image
    variation.

liver slice is held in three slightly different positions within the acoustic field, and each demonstrates a variety of internal structures, including blood vessels and bile ducts. These images also point out the variable appearance of a tissue interface such as the beveled edge, which is seen on the right side of the images. While the internal structures can be readily seen in certain positions, their visibility is highly dependent upon their position in the focal plane, as well as the angle at which the impinging energy strikes.

This is demonstrated in Fig. 9, where a large fibrotic uterus is imaged, first from one direction and then, after a 180° rotation, from the other (Figs. 9b and 9c). As in Fig. 7, if contrasting images were not available, one's impression would depend on which image one viewed. One could easily conclude that uterine fibroids exhibit a high degree of acoustic absorption or, on the other hand, that they exhibit the same ultrasonic transparency that many other tissues do. Either conclusion would be incorrect and over-simplified.

The fact that the same specimen can present widely dif-fering images even when imaged in the same conditions implies that opacity is due to interface effects, i.e. critical angle, etc., rather than to absorption. These figures, using relatively simple tissue specimens, demonstrate the high risk for interpretive error. With complicated images, as one would obtain in the through transmission imaging of a living abdomen, ignorance of the overwhelming role of interface effects vis à vis absorptive effects could lead to major interpretive errors.

Pursuing the problem of absorption vs. interface effects further, the relative lack of ultrasonic absorption seen with through transmission imaging is also demonstrated in Fig. 10. From theoretical considerations one would ex-pect a significantly greater absorptive effect as one imaged with higher frequencies. However, this absorptive effect, while present, turns out to be minimal.

The primary difference in the two images of Fig. 10 is increased resolution and with it, increased clarity, derived from the fact that edge effects, which make up the major part of the image, are crisper, and real tissue details can

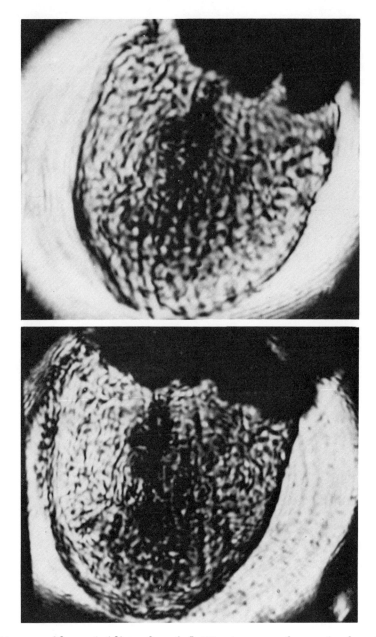

Figures 10a and 10b: 3 and 5 MHz images of surgical spleen
    demonstrating increased resolution seen with higher fre-
    quencies and the minimal absorptive effect at either
    frequency.

be more easily differentiated from artifacts. The hilar region of the spleen, which is thicker than the surrounding tissue and contains large amounts of fibrous tissue, fat, and vascular structures, appears relatively opaque. This is due to a variety of effects, both absorptive and reflective, as well as to dynamic range problems inherent in static photography. When the image was viewed in real time, the improved resolution seen at higher frequencies and the relatively minimal increase in ultrasonic absorption were much more apparent.

Further examples illustrating the absorptive vs. the interface effects are seen in Figs. 11 and 12. Fig. 11 clearly demonstrates the critical role of tissue positioning with a system such as the one we used in our experiments. This freshly excised multiloculated breast cyst is shown in two slightly different positions. The arrows in Fig. 11b point out several small cysts and the edge of a larger one. As can be seen from the corresponding acoustic images, these smaller cysts appear either opaque or transparent, depending on their relative position in the impinging illuminating ultrasonic field (Fig. 11a). In Fig. 11c the edge of one of the larger cysts in this breast mass is indicated by the arrow; note that this larger cyst is not at all visible in Fig. 11a, while the smaller cysts are not visible in Fig. 11c. This is primarily due to the high dependence of the image formation on critical angle. There was almost unlimited variation in the images obtained, and this tremendous variation is, again, very obvious in the real-time or dynamic imaging situation.

Also of interest in Fig. 11 is the almost complete transparency of this cystic specimen, as well as the presence of clearly visualized and degrading system artifact.

Fig. 12 again demonstrates the relative lack of acoustic absorption seen with tissues of different types. This lack of absorption is both absolute and differential. In these images of esophageal carcinoma, colonic carcinoma and the appendix, the differentiating features of the image are attributable to interface effects derived from the shape of the specimen itself. While these shapes are helpful in differentiating the features of the specimen, in the more complicated real-life situation, with large amounts of overlying tissue, differentiation by shape alone may

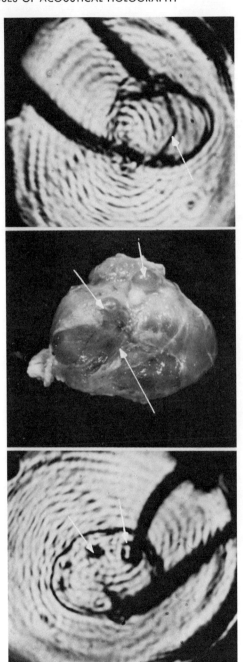

Figures 11a and 11c: 5 MHz images of multiloculated breast cyst. Figure 11b: Photograph of the specimen. Note almost total acoustic transparency except for edge or interface effects. Cysts (arrows) show up as dark or bright regions depending on position and critical angle of cystic interface.

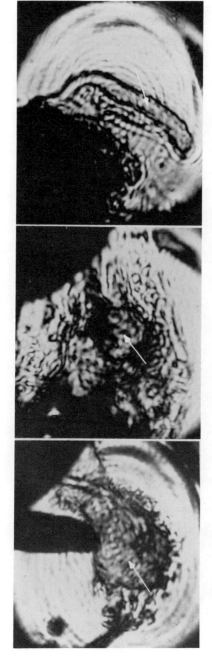

Figure 12a: 5 MHz image of 3 cm diameter esophageal carcinoma. Figure 12b: 5 MHz image of 2 cm diameter large bowel carcinoma. Figure 12c: 5 MHz image of fluid-filled appendix and suspending tissues. All three images illustrate the relative acoustic transparency of the solid tissue mass and the minimal effect of absorption in transmission image formation.

become a difficult and persistent problem. When these images are viewed in real time and at higher frequencies, some differential acoustical absorption is seen. The use of higher frequencies, together with imaging systems capable of greater variation in the parameters involved in this type of image formation, may result in images utilizing the phenomena of absolute and differential acoustic absorption to a greater extent. Several directions that might be investigated with regard to improving the absorption aspects of these images are the use of a shorter pulse length, non-coherent illumination, including frequency sweeping, and detection systems with greater dynamic range.

Imaging Fat Tissue

Of all the tissues imaged, fat tissues presented the most difficult problems and the most characteristic image. As indicated in previous papers,[4, 5] and from the material presented in this paper, we have concluded that fat has a highly characteristic ultrasonic pattern when imaged in the through transmission manner. The image formed is extremely complicated, due to a variety of effects. Interestingly, in the case of fat, absorption plays an increased role.

Fig. 13 demonstrates this increased absorption when fat is imaged. When this beef kidney was imaged intact (Fig. 13a) or in a dissected, lobulated form (Fig. 13b), the solidified fat of the hilar region was acoustically opaque, regardless of position. While this unfixed beef kidney specimen was older than most of our surgical specimens, this phenomenon of apparent increased acoustic absorption was seen in all fatty tissues, regardless of whether they were fresh or fixed. Fig. 13 also demonstrates the glass-like appearance of normal kidney tissue and the visualization of this tissue primarily by interface or edge effect. This is quite apparent in the left side of the image in Fig. 13a. The contour of the kidney can be seen, primarily from the shadow created by partial deflection of the illuminating ultrasonic beam.

Another highly characteristic pattern found in fat tissues is seen in Fig. 14b, where a freshly excised human kidney, surrounded by its perinephric fat, was imaged at

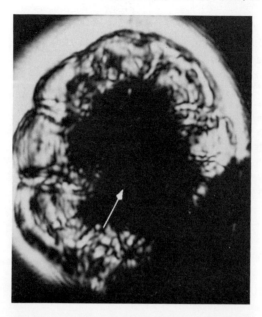

Figure 13a: 3 MHz image of lobulated beef kidney.

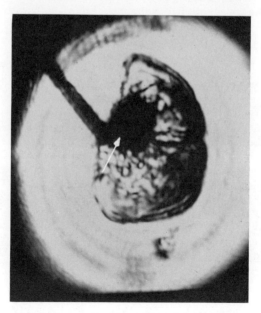

Figure 13b: 3 MHz image of beef kidney lobule.

3 MHz. The "wormy" appearance seen in the left part of
the image is typical of fat images throughout the body and
is probably due to the large number of interfaces present
in fat tissue, which is made up of islands of fat cells
intermingled with a fibrous stroma, as well as interdigi-
tating vascular structures. This tissue, with its highly
convoluted mass of interfaces, predictably presents a confusing image. The darker mass to the right of the image
is tumorous kidney and overlying fat layers. The relative
opacity of this region is due to the scattering and absorption of the sound as it attempts to pass through this relatively thick specimen.

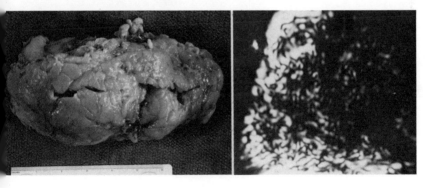

Figure 14a: Fresh surgical specimen of human kidney with
large renal cell carcinoma surrounded by perinephric fat.

Figure 14b: 5 MHz image through the kidney and perinephric
fat.

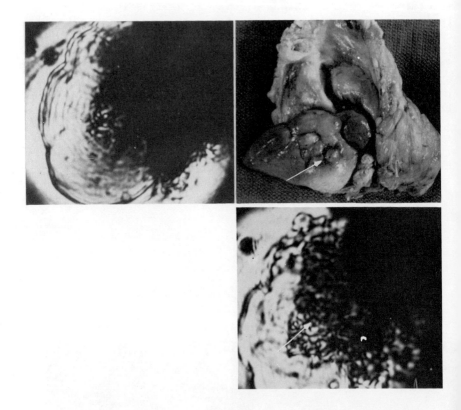

Figures 15a and 15c:　3 MHz images of human kidney with
carcinoma.　Note transparency and glass-like appearance of normal kidney tissue, marbled appearance of
underlying tumor, and relative opacity of surrounding
fat (right).　There is some gray-scale shadowing in
both images.　Figure 15b: Photograph of same kidney
with perinephric fat folded back.

This same specimen was dissected, the perinephric fat stripped back from the underlying tumorous kidney, and it was then re-imaged (Fig. 15). The glassy appearance of the normal kidney tissue can be seen in Fig. 15a, together with the opaque withdrawn perirenal fat tissue to the right of the image. In Fig. 15b and 15c the arrows indicate the cancerous kidney tissue, which can be easily differentiated from the normal kidney. The cancerous tissue appears to absorb or deflect more acoustic energy than the normal tissue, but less than the fat tissue. Note that within this cancerous kidney tissue numerous edges or tubular structures are seen, probably because of the more vascular nature of the kidney tumor. One can conclude from Figs. 14 and 15 that, while differentiation is possible when a structure is relatively simple and there are few intervening tissues of different types, if there is a great deal of overlying tissue present, deciphering these images may be extremely difficult and will be very much a function of learning and experience.

In contemplating the possible clinical applications for this imaging modality, one problem that will arise in intra-abdominal imaging is the presence of gas within the bowel lumen. In order to demonstrate that this problem is real and worthy of serious consideration, a simple experiment was undertaken using an excised surgical colon together with its attached mesentery (Fig. 16). As indicated by the arrows, the bowel was gradually injected with air, and the resulting images recorded. Fig. 16a shows the bowel lumen as it was received from Surgery, with small pockets of gas present within it; in Fig. 16b, the lumen is partially filled with injected air; and in Fig. 16c, completely filled and distended. The gas or air, as one would expect, causes a total reflection of the impinging ultrasonic energy, resulting in opacity of the image. This phenomenon, while creating difficulties in image interpretation in some instances, may in others be very useful in the visualization of a variety of intra-abdominal pathologies. With appropriate technical development, it is conceivable that transmission ultrasonic imaging might serve as a strong complement to barium X-ray fluoroscopy.

Of note in Fig. 16b is the effect of the vascular structures within the mesentery. The mesentery in general appears acoustically transparent, except for the edge

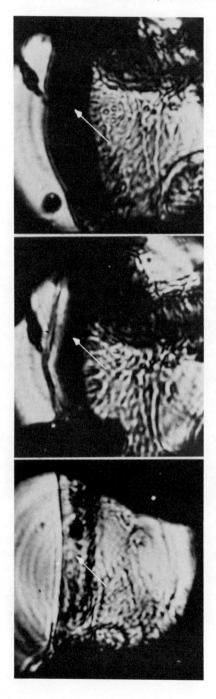

Figures 16a, 16b and 16c: 5 MHz images of large bowel segment with mesentery, showing effects of gas injected into bowel lumen.

Tissue Age

One of the more interesting phenomena observed in our investigations with tissues of various types was the marked effect of tissue age on the acoustic image. It has been consistently observed that with younger tissues, such as in fetal imaging (Fig. 3), there is an almost complete lack of soft tissue differentiation and detail. With older tissues, such as the appendages of children and teenagers, an increasing number of internal soft tissue structures are visualized; in the adult the largest number are seen. This effect is true even in adults of differing ages. While we have not done sufficient definitive studies to fully elucidate the operative factors, the implication is that the older tissue interfaces have a greater ultrasonic impedence mismatch. This is probably due to the relative increase in elastic or connective tissue as well as the amount of fat present.

A very clear example of this is seen in Fig. 17, showing the in vitro imaging of adult and neonatal livers. Fig. 17a shows a complicated image of the adult liver, containing a large number of tubular structures, which are the highly convoluted vascular and bile drainage systems. The amount of acoustic absorption and interface effect is much greater than that seen in the neonate's liver (Fig. 17b.) This liver, imaged at 3 MHz, appeared glass-like and almost completely transparent, being visible primarily by its edges and by the shadings produced from partial deflection due to critical angle. This age phenomenon has held true for all tissues imaged, regardless of type. When this kind of imaging comes into more routine clinical usage, this will probably work to the ultrasonographer's advantage, since the pathology that he/she will be most interested in in the newborn is probably that of the joints and abdomen. The fact that soft tissues are relatively transparent in the young child will permit better visualization of the highly cartilaginous portions of the pelvis, knees, shoulders, and other joints of interest to the pediatrician. In the adult, there is a greater need to visualize vascular structures, foreign bodies, tumor masses, and other non-skeletal structures.

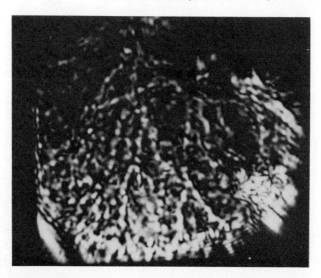

Figure 17a: 3 MHz image of surgically removed adult liver
   showing vascular and bile drainage structures and shadings produced by tissue inhomogeneities and interfaces
   within the liver.

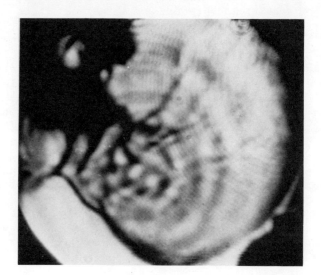

Figure 17b: 3 MHz image of in virto liver of one-day-old
   baby. Note almost complete lack of structural detail.
   Dark objects at left are fingers of specimen holder.

One of the more obvious application areas of through transmission imaging may be obstetrics. The fluid-filled uterus, with its "glass man" inside, should lend itself well to such a high resolution, non-invasive and, from the evidence to date, non-injurious imaging technique. In several of our previous papers we have dealt at length with this possibility. discussing the results of a number of in vitro studies. We have included several uterine images here, both to illustrate the problems facing this application and to demonstrate its potential.

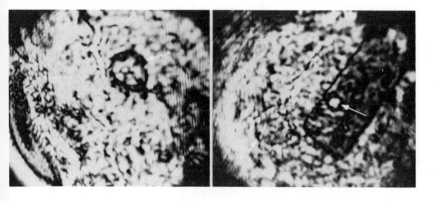

Figure 18a: 3 MHz image of excised postpartum uterus with plastic intrauterine contraceptive device (Dalkon shield) within uterine cavity (arrow). Diffraction artifacts to right of image are due to system misalignment. Figure 18b: Same uterus with ¼" thick plastic block inserted into the uterine cavity alongside Dalkon shield (arrow to left). Note apparent increased absorption through plastic strip and excellent resolution of ¼" holes drilled through plastic strip (arrow to right). Uterine wall is approximately 6-8 cm thick.

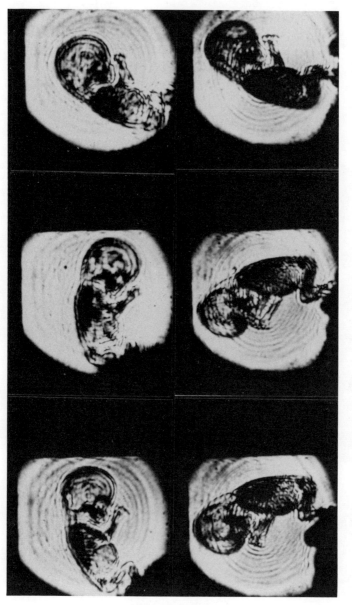

Figure 19: 5 MHz images of in vitro 14-week fetus demonstrating apparent differences in acoustic absorption as specimen is moved about in ultrasonic field. Note especially the abdominal area; in some images the liver appears translucent and in others, opaque, depending on interactive effects of the impinging ultrasound.

As seen in Figs. 3, 19 and 20, the fetus and a number
of its internal structures can be clearly visualized in the
extrauterine situation. The question immediately arises as
to how much degredation would be produced by overlying uterine and abdominal tissue when fetuses are visualized in
the in vivo, real-life situation.

Using a large postpartum uterus as an example, Fig. 18
illustrates that foreign bodies such as intrauterine contraceptive devices (in this case a Dalkon shield) can be clearly
visualized, with excellent resolution. In Fig. 18a the Dalkon shield resting within the uterine cavity (arrow) is
visible, primarily by its edge or interface effects. In
Fig. 18b the positioning is slightly different, and although
the Dalkon shield is not as readily visible, it it still discernible. Fig. 18b also illustrates the relatively small
amount of distortion caused by overlying tissues in uterine
imaging.

In addition to the Dalkon shield, this uterine specimen
had a plastic block inserted in the uterine cavity. This
block had a ¼" hole drilled through it, which is seen in the
right-hand side of the image in Fig. 18b (arrow). Note that
this plastic block appears to be more acoustically absorptive than the surrounding uterine tissues, but in the real-
time situation this was seen to be due, again, to the critical angle effect, which caused partial deflection of the impinging ultrasonic energy.

In other studies we have found that we are able to
image transversely through a 30-week pregnant abdomen and
still retain resolution comparable to Fig. 18b.[5] Furthermore, when the patient was moved to a position within the
focal plane, the living intrauterine fetus was easily visualized including, in the real-time situation, the movement of
the fetal heart within the chest. While the system we are
presently using is somewhat deficient in sensitivity and
illuminating power, the unpublished results we obtained in
the in vivo imaging, together with the more extensive studies of the in vitro excised pregnant uterus, indicate a
very high probability of early application in the obstetri-
cal area. This potential is suggested by Fig. 21, in which
a 12-week in vitro intrauterine fetus is visualized. The
ossified regions of the facial and frontal bones, as well as

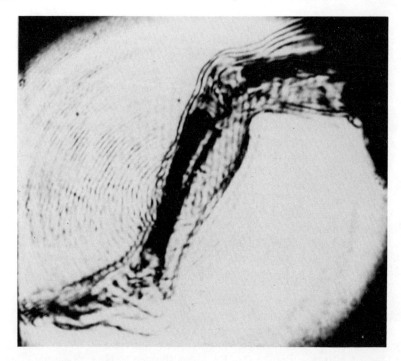

Figure 20: 7 MHz image of 26-week _in vitro_ fetus, demon-
    strating complete opacity of the ossified regions of
    the skeletal structure of the leg, translucency of the
    cartilaginous portions, and the delineation of the
    soft tissues by their edges.

those of the ribs and spinal column, are readily seen, as
are the slightly out-of-focus arms, umbilical cord and
legs of this fetus. In other papers we have demonstrated
the visualization of the developing intracranial neural
structures, as well as some of the intra-abdominal fetal
tissues.[4, 5] Once again, it must be emphasized that
none of these images gives a true picture of just how
clear these images can be, since this modality, to be
fully utilized, must be used in a real-time or dynamic
mode.

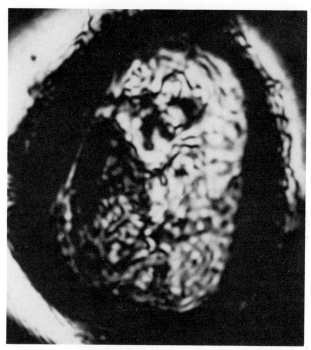

Figure 21: 3 MHz image of excised intact pregnant uterus, demonstrating 12-week <u>in utero</u> fetus. Note the clear delineation of the uterine cavity, together with the ossified portions of the developing fetus within this cavity, especially those of the maxilla, mandible, frontal bones of the cranium, and vertebral column.

The results presented here are but a cross-section of far more extensive studies that we have undertaken, and they are chosen to illustrate the nature of the through transmission phenomenon, the types of problems that can be expected in exploring this application, and to give some idea of what we feel is the tremendous potential that is waiting to be developed in this area of ultrasonics.

CONCLUSIONS

From the body of data now accumulated, a number of conclusions can be drawn regarding the interactive biological effects to be expected using this imaging modality, as well as engineering developments that will be required in order to fully utilize the clinical potential inherent in transmission imaging.

Specific conclusions regarding biological effects using through transmission imaging are:

1.   The predominant interactive effects seen with through transmission imaging are interface or edge effects rather than absorptive effects. In the system used in our experiments, diffraction was heavily involved in image formation, due to the high coherency of the transilluminating ultrasonic beam. This diffractive effect in many instances overshadowed or obscured the other interface effects, refraction and reflection, and made it even more difficult to visualize what little absorptive effect there was.

2.   This relative lack of absorptive effect was apparent regardless of tissue type, i.e. whether the tissues were from different organ systems, such as bowel, thyroid, breast, muscle and kidney; or from the same organ system, such as solid or cystic breast masses. This may in part be due to the parameter limitations of presently available equipment. Further investigations are necessary using equipment with higher imaging frequency, increased sensitivity, narrower pulse width, and possibly frequency sweeping. Some evidence to date indicates that imaging frequencies greater than 7 MHz may permit visualization of those absorptive effects that are present. In our experiments we were handicapped by considerably lower power levels at the higher 7 MHz frequencies, so that definitive investigations could not be undertaken.

3.   Greatly increased resolution is seen with higher frequencies, with a disproportionate increase in image clarity and ease of interpretation. While theoretical calculations would lead one to expect improved resolution at higher frequencies, the marked increase in image clarity appears to be primarily attributable to a lessened effect of artifactual information, such as diffraction edges.

The fact that absorption appears to play a lesser role than initially expected, coupled with improved images at higher frequencies, probably indicates that higher frequencies should be used as soon as the technology permits.

4. The use of a focused image plane in transmission imaging systems presents a variety of problems, especially in systems using both the focused image plane approach and coherent illumination. The very strong diffractive effects in these systems lead to erroneous interpretation, due to artifactual information created in the zone immediately on either side of the focal plane. As demonstrated in our experiments, it becomes very easy to confuse solid and tubular structures. It is not yet clear whether the solution lies in shallower or in thicker focal planes. Systems are now under development which have a variable thickness focal plane, and this may help in solving this problem. Alternatively, the minimization of diffractive effect through the use of broader band, less coherent ultrasonic illumination may help solve the problem.

5. The amount and nature of soft tissue visualization using transmission ultrasonography is, to a considerable degree, a function of tissue age. We have consistently found that younger tissues, especially those of fetuses and young babies, appear acoustically transparent, with few absorptive or interface effects, except in ossified portions of the skeletal system. With increasing tissue age, progressively more interface effects are observed, with improved visualization of numerous soft tissue structures, including tendons, arteries and veins. The nature of the age dependency has not been fully elucidated and is probably a function of several parameters; the most important appear to be the amount of connective or elastic tissue and the amount of fat tissue present.

6. Fat tissue appears to present a major impediment to the early application of through transmission imaging to the clinical situation. Of all the tissues examined, fat creates the most confusing images, especially in a system such as the one used in our experiments, with its strong diffraction effects. When a focused image plane system is used, fat tissue has a characteristic "wormy" appearance, making it almost impossible to determine the nature of any underlying non-fatty tissues. This is

presumably due to the highly loculated nature of fat and
the convoluted vascular structures penetrating it. These
interpretive difficulties with fat tissue are somewhat
minimized by the use of a real-time or dynamic imaging
modality, together with higher frequencies, which lessen
the effect of artifact. Nevertheless, the problem with
fat tissue imaging remains a serious impediment, and further work is necessary, both in biology and engineering.

7. Real-time or dynamic imaging capability is essential in any clinically practicable system using through
transmission ultrasonography. As has been demonstrated
repeatedly in our experiments, the image-degrading effect
of the critical angle phenomenon and other edge effects
can be relatively easily overcome by using real-time imaging. Static image recording alone will almost certainly
lead to gross image misinterpretations.

Insofar as necessary engineering developments in this
field are concerned, three areas are suggested:

1. The development and application of improved detection methods, such as the piezoelectric phenomenon. This
should produce systems with much greater sensitivity and
dynamic range, while at the same time requiring lower illuminating power levels. The growing concern about possible
safety hazards associated with ultrasonic equipment presently in routine clinical use adds considerable pressure
for development of these improved detection methods. Several systems using more sensitive detection methods are
nearing completion. The foremost of these is the unit
presently under development at the Stanford Research Institute, which uses a linear array of piezoelectric transducers to produce near real-time transmission images.

2. The development of systems capable of higher frequency illumination and detection, with levels of illuminating power and detection comparable to those with operating
ranges in the lower frequencies of 1-3 MHz. The suggested
operating range for such equipment is between 5 and 10 MHz.
The use of higher frequencies will result in improved resolution, fewer problems with artifact differentiation, and
more extensive ability to use the absorptive phenomenon for
tissue differentiation.

3. If focal plane imaging is to be used, a solution must be found to the problems associated with presently available systems that have a focal plane thickness of 5 to 10 mm. This difficulty may be alleviated with systems using broader band illumination, thereby minimizing diffractive effects. Should this not prove to be the case, there must be investigation of whether to move toward a thinner or a thicker focal plane. The system presently being developed by Actron Corporation, which uses a true holography-type detection method and has no acoustic lens train, may provide some of the necessary answers to this problem.[11]

In summary, much further work remains in this field, both in biological investigations and in engineering development. There must be a better understanding and delineation of the basic interactive effects of high frequency ultrasound with various tissue types found in the human body. This will require extensive quantitative evaluation using selected in vitro specimens under a spectrum of imaging conditions. These results will have to be correlated with results obtained in actual clinical practice. Images obtained with presently available equipment in the clinical situation are complex and difficult to interpret. Without a thorough knowledge and understanding of the basic effects involved with the specific tissues being imaged, accurate interpretation will continue to be difficult. Early widespread application of through transmission imaging in clinical medicine without this necessary foundation might well result in premature discouragement with the clinical possibilities of the field as a whole. This would indeed be unfortunate.

As we have stated here and elsewhere, we are most encouraged by our preliminary results, but there are many problems that must be overcome before the true worth of this imaging modality can be estimated. We urge an intensified engineering effort to produce equipment with which the medical profession can undertake the extensive investigations necessary to bring transmission fluoroscopy-type ultrasonic imaging to its maximum potential.

ACKNOWLEDGEMENTS

The authors gratefully acknowledge the invaluable help given by Ms. Kathy Ward and Ms. Martha Sue McConnell and the technical assistance of Mr. William Pitt in the preparation of this paper. This research was made possible through the kind support of the Surgical Research Fund of Children's Hospital of San Francisco.

REFERENCES

1. Dussik, K.T., "Über die Möglichkeit hochfrequente mechanische Schwingungen als diagnostisches Hilfsmittel zu verwenden," Z. ges. Neurol. Psych. 174: 153 (1942).

2. Weiss, L. and E.D. Holyoke, "Detection of Tumors in Soft Tissues by Ultrasonic Holography," Surg. Gyn. Obst. 128(5):953 (1969).

3. Holbrooke, D.R., H. Shibata, B. Hruska, E.M. McCurry, and E. Miller, "Diagnostic Holography - a Feasibility Study," in Acoustical Holography, Vol. 2, A.F. Metherell and L. Larmore, eds., Plenum Press, New York, pp. 251-263 (1969).

4. Holbrooke, D.R., E. McCurry, V. Richards, and H. Shibata, "Acoustical Holography for Surgical Diagnosis," Annals of Surgery (Oct. 1973, in press).

5. Holbrooke, D.R., E.M. McCurry, V. Richards, and H. Shibata, "Through Transmission Ultrasonic Imaging of Intrauterine and Fetal Structures Using Acoustical Holography," Proceedings, 2nd World Cong. of Ultrasonics in Medicine, Excerpta Medica, Amsterdam, The Netherlands (in press).

6. Smith, Moody, "Imaging of Biological Structures Using the Principles of Holography," in Proceedings, Conf. on Engineering in Medicine and Biology, Vol. 9, Paper No. 10.7, Plenum Press, New York (1967).

7. Green, P.S., L.F. Schaefer and A. Macovski, "Considera-
    tions for Diagnostic Ultrasonic Imaging," in Acousti-
    cal Holography, Vol. 4, Glen Wade, ed., Plenum Press,
    New York, pp. 97-111 (1972).

8. Brenden, B.B., "Ultrasonic Holography: a Practical
    System," in Proceedings of a Japan-U.S. Seminar on
    Pattern Information Processing in Ultrasonic Imaging,
    Plenum Press, New York (in press).

9. Hildebrand, B.P. and B.B. Brenden, An Introduction to
    Acoustical Holography, Plenum Press, New York (1972).

10. Acoustical Holography, Vol. 1 (1969), Vol. 2 (1970),
    Vol. 3 (1971), Vol. 4 (1972), Plenum Press, New York.

11. Metherell, A.F., S. Spinak and E.J. Aisa, "Temporal
    Reference Acoustical Holography," Appl. Optics 8(8)
    1543 (1969).

A MEDICAL IMAGING ACOUSTICAL HOLOGRAPHY SYSTEM USING LINEARIZED SUBFRINGE HOLOGRAPHIC INTERFEROMETRY

A.F. Metherell, K.R. Erikson, J.E. Wreede,
R.E. Norton, Jr., and R.M. Watts

Actron Industries, Inc.
A Subsidiary of McDonnell Douglas Corporation
700 Royal Oaks Drive
Monrovia, California 91016

ABSTRACT

Linearized subfringe interferometric holography, described in Chapter 4, has been used for detection and recording in an acoustical holography system which is presently under development at McDonnell Douglas Corporation. This paper describes the laboratory system used to test the feasibility of the concept and to work out some of the problems associated with this technique. Acoustical hologram movies have been recorded at 16 frames per second, with each frame being recorded in 6 microseconds from a 100-microsecond burst of sournd. Each movie frame is the recording of a complete acoustical hologram of 30-cm aperture at a sound frequency of 1 MHz. With a sound wavelength of 1.5 mm we have resolved about 2.5 mm and imaged objects less than 1 mm. The three-dimensional image volume is about a 30-cm cube. There is no acoustic lens in the system and thus the resolution is limited by the 30-cm aperture of the acoustic hologram. We have resolved about 1.5 times the Rayleigh limit.

We are presently constructing a new version of this system which is intended to be used clinically to obtain images of soft tissues and structures and organs in the human body.

INTRODUCTION

Over the past seven years many different approaches have been proposed for recording acoustical holograms suitable for medical imaging.[1-5] The

purpose of the work described here is to develop a practical system suitable
for producing acoustical holographic images of the soft tissue organs and
structures in the human body (primarily the abdomen).

This work is a continuation of the program which began at the Douglas
Advanced Research Laboratories in Huntington Beach and which was first
referred to as the linearized subfringe interferometric holography technique
at the Second International Symposium on Acoustical Holography.[6] This
paper reports on the development of this technique as it relates to the
medical imaging objective. Some preliminary images produced on an experimental laboratory system are presented but its ultimate usefulness cannot be
evaluated until it has been adequately tested in a clinical environment.

THE EXPERIMENTAL LABORATORY SYSTEM

The linearized subfringe interferometric holography technique for
recording acoustical holograms has been fully described in sections 3, 7 and 8
of Chapter 4 in this book. An experimental laboratory system using these
principals was constructed and is illustrated in Figure 1 and shown in the
photograph in Figure 2. A 12-inch square, one megahertz sound source is
placed in the bottom of a water filled tank, the test object is placed in the
tank, the thin plastic acousto-optic membrane 30 cm in diameter is stretched
across the top of the water tank. The sound source is turned on for 100 μsec
acoustically illuminating the object and the scattered sound waves impinge on
the acousto-optic membrane. Everything above the acousto-optic membrane
is the linearized subfringe holographic optical detection and recording system. The pulsed xenon laser on the right is turned on for 6 μsec and the light
beam bounces off one mirror, goes through a beamsplitter, and then passes
through a lens and pinhole to optically illuminate the entire acousto-optic
membrane. The large lens placed over the acousto-optic membrane acts like a
condensing lens to concentrate the reflected light back onto the photographic
film in the movie camera body at the top of the system. The reflected beam
from the beamsplitter passes through the optical modulator and around the
system to form the optical reference beam that is superimposed on top of the
beam reflected from the membrane. One frame from the movie film then consists of an optical hologram of the acousto-optic membrane. As described in
Chapter 4, the reconstructed optical image of the membrance is a linearized
subfringe inferometric hologram image which is itself an acoustic hologram

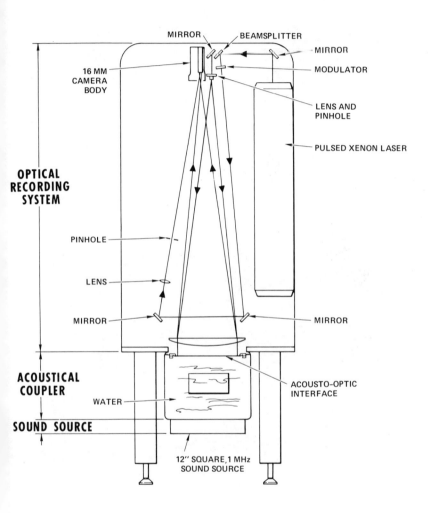

Figure 1. Layout used in the laboratory experimental apparatus.

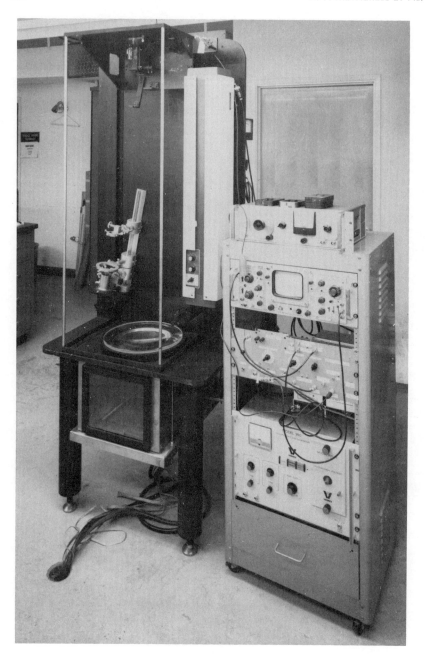

Figure 2. Photograph of the laboratory experimental apparatus.

of the sound wave impinging on the membrane. Therefore, when the hologram film in the movie camera is processed the reconstructed image of the membrane is then put on a second movie film which becomes the acoustic hologram movie which in turn can be reconstructed later to obtain the acoustical image of the object. A typical frame of the acoustical hologram movie is shown in Figure 3.

Figure 3 is the image of the acousto-optic membrane and the fringe pattern contains the acoustical hologram information. The optical image of the object is then obtained by placing this movie film in a coherent laser beam and viewing the reconstructed image on a TV vidicon or a ground glass viewing screen.

The top left picture in Figure 4 shows a molecular model consisting of ¾-inch diameter and 1½-inch diameter Styrofoam balls connected with wooden toothpicks. The other three pictures show the reconstructed image focused at

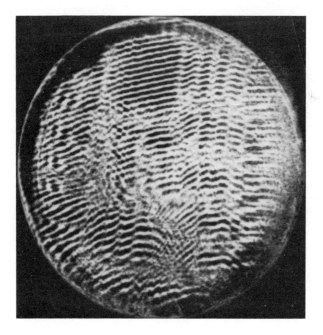

Figure 3. Image of the acousto-optic interface showing the acoustical hologram fringes obtained from one frame of the optical linearized subfringe interferometric hologram movie film.

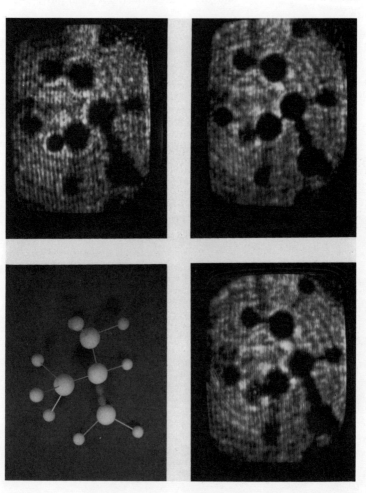

Figure 4. (upper left) Three-dimensional model made from 1.5-inch diameter and 0.75-inch diameter balls connected by wooden toothpicks. (upper right) Reconstructed image focused on closest ball; (lower right) focused on center; (lower left) focused on farthest ball, all from the same hologram.

three different depths, all obtained from the one acoustical hologram. The top right shows the viewer focused on the closest Styrofoam ball with the other balls out of focus. The lower right image is focused about midway down where the center balls and the two lead sinkers along with the connecting toothpicks are in focus. The lower left picture is focused on the farthest balls with the nearest ball completely out of focus. The diameter of the connecting toothpicks is equal to about one wavelength of sound. We have imaged structures less than one wavelength and resolved about 1½ wavelengths. As can be seen, the depth of focus of any focal plane in the reconstructed image is only a few millimeters. This is because the imaging is accomplished with an equivalent numerical aperture of 1 or less.

Figure 5 shows two reconstructed images of a single hologram of a hand made at one megahertz and wavelength 1.5 millimeters. On the left the image is focused on the tips of the fingers and the back of the hand is out of focus. Notice that the sound is completely absorbed by the bone; consequently, there isn't very much soft tissue structure to see at this wavelength. On the right the image is focused on the back of the hand where the metacarpal bone is clearly in focus. There is a structure between the first and second metacarpals that resemble the metacarpal artery which then branches into the first and second digital arteries.

A nineteen-week fetus is shown in Figure 6. It should be pointed out that the frequency used in these experiments was chosen for penetrating through a full sized adult abdomen and was not intended for viewing such a small subject. As a result most of the soft tissue structures are beyond the resolution limit of this imaging system.

Figure 7 shows the two positions of focus obtained from one frame of a holographic movie which was made by suspending the fetus on a wire frame and rotating it in the water tank to obtain a holographic movie. The picture at the right is focused about midway through the fetus and the wire frame to which the specimen was tied can be clearly seen. These wires were slightly over one wavelength in diameter. The facial bone structure can be seen through the back of the head. The image on the left is focused at a slightly different depth and two dark structures in the top of the head are visible which may possibly be the lateral ventricles of the brain. These two structures were consistently seen when viewing the specimen from this direction.

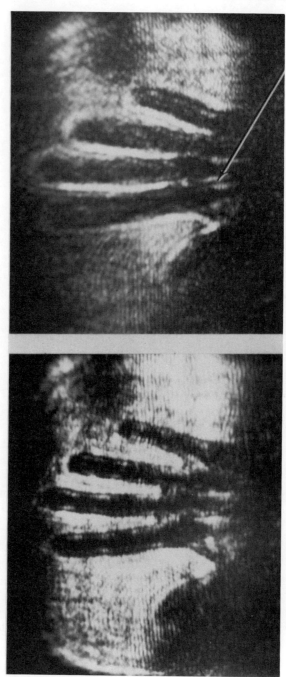

Figure 5. Two reconstructed images obtained from one acoustical hologram of a hand *in vivo*.

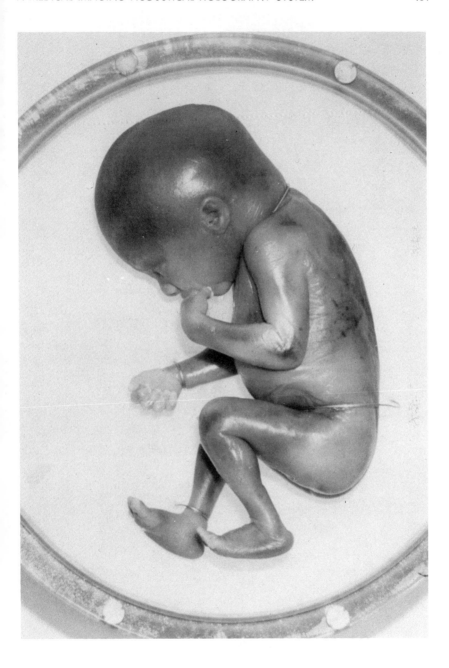

Figure 6. Photograph of a 19-week human fetus *in vitro*.

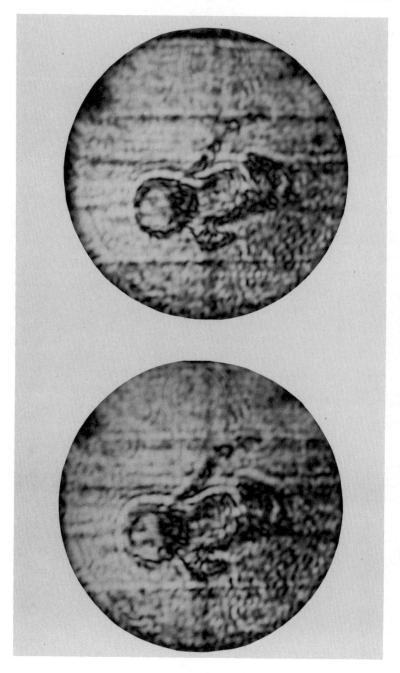

Figure 7. Two reconstructed images obtained from one frame of an acoustical hologram movie of the 19-week fetus.

With the fetus placed in a position similar to that shown in the photograph in Figure 6, an acoustical hologram was recorded which produced the reconstructed image shown in Figure 8. The image is focused on the median sagittal plane of the fetus. The structures such as the arms are out of focus since they are out of the focal plane. In this case the specimen was acoustically illuminated with a single fifteen centimeter square sound source operating at one megahertz. The near field pattern of the sound source is creating a strong noiselike pattern surrounding the subject. Nevertheless, some internal detail can be seen. Figure 9 shows a diagram reproduced from Gray's Anatomy of a sixteen-week fetal brain. Although three weeks younger than the specimen actually tested, there are some close similarities. In Figure 8 the overall shape of the head and the back of the fetus including the spine are clearly visible. The facial bone structure appears dark in the acoustic image. A number of the important structures are identified. We cannot be certain that all the structures indicated in this picture actually appear in the reconstruction, but they do show where the structures ought to be from the anatomy. The most prominent structure in the reconstruction appears to be the bright structure which corresponds with the position of the pons. From the pons we have the medulla oblongata which then passes into the brain stem. The dark structures behind the medulla oblongata should be the fourth ventricle and the cerebral aqueduct leading into the third ventricle. The third ventricle structure appears to be well defined and the front edge of the cerebral hemisphere is very clear. The cerebral hemisphere becomes indistinct at the top back and back of the head. The inability to resolve the top back and back of the cerebral hemisphere as well as the cerebellum may be due to the presence of a massive cerebral hemorrhage which we discovered later when the specimen was dissected.

Some of the disadvantages associated with acoustical holography are apparent in these images. Notice that when the edge of the head goes out of focus, instead of becoming blurred, there is an edge ringing effect. Also the background tends to be very blotchy which is caused mainly by the near field pattern of the illuminating sound source as well as acoustic speckle effects. These characteristics are the result of using coherent radiation which is necessary for the recording of a hologram. Certain incoherent imaging systems can avoid these effects but they lose the capability of recording a three-dimensional volume on a single movie frame.

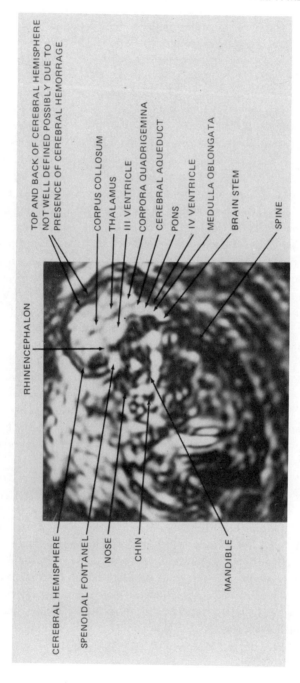

Figure 8. Reconstructed image focused on the median sagittal plane of the 19-week fetus.

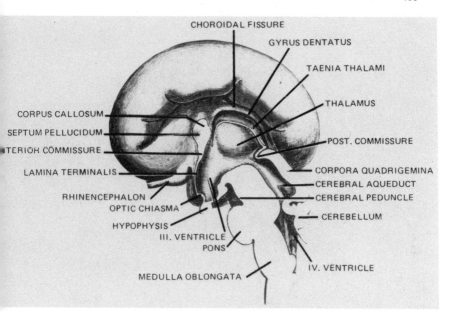

Figure 9. Median sagittal section of a 16-week fetal brain
(from Gray's Anatomy).

THE CLINICAL PROTOTYPE

The physical arrangement of the laboratory system used for the experiments just described is not convenient for a clinical application. Figure 10 shows the basic arrangement that we are intending to use in the clinical system that is presently under development. The camera system is basically an inverted version of the original laboratory system where we now have the sound source on top of the subject and the water tank containing the acousto-optic membrane below. The optical recording system is contained in the cabinet below the patient. Rather than have the patient immerse themselves in a water bath the patient is acoustically coupled through a water filled membrane which forms the center cushion of the examination table. The sound source is also coupled via a water pillow. After the movie film is recorded it is removed from the camera system and developed in the processor. The second acoustical hologram movie is then produced and is then available for viewing at a later time in the viewing system.

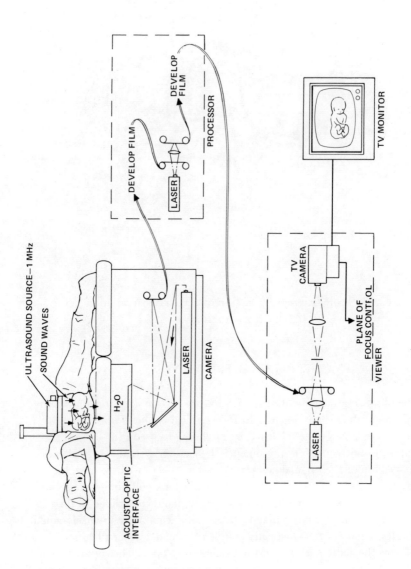

Figure 10. Schematic of the acoustical holography medical imaging system.

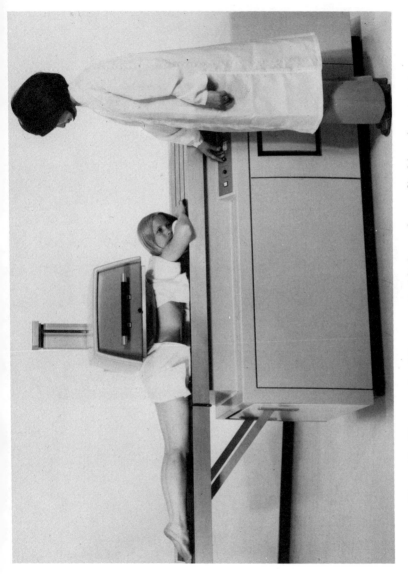

Figure 11. Photograph of the acoustical holography medical imaging camera.

Figure 12. Photograph of the acoustical holography medical imaging viewer.

The clinical prototype system which is presently being prepared for system testing is shown in Figure 11. The end cushion can be folded down and the sound head swung back to allow the patient to position herself on the table. The machine is intended to be operated by an X-ray technician.

The patient is positioned on the water cushion with the sound head contacting her back. This system is primarily intended for viewing the internal soft tissue structures and organs in the abdomen. For an advanced pregnancy the patient can be positioned on her side to allow transverse acoustic illumination of the pregnant uterus.

In the back of the machine there is a small window equipped with a mirror which allows the camera operator to look up through the bottom of the lower sound coupling cushion to check that the patient's skin is making good contact with the plastic membrane. In practice the patient's skin which is in contact with both the upper and lower coupling cushions is coated with a sound coupling fluid similar to that used with present diagnostic ultrasound units.

Figure 12 shows the viewer which can be placed on a table in the physician's office. The results of each patient examination comes to the physician in the form of a short spool of Super 8 mm movie film. Each frame of the movie film is an individual acoustical hologram of the patient's abdominal volume. The physician then inserts the film in the automatic projector. He can then run the film forward or backward while viewing the image on the TV monitor. When he gets to a particular point of interest, he can stop it on a single frame. Because the hologram produces a three-dimensional image he can vary the focus so that he can view any plane through the entire abdominal volume.

CONCLUSIONS

This paper has presented some preliminary results from a new acoustical holography imaging system which demonstrates a capability for soft tissue imaging. Conclusions as to its clinical usefulness must await results of the clinical testing which is due to begin before the end of 1973.

REFERENCES

1.   B. P. Hildebrand and B. B. Brenden, *An Introduction to Acoustical Holography*, Plenum Press, New York, 1972.

2.   A. F. Metherell, H. M. A. El-Sum, and L. Larmore, editors, *Acoustical Holography Volume 1*, Plenum Press, New York, 1969

3.   A. F. Metherell and L. Larmore, editors, *Acoustical Holography Volume 2*, Plenum Press, New York, 1970

4.   A. F. Metherell, editor, *Acoustical Holography Volume 3*, Plenum Press, New York, 1971.

5.   G. Wade, editor, *Acoustical Holography Volume 4*, Plenum Press, New York, 1972.

6.   A. F. Metherell, S. Spinak, and E. J. Pisa, "Temporal Reference Acoustical Holography," *Applied Optics* 8(8), 1543–1550, August, 1969.

DIGITAL PROCESSING OF ACOUSTICAL HOLOGRAMS

C. S. Clark and A. F. Metherell

Actron Industries, Inc.
700 Royal Oaks Drive
Monrovia, California 91016

ABSTRACT

Acoustical holograms recorded at 1 MHz using the McDonnell Douglas medical imaging system were digitized into 1024 x 1024 elements and 64 grey levels (6 bits). Digital processing, similar to that used in conventional digital image processing, was employed in the acoustical hologram plane rather than the reconstructed image plane to improve the image signal-to-noise ratio and to perform other image manipulations. The analysis and experimental results obtained comparing processed and unprocessed images are presented.

INTRODUCTION

The development of high-speed general-purpose digital computers has recently led to the emergence of a relatively new field of study. This, along with the discovery of the Fast Fourier[1] and Fast Hadamard[2] transform algorithms, has made it feasible to digitally process images of natural photographic quality. The expansion of digital image processing technology has been rapid and widespread with many diverse types of images being processed, an image being defined by a brightness function of two space variables (x,y) where the variations in gray scale describe pictorial information.

Digital image processing, a broad field with many applications, can be divided into three categories. The first category is the study of pattern recognition. This includes feature extraction, classification of images, matched filtering and maximum likelihood theory. Pattern recognition is used in radar and communications systems.

The second type of image processing is image coding or bandwidth reduction. The objective of image coding is the efficient representation of images by removal of redundancy. One method is the coding of differences in adjacent samples rather than the coding of the samples themselves. This usually requires fewer bits for transmission since the differences are usually very small. As a result, an overall bandwidth reduction occurs. Another method of bandwidth reduction is transform image coding.[3] In this method a two-dimensional transform is taken, usually a Hadamard or Fourier transform, and the transform is transmitted rather than the image itself. Since most of the image information is concentrated in the low frequency coefficients for the Fourier transform and in the low sequency coefficients for the Hadamard transform, only a few coefficients of the transform need be transmitted in order to transmit most of the information. Transmitting less information results in a bandwidth reduction.

The third and most interesting category of digital image processing is image enhancement and image restoration. Image enhancement is defined as the manipulation of the image to improve image quality. Image restoration attempts to restore the image to an approximation of some ideal (original) image which has been degraded in the formation of the image, for example degradations caused by an optical imaging system. The computer is used to mathematically "undo" or invert the degrading process. Inverse and linear least-square filters are typically used for image restoration. Image enhancement techniques such as contrast enhancement are more subjective in that they attempt to make the image more pleasing to the human visual system.

MATHEMATICAL BACKGROUND

Many one-dimensional analysis techniques already developed in areas such as signal processing and communications systems can be easily extrapolated to two-dimensional techniques, for example, the two-dimensional Fourier transform.

$$F(u, v) = \tfrac{1}{2}\pi \int\int_{-\infty}^{\infty} f(x, y) \exp\left[-i(ux + vy)\right] dx\,dy. \tag{1}$$

The discrete form of the two-dimensional Fourier transform is[4]

$$F(u, v) - 1/N \sum_{x=0}^{N-1} \sum_{y=0}^{N-1} f(x, y) \exp\left[-2\pi i/N(ux + vy)\right]. \tag{2}$$

For linear space invariant systems the image can be expressed as a convolution of the object and the point spread function of the system.[5]

$$f(x, y) = \int\int_{-\infty}^{\infty} g(\xi, \eta)\, h(x - \xi, y - \eta)\, d\xi\, d\eta \tag{3}$$

It is well known that the convolution of two functions is equivalent to the multiplication of their transforms. Thus,

$$F(u, v) = G(u, v)\, H(u, v) \tag{4}$$

An inverse filter which will mathematically invert the degradation of the optical system can be constructed by multiplying the transform of the image by the inverse of the transform of the point spread function to recover the original object transform

$$G(u, v) = F(u, v)/H(u, v) \tag{5}$$

and then taking the inverse transform to obtain $g(x, y)$.[6] This method is easily implemented via digital computer providing the point spread function is known. However, when the image contains an additive noise component, the noise in the image can be greatly amplified since at high spatial frequencies $1/H(u, v)$ is usually very small. Thus, this type of filtering works well only on images with low noise components.

Other types of filters which are useful for image restoration and enhancement are low-pass filters for contrast, high-pass filters for edge enhancement and digital Wiener filters[7] and image estimation techniques for noise reduction. Some image enhancement techniques include contrast stretching and histogram equalization.[8]

COMPARISON OF CONVENTIONAL IMAGES
TO ACOUSTICAL HOLOGRAMS

Holograms are not images in the true sense because they do not contain pictorial information but instead are an interference pattern of coherent wave fields. They are similar to imagery in that they are two-dimensional in nature and can be represented by a brightness function of a range of gray levels. However, although holograms exist in a two-dimensional plane, three-dimensional images can be reconstructed from them. Thus, information about a three-dimensional image is contained in the two-dimensional hologram.

Holograms have different spatial properties than the pictures for which image processing techniques have been developed. Most of the information in conventional images is contained in the low frequency portion of the spatial frequencies and a carrier frequency forming the interference fringes. Because of the high spatial frequency content a much larger grid is required to accurately represent the acoustical hologram than is usually required for a conventional image. Since many more samples are required for an acoustical hologram than for images much more computer time is needed for the computer processing of holograms.

Acoustical holograms have different types of degradations than do ordinary pictures, x-rays, and other types of pictures which typically have been digitally processed. Degradations in conventional imaging systems might include diffraction effects, aberrations, atmospheric turbulence, motion of object or camera, recording system nonlinearities, and various types of noise. On the other hand, degradations in the acoustical holographic system include film-grain noise, noise in the optical system, noise in the acoustical system, nonlinearities in the recording system which cause the fringes not to be perfectly sinusoidal, and speckle due to the coherent property of the source energy. Speckle is the most difficult type of noise to remove because it is nonlinear and is multiplicative noise rather than additive noise.[9, 10]

Digital processing of acoustical holograms is a new application of digital image processing technology. This problem differs from conventional digital image processing since the acoustical hologram is operated upon in order to improve the quality of the image which is obtained by reconstructing the acoustical hologram rather than operating on the image itself. Thus the image is manipulated indirectly by computer processing on the intermediate stage, that is the acoustical hologram, rather than directly by operating upon the image itself. By processing the two-dimensional hologram rather than the reconstructed image the entire three-dimensional image is improved. Since it is possible to focus on any plane of the reconstructed image, many images can be improved by processing one hologram.

APPLICATION OF DIGITAL IMAGE PROCESSING TO ACOUSTICAL HOLOGRAMS

For the following discussion refer to Figure 1. In order to perform digital processing, the acoustical hologram must first be converted into a format compatible with the digital computer. This is accomplished by the process of digitization or the conversion of the analog hologram to gray levels corresponding to densities and the creation of digital information representing the hologram on magnetic tape. Acoustical holograms recorded at 1 MHz were scanned on a grid of 1024 by 1024 elements with six bits per element or 64 gray levels. Such a large number of samples was necessary because of the high spatial frequency content of the holograms.

After the information is available to the computer on magnetic tape the holograms are input into the computer, the computer operations are performed, and a new tape is output containing the digitally processed holo-gram. An IBM 370/165 was used for the computer processing. The next step is the playback of the hologram or the digital to analog conversion. The

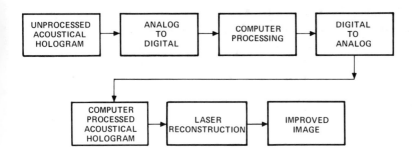

Figure 1. System flow chart for computer processing of acoustical holograms.

digital numbers are converted back to densities and the computer-processed hologram is generated from the computer tape. This hologram is then photo-reduced to the proper size and reconstructed by laser to obtain the improved image.

RESULTS OF PROCESSING

Some preliminary results have been attained in a first attempt at digital processing of acoustical holograms. A smoothing operation was performed in order to eliminate high frequency noise. Each sample of the hologram was replaced by a weighted average of all the samples in a square neighborhood of which the selected sample is the center. For example, for a 3 by 3 neighborhood smoothing, a total of nine sample points are averaged to replace each point of the hologram.

On the left side of Figure 2 is an original acoustical hologram where the object is a pig's heart. On the right is the reconstruction of the original hologram without any computer processing. A particularly noisy hologram was chosen in order to demonstrate a more dramatic improvement.

Figure 3 demonstrates the results of computer processing on the hologram in Figure 2. The upper left image was reconstructed from the hologram which had been digitized and played back only. Since the image seems to have been improved it can be assumed that digitizing alone can improve the reconstructed image. This is probably due to the averaging of the high frequency noise and speckle since a finite area of the hologram is averaged to obtain each sample in the digitizing process.

The upper right image in Figure 3 is a reconstruction from a hologram which was averaged with a 3 by 3 neighborhood smoothing. The width of the neighborhood was about two tenths of the width of the fringes measured from peak to peak. It can be seen that the image has been improved as the detail has become clearer and the inside wall of the heart is made more visible.

The lower left image in Figure 3 is a reconstruction from a 5 by 5 neighborhood smoothing of the same hologram. The width of the neighborhood was approximately 0.36 of the width of the fringes. The image seems to be improved still more.

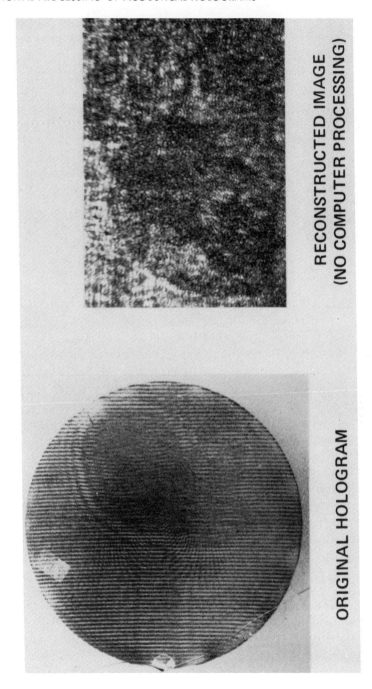

Figure 2. Acoustical hologram and reconstructed image.

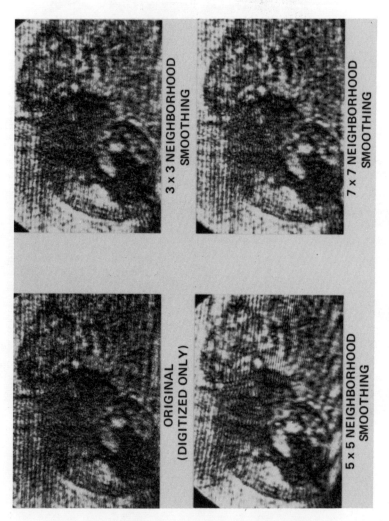

Figure 3. Reconstructed images from holograms processed by neighborhood smoothing.

The lower right image in Figure 3 is a reconstruction from a 7 by 7 neighborhood smoothing. The width of the neighborhood was approximately one half the width of the fringes. This image appears to be slightly more degraded than the one with the 5 by 5 smoothing. If the size of the neighborhood were made larger and larger, eventually all of the information would be lost and the image would be destroyed.

SUMMARY

The results just shown indicate the validity of digital processing of acoustical holograms. This is only a first look at the application of digital image processing techniques to holograms. Many other techniques besides smoothing can be applied such as any of the various types of filtering. Thus far no algorithms have been developed for the removal of speckle. Image processing by computer methods might also be applied to the reconstructed images. The holograms could be reconstructed digitally by the computer as well.

Digital processing of acoustical holograms is important to the development of medical imaging systems as it offers a great potential and flexibility. It can be used to learn about and to evaluate the system. Furthermore, digital methods proved to be useful could be implemented via optical, electronic or analog means. As demonstrated by the pulse echo technique used in diagnostic ultrasound a very small improvement in image quality could lead to greatly increased usefulness. Image features which might otherwise not be apparent would be made visible. Thus the development of digital processing for acoustical holograms could greatly improve the validity and usefulness of the resulting images.

REFERENCES

1. J. W. Cooley and J. W. Tukey, "An Algorithm for the Machine Calculation of Complex Fourier Series," *Math Computation,* Vol 19, pp. 297–301, April 1965.

2. W. K. Pratt, J. Kane, and H. C. Andrews, "Hadamard Transform Image Coding," *Proc. IEEE,* Vol 57, No. 1, pp. 58–68, January 1969.

3.   H. C. Andrews and W. K. Pratt, "Transform Image Coding," Polytechnic Institute
o   of Brooklyn, *Internat. Symp. Computer Processing in Communications*, April 1969.
      Polytechnic Instit. Brooklyn, New York, 1969.

4.   H. C. Andrews, A. G. Tescher, Richard P. Kruger, "Image Processing by Digital
      Computer," *IEEE Spectrum*, Vol 9, No. 7, July 1972, pp. 20–32.

5.   J. W. Goodman, *Introduction to Fourier Optics*, New York, McGraw-Hill, 1968.

6.   H. C. Andrews, and W. K. Pratt, "Ditial Computer Simulation of Coherent Optical
      Processing Operations," *IEEE Computer Group News*, Vol 2, pp. 12–19,
      Nov. 1968.

7.   W. B. Davenport, Jr., and W. L. Root, *Random Signals and Noise*, New York,
      McGraw-Hill, 1968.

8.   E. L. Hall, R. P. Kruger, S. J. Dwyer, III, D. L. Hall, R. W. McLaren, and
      G. S. Lodwick, "A Survey of Preprocessing and Feature Extraction Techniques
      for Radiographic Images," *IEEE Transactions on Computers*, Vol C-20, No. 9,
      pp. 1032–1044, Sept 1971.

9.   S. Lowenthal and H. Arsenault, "Image Formation for Coherent Diffuse Objects:
      Statistical Properties," *J.O.S.A.*, pp. 1478–1483, Nov 1970.

10.   A. V. Oppenheim, R. W. Shafer, and T. G. Stockhom, Jr., "Nonlinear Filtering of
      Multiplied and Convolved Signals," *PROC. IEEE*, Vol 56, pp. 1264–1291,
      Aug 1968.

AN ULTRASONIC HOLOGRAPHIC IMAGING SYSTEM FOR MEDICAL APPLICATIONS

Wayne R. Fenner and Gordon E. Stewart

The Aerospace Corporation
2350 E. El Segundo Blvd.
El Segundo, California 90009

INTRODUCTION

The application of ultrasonic holography to medicine has been suggested by many researchers as an alternative to the use of pulse-echo systems for medical diagnosis.[1,2] This program has as an ultimate objective the development of a real-time holographic imaging system with digital image reconstruction and its evaluation for medical research. The present system configuration is such that, although in actual operation it does not constitute a real-time imaging system, in every other way one can evaluate its capabilities. Data acquisition, recording, and image reconstruction are performed under control of an Alpha-16 minicomputer. Image reconstruction is performed, however, off-line by a CDC 7600 computer system. The system offers a variety of modes of operation. Holographic data may be collected by means of a piezoelectric sensing array, or may be modeled by means of digital computation. Although the system is being built with a particular array sensor, through the use of modeling, the performance of arrays with different elements and different spacings may be readily evaluated. Additionally, one of the most difficult problems in ultrasonic diagnosis is the interpretation of images. Modeling allows the system to display images corresponding to computer generated holograms from a series of different boundary value problems to aid in the interpretation of real data.

IMAGING TECHNIQUE

Of the ultrasonic pressure distribution measurement techniques, piezoelectric devices offer the lowest threshold sensitivity.[3] It has been shown by Vilkomerson[4] and other investigators that for tissue imaging deep inside the body, only piezoelectric array detection has adequate sensitivity. In addition to sensitivity, piezoelectric array detection allows the direct recording of amplitude and phase without the addition of an external reference beam. The removal of this external reference beam reduces the spatial bandwidth required for the sensing array, which in turn allows a greater spacing and a higher sensitivity for the array elements. Once sensing of the incident ultrasonic radiation has been accomplished, a spatial carrier can be introduced producing a truly holographic system. However, as has been noted by Mueller[5], the addition of this spatial carrier introduces the problem of carrying a conjugate image as well as increasing the noise in the system.

Reconstruction of the image without the introduction of a spatial carrier can be viewed in several ways. One group of investigators[6] considers this reconstruction to be basically a problem in backward propagation. Alternately, one can consider the process of reconstruction directly from the amplitude and phase data as computer modeling of an acoustic lens. We have taken the former point of view since it is somewhat simpler conceptually. One has several advantages in the use of such an imaging system. System aberrations may be readily compensated, reconstruction can be performed from the data at later times corresponding to any object depth desired, and images corresponding to different illumination angles may be superposed.

Assuming the scatterers to be located in a plane at $-d_0$ as shown in Fig. 1 producing a pressure distribution $p_1(x_1, y_1)$, for the scattered wave, one can write the corresponding amplitude at $z=0$ in the Fresnel approximation as a two-dimensional convolution.[7]

$$p_2 = p_1 \circledast h \tag{1}$$

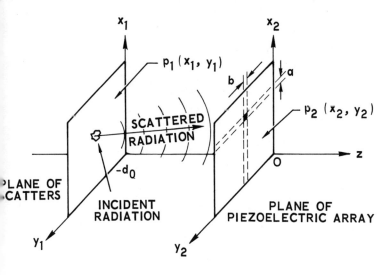

Fig. 1.   Scattering and Array Sensing Configuration.

where

$$h(x_2, y_2; x_1, y_1) = \frac{e^{-jk_0d_0}}{j\lambda d_0} \exp\left\{ -\frac{jk_0}{2d_0}\left[(x_2-x_1)^2 + (y_2-y_1)^2 \right] \right\}$$

Fourier transforming Eq. (1) and expressing the transform of the convolution as the product of Fourier transforms

$$P_2(k_x, k_y) \propto P_1(k_x, k_y) \exp\left\{ +j\left(\frac{k_0d_0}{2}\right)\left(\frac{k_x^2+k_y^2}{k_0^2}\right) \right\} \quad (2)$$

where we have dropped extraneous phase factors.

Since each piezoelectric element averages the pressure over its surface, the effective pressure on an element of height a and width b is

$$p_e(x_2, y_2) = \int\limits_{x_2 - \frac{a}{2}, \ y_2 - \frac{b}{2}}^{x_2 + \frac{a}{2}, \ y_2 + \frac{b}{2}} p_2(\xi, \eta) d\xi d\eta$$

$$= \frac{1}{ab} \int\limits_{-\infty}^{+\infty} p_2(\xi, \eta) \ u(x_2 - \xi, y_2 - \eta) d\xi d\eta \qquad (3)$$

where $u(x, y) = 1 \qquad |x| \leq \frac{a}{2}, |y| \leq \frac{b}{2}$
$\qquad\qquad\qquad 0 \qquad$ elsewhere.

The Fourier transform of the convolution is the product of two Fourier transforms

$$P_e(k_x, k_y) = P_2(k_x, k_y) \ \text{sinc}\left(\frac{k_x a}{2}\right) \ \text{sinc}\left(\frac{k_y b}{2}\right) \qquad (4)$$

where $\text{sinc}(x) = \sin(x)/x$. Thus, omitting extraneous phase factors, we can write from Eqs. (2) and (4);

$$P_1(k_x, k_y) = \frac{P_e(k_x, k_y) \ \exp\left[\left(\dfrac{-jk_o d_o}{2}\right)\left(\dfrac{k_x^2 + k_y^2}{k_o^2}\right)\right]}{\text{sinc}\left(\dfrac{k_x a}{2}\right) \ \text{sinc}\left(\dfrac{k_y b}{2}\right)} \qquad (5)$$

Image reconstruction proceeds by Fourier transforming Eq. (5). The resultant image is a "corrupted" estimate[8] of $|p_1(x_1, y_1)|^2$ due to the finite aperture and discrete sampling used in the transformation.

SYSTEM DESIGN

The system has been designed in a manner to provide maximum flexibility during the development phase. A block diagram is shown in Fig. 2. Signals from an array

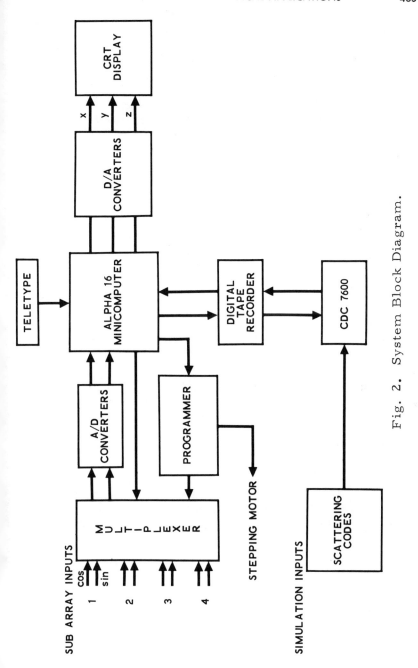

Fig. 2. System Block Diagram.

of piezoelectric elements are amplified and coherently
detected to provide full amplitude and phase information
about the pressure distribution. The detected signals are
then multiplexed, digitized, and stored in core memory
of the Alpha-16 minicomputer.

A digital programmer allows the operator to select
gate times for the transmitter and receiver and to select
the operating frequency.

The minicomputer controls data acquisition, digital
tape recorder, and CRT display functions of the system.
As presently implemented, the data is written on magnetic tape for subsequent processing by a CDC-7600 computer. After reconstruction, the image information is then
loaded on tape and read back into the minicomputer for
display on the CRT. Data may also be synthesized on the
CDC-7600. At present, this synthesized data is being
used to test the reconstruction codes and the CRT display.
However, it is felt that future computer modeling will be
a valuable tool in interpreting actual data from complex
targets such as occur in the human body.

The array of mechanically scanned piezoelectric elements is shown in Fig. 3. A stepping motor drives the
array across a 4-inch by 4-inch aperture. The array itself is linear, 4 inches long, and composed of 64 elements
of PZT-5A.

The elements are cut to a resonant frequency of 1 MHz.
The array is cast in epoxy and mounted in a lucite plate.
The array assembly is scanned across a mylar film which
is in contact with the surface of the subject to be imaged.
The output of each piezoelectric element is individually
amplified by an RF amplifier of 40 dB gain and 2.5 MHz
bandwidth. The outputs of the 64 amplifiers are multiplexed
into 4 channels. The signal from each channel is multiplied with two reference signals, one in-phase and one at
quadrature, resulting in 8 outputs which characterize the
full amplitude and phase information from 4 elements.
This data is then sampled by an 8-channel A/D converter
controlled by the minicomputer. Total system gain is
approximately 65 dB when multiplier gain and subsequent
DC amplification are included with RF amplification. Sensitivity (associated with the LSB on the A/D) is approximately 11-μV(rms) referred to the input. The maximum

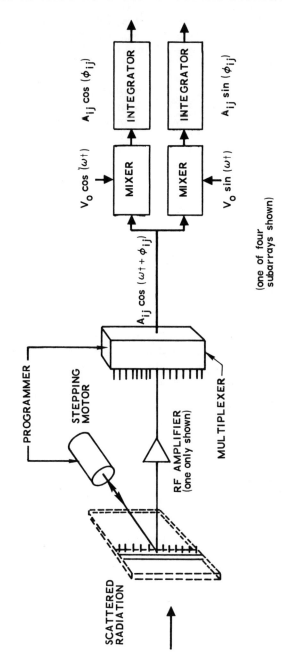

Fig. 3. Mechanically Scanned Linear Array.

signal referred to the input is about 5.5 mV(rms) resulting in a dynamic range of over 50 dB. Compensation for gain and phase differences between channels will be included in the data processing.

As previously stated, the digitized data is temporarily stored in the core memory of the minicomputer and then written on magnetic tape. The tape is then processed in a CDC-7600 computer system. Image reconstruction involves fast Fourier transforming the complex pressure distribution sampled at the aperture, removal of the quadratic phase term in the Fresnel approximation, and then retransforming to obtain the pressure distribution of scattered wave in the plane of the reconstruction. Aberrations are encountered in the reconstruction in the Fresnel approximation due to the large aperture(f/1.0 for an object at a depth of 4.0") of the system. The focal-plane for reconstruction of a point scatterer has been arbitrarily chosen as the plane containing the circle of least confusion.

The intensity data in this assumed focal plane is then normalized to the maximum intensity in the plane, converted to fixed point numbers corresponding to the 6-bit intensity levels carried through the display system. To compensate for CRT characteristics, the data is then shaped to produce an approximately linear variation in perceived CRT intensity. Finally, a tape is written and then displayed on the CRT under minicomputer control.

MODELING STUDIES

The system is constructed in such a way that computer modeled as well as real holographic data may be used as inputs to the display system. There are two basic reasons for this: First, the modeling code allows the testing of the system with a variety of array parameters to check its imaging capabilities. Second, a major problem in medical diagnosis with ultrasonics is the interpretation of image data. The use of modeling codes allows one to determine how semi-transparent bodies of various size will image for given system parameters. Of particular interest are objects which might be relatively smooth in terms of wavelengths. To date modeling work has been conducted primarily with point scatterers and impedance spheres although there are numerous other boundary value problems which admit analytic solutions which should prove to be of interest.

A series of tests of the imaging characteristics of different array configurations were performed using point scatterers. For the initial system parameters selected, a 64 x 64 element array with an interelement spacing of $1/16"$ operating at a frequency of 1 MHz, the reconstructed image of a letter A located at a depth of 3.50" is displayed in Fig. 4. This corresponds to $k_o d_o = -400$. The reconstruction plane corresponding to the circles of least confusion is found to occur at 420 which is between the Gaussian focus and the marginal focus. This difference occurs because of the use of the Fresnel approximation in the backward propagation. The position of the plane of reconstruction, measured in terms of $k_o z$, is displayed in the lower lefthand corner of the reconstruction.

As an example of imaging more complex objects, Fig. 5 shows the reconstruction at selected planes of the scattered pressure due to a sphere of acoustic impedance differing from that of the water in which it is assumed to be located. The phase velocity in the sphere is 0.9 times that of the water and the compressibility is taken to be the same. The wave is assumed to be incident with its wavefront parallel to that of the array. The sphere has a radius of about 8λ.

Reconstructed images of the sphere display primarily scattering from the front and rear surfaces. Reconstruc-

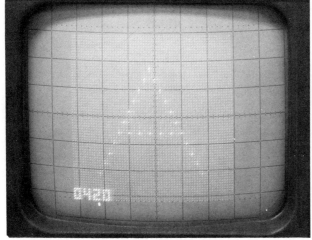

Fig. 4. Reconstruction from Calculated Pressure Distribution of Point Scatterers Arranged to form an A at a Depth of 3.5".

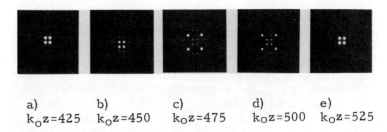

a)   b)   c)   d)   e)
$k_o z=425$ $k_o z=450$ $k_o z=475$ $k_o z=500$ $k_o z=525$

Fig. 5.   Reconstruction from Computed Pressure Distri-
          bution of an Impedance Sphere at Various Depths.

tion at the front tangent plane, three equally spaced intermediate planes, and the rear tangent plane, are shown in
Figs. 5a through 5e. Varying the angle of incidence of
the wave causes the specular points to move across the
surface of the sphere producing, as would be expected, a
change in depth of the specular point.

CONTACT SCANNER

The system under development has been designed primarily for imaging of the abdomen. Of particular interest
will be the visualization of the pancreas, the liver, the
aorta, and kidneys.[9] An array in contact with the skin
would appear to be the optimum choice for such imaging
for two reasons. First, by locating the array as close as
possible to the scattering centers, the optimum resolution
can be obtained for given aperture dimensions. Second,
the use of a waterbath is eliminated.

The ideal choice for sensing the ultrasonic radiation
would be an MxN array of elements. However, the cost of
electronics and packaging considerations weigh against
present construction of a large MxN array. As a compromise, a scanner shown schematically in Fig. 6 is being
developed. A linear array will be moved behind a stationary mylar window. Amplifying and multiplexing functions
would be performed by integrated circuitry built into the
head of the scanner.

A prototype of this scanner, shown in Fig. 7, is now
being tested in conjunction with a waterbath. The 64 individual RF amplifiers are mounted on PC boards directly
behind the array. Only one of eight such cards is shown
installed. A complete scan of the aperture now requires

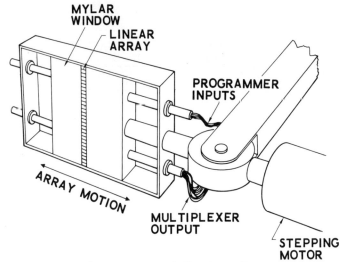

Fig. 6. Proposed Contact Scanner.

several seconds. Future versions of the scanner will take
the order of one second to sample the pressure distribution
across the aperture, however. Data processing on the
CDC-7600 computer can be done in less than a second per
reconstruction depth after transmittal of the data to the
computer center. It is estimated that near real-time reconstructions can be obtained using a hard-wire fast Fourier
transform processor.

Fig. 7. Prototype of Contact Scanner.

REFERENCES

1. F. L. Thurstone, "Ultrasound Holography and Visual Reconstruction,"Proceedings of the Symposium on Biomedical Engineering, Vol. 1 , Marquette University, Milwaukee, Wisconsin, (1966) pp. 12-15.

2. D. R. Holbrooke, H. R. Shibata, B. B. Hruska, E. McCurry, and E. R. Miller, "Diagnostic Holography- A Feasibility Study", Acoustic Holography Vol. 2 , Plenum Press, New York (1970) pp. 243-263.

3. H. Berger, "A Survey of Ultrasonic Image Detection Methods, " Acoustic Holography Vol. 1 , Plenum Press, New York (1969) pp. 27-48.

4. D. Vilkomerson, "Analysis of Various Ultrasonic Holographic Imaging Methods for Medical Disgnosis, " Acoustic Holography Vol. 4 , Plenum Press, New York (1972) pp. 401-429.

5. R. K. Mueller, "Acoustic Holography, " Proceedings of the IEEE Vol. 59 , September (1971) pp. 1319-1335.

6. A. L. Boyer, P. M. Hirsch, J. A. Jordan, Jr., L. B. Lesem, and D. L. VanRooy, "Reconstruction of Ultrasonic Images by Backward Propagation, " Acoustic Holography Vol. 3 , Plenum Press, New York (1971) pp. 333-348.

7. J. W. Goodman, Introduction to Fourier Optics , McGraw-Hill, San Francisco (1968) p. 60.

8. G. D. Bergland, "A Guided Tour of the Fast Fourier Transform, " IEEE Spectrum , July (1969) pp. 41-52.

9. E. R. Miller, "Immediate Aims of Acoustical Imaging in Medical Practice, " Acoustical Holography Vol. 3 , Plenum Press, New York (1971) pp. 19-22.

A NEW, HIGH-PERFORMANCE ULTRASONIC CAMERA

Philip S. Green, Louis F. Schaefer,
Earle D. Jones, and Joe R. Suarez
Stanford Research Institute
Menlo Park, California   94025

ABSTRACT

By virtue of their true-perspective presentation, focused and holographic methods of ultrasonic imaging hold great promise for soft tissue visualization. However, none of the numerous methods demonstrated to date combine the features required of a clinically useful instrument.

A new ultrasonic camera for real-time imaging of soft tissue is described. This system produces high resolution images of a 15- × 15-cm field. Its high sensitivity (10^{-11} W/cm^2 threshold) permits penetration of a considerable depth of tissue. Both transmission and reflection imaging are provided, with pulse/range-gating operation in both modes. The image is displayed on a television monitor and may be recorded on video tape.

The ultrasonic camera incorporates a unique wavefield deflection system for scanning the image past a linear array of 192 piezoelectric detectors. Linear and logarithmic amplification is provided in each signal channel.

Ultrasonic images of test objects and of human extremities are presented.

INTRODUCTION

Ultrasound, because of its nontoxic and noninvasive nature, and because of its ability to discriminate between soft tissues, is finding increasing use in diagnostic medicine. Although existing echographic methods have established the potential effectiveness of ultrasonic visualization in diagnosis, they have not found wide clinical acceptance. Most of the equipment now available provides inadequate lateral resolution, has limited gray scale capability, and requires a lengthy period to build up a single image.

Although research is under way to overcome these deficiencies, the most serious drawback of the echographic methods is the image presentation itself. Physicians are accustomed to the lateral and A-P (anterior-posterior) radiographic views; the cross-sectional echographic images are not only unfamiliar, but difficult to relate to the anatomy. A focused ultrasonic system that produces images similar to those obtained by radiographic techniques would help bridge this interpretation gap.

In focused ultrasonic imaging, the entire area to be visualized is flooded with ultrasonic waves. The waves scattered by the "object," which may be the internal organs of the patient, are collected by an ultrasonic lens and focused onto an image plane. Detectors then convert the ultrasonic energy to electrical signals for television display.

SYSTEM FEATURES

In order to be useful in diagnostic medicine, a focused ultrasonic imaging system must incorporate certain performance features. Although many ultrasonic imaging techniques have been devised over the years, none have combined the features required of a clinically useful instrument. The SRI ultrasonic camera system was designed specifically for clinical application. It incorporates a set of "illuminating" transducers, ultrasonic lenses, and image deflectors; a linear array of receiving transducers; and electronics for displaying

the image on a TV monitor. Some of the system's important
features are:

- Real Time Image Presentation--The laboratory system
  clearly demonstrates the importance of the presentation of live images. Normal motion of internal
  organs is readily observed. Additional information
  can be gained by palpating or stressing the area examined. Subtle details often become more visible or
  identifiable with small movements. The examiner's
  instantaneous view of the image allows him to position the patient as desired.

- High Sensitivity--The sensitivity of the receiving
  array is of the order of 10^{-11} W/cm^2. This has prov-
  en adequate, using 2-MHz ultrasonic waves, for penetration of an average abdomen at power levels well
  below those employed in current diagnostic practice.

- High Resolution--The resolution attained with 3.5-MHz
  soundwaves is approximately 0.75 mm. This frequency
  should be useful for visualizing the extremities,
  breasts, and thyroid. The addition of a magnifying
  lens further improves the resolution.

- Variable Focal-Plane Position--The examiner can focus
  the camera to any depth within the patient. The zone
  of focus is in a plane perpendicular to the direction
  of propagation of the ultrasonic waves and is about
  1 cm deep.

- Transmission or Reflection Imaging--A transmission
  image is made if the illuminating sound field passes
  through the patient to the receiving array. In some
  cases, an area of interest may be screened on one
  side by a natural barrier such as the air-filled lungs
  or gas in the intestines. Reflection-mode imaging can
  be used in these regions.

- **Pulsed and Range-Gated Operation**--The use of pulsed (rather than continuous) ultrasonic illumination has several advantages. The low duty cycle--here less than 2 percent--reduces average power, minimizing patient exposure. In addition, the long time between pulses allows reverberations to die out. Range gating is accomplished by turning on the receiving amplifiers only during the short interval in which the image sound field arrives at the receiving array. This discriminates against unwanted multiple reflections that would degrade transmission images. In the reflection mode, range gating allows the selection of only those reflections from the zone of focus.

- **TV Display and Recording**--The images are presented on a standard television monitor, and can be conveniently recorded on video tape for further study and patients' records. The video signals can be processed in many ways for image enhancement.

DESCRIPTION OF EXPERIMENTAL SYSTEM

The experimental ultrasonic camera now operating in the laboratory incorporates a large water tank in which the subject under examination is partially immersed. Referring to Figure 1, the operation of the system can be explained as follows: The variable-power RF amplifier transmits a pulse of energy to the selected illumination transducers. The transducers convert the electrical signal to a pressure wave that propagates through the tank of water toward the object. The pressure wave is scattered by the object; the plastic lenses collect and focus the scattered waves to form an ultrasonic image. This image is detected by a line array of 192 small receiving transducers. The receiving transducer array detects only one picture line of the image. A complete picture of about 400 interlaced lines is obtained by repeatedly sweeping the focused ultrasonic image past the line array. This is accomplished by counterrotating the plastic ultrasonic prisms located between the lenses.

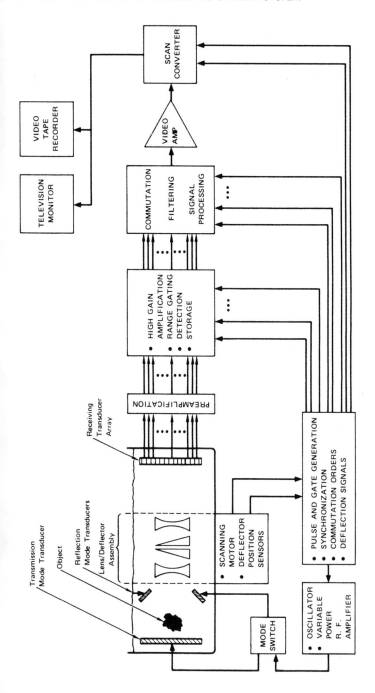

FIGURE 1   BLOCK DIAGRAM OF THE SRI ULTRASONIC CAMERA SYSTEM

The detected signals are first amplified by the preamplifiers, then by the range-gated high-gain amplifiers. The
192 discrete signals comprising a picture line are simultaneously gated, detected, and stored. The stored information
is then commutated--read out serially--to form the video
signal, which is further amplified. The video and deflection signals are synchronously applied to the scan converter.
The original picture of 15 fields per second is then converted to the standard television format of 60 fields per
second and displayed on a television monitor. The normal
field of view is about 15 cm on a side. Proper synchronization of all the pulse, gating, and deflection waveforms
is based on signals sent from the prism position sensors
located in the lens/deflector package.

The ultrasonic components are currently housed in a
1200-liter Plexiglass aquarium. The various components of
the system are shown in Figure 2. The receiving array and
preamplifier assembly are mounted at the left end of the
tank. The lens/deflector assembly is positioned about 30 cm
to the right. Illumination transducers and lenses are sus-
pended from a rail mounted to the top of the tank. Monitors
and other electronic equipment can be seen on the tables to
the left of the tank. The low-angle view in Figure 3 shows
the receiving array and the rack containing the 192 high-gain
amplifiers.

EXPERIMENTAL RESULTS

Various objects have been used to test the system's sensitivity, resolution, and dynamic range. Images of two frequently used test objects are seen in Figure 4. An auxiliary
ultrasonic lens has been inserted, yielding a magnification
of about 2.5X. The image on the left is that of a perforated aluminum sheet, 0.5-mm thick. The holes in the sheet
are about 1.5 and 3 mm in diameter. The view of the epoxy-
glass printed circuit board on the right reveals the pattern
of the glass cloth matrix as well as the metallic conductors.
The smaller holes are less than 1 mm in diameter. Transmissio

FIGURE 2   LABORATORY ULTRASONIC CAMERA SYSTEM

FIGURE 3   RECEIVING ARRAY ASSEMBLY AND ELECTRONICS

 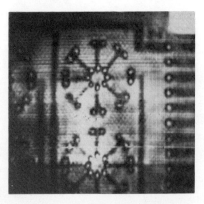

FIGURE 4   TRANSMISSION IMAGES OF A PERFORATED ALUMINUM
SHEET (LEFT) AND AN EPOXY-GLASS CIRCUIT BOARD
(ACOUSTIC MAGNIFICATION, 2.5X)

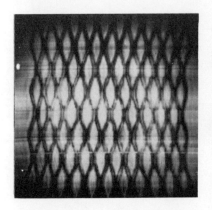 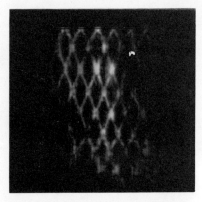

FIGURE 5   TRANSMISSION (LEFT) AND REFLECTION IMAGES

and reflection images of an expanded aluminum sheet are   shown
in Figure 5 (note that the images are of opposite polarity.)

Images of the hand and arm are shown in Figure 6.   Bones,
tendons, blood vessels, and muscle fascia are clearly visible.
Observation of the images in real time provides a far greater
appreciation for the spatial relationships between anatomical
structures.

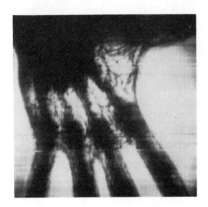 

FIGURE 6   TRANSMISSION IMAGES OF A HUMAN HAND AND FOREARM

CLINICAL SYSTEM

The laboratory imaging system described in this paper
requires that the area of the body to be visualized be immersed in water. This is quite convenient for viewing the
extremities, and has been used successfully for abdominal
and pelvic imaging. However, whole body immersion is somewhat inconvenient and time-consuming, and is probably ill-
suited to many applications, including mass screening. A
nonimmersion camera is now under construction for subsequent
clinical evaluation. In this unit, water-filled flexible
chambers provide ultrasonic coupling to the patient. An
artist's conception of the clinical camera is shown in Figure 7.

FIGURE 7   ARTIST'S CONCEPTION OF THE CLINICAL VERSION OF THE
ULTRASONIC CAMERA

CONCLUSIONS

In a brief preclinical trial in our laboratory, the ultrasonic camera has been used successfully to image abdominal
and pelvic structures as well as extremities. The images obtained were of a quality that suggest many possible clinical
applications. Additionally, diagnostically significant ultrasonic pictures of the brain, breast, thyroid, and circu-
latory system should be possible. The camera is also expected
to be effective in obstetrics and for many pediatric and or-
thopedic applications.

The results of these experiments emphasize the importance of the live image presentation. The area of interest is most readily brought into focus by positioning the subject while watching the monitor. Many easily recognized anatomical features (e.g., spinal column, ribs, pelvic bones) aid the examiner in locating specific soft tissue structures. Regions of the body shielded from view by ultrasonically opaque substances (e.g., bone or gas-filled intestines) can often be observed by changing the viewing aspect or by focusing past the obstruction.

The system currently being tested in our laboratory was designed for general usage. Our initial tests indicate other configurations may be desired for specific diagnostic applications.

ACKNOWLEDGMENT

The development of the SRI ultrasonic camera was supported in part by Public Health Service Grant GM18780-01.

POTENTIAL MEDICAL APPLICATIONS FOR ULTRASONIC HOLOGRAPHY

R. E. Anderson, M. D.*

Assistant Professor of Radiology

University of Utah, Salt Lake City, Utah

INTRODUCTION

The term "non-destructive testing" is usually used in the context of an industrial testing program. But in fact, many medical diagnostic procedures could be placed in this same category, especially those in the field of radiology.

Many of the important advances in diagnostic radiology in recent years have been made through the introduction of artificial contrast agents into the body by invasive techniques. Some of these procedures unfortunately carry with them a hazard for the patient, ranging from discomfort to death. The reason that they continue to be done in spite of the risks involved is that information can be gained about serious illnesses which at present can be obtained in no other way. New techniques for non-invasive testing of the human body would be most welcome.

While conventional pulse-echo ultrasound scanning methods have provided a new means of examining some of the internal structure of the human body, current images leave much to be desired. We feel that the development of practical, real time ultrasonic imaging equipment will greatly expand the usefulness of ultrasound for medical purposes.

*Scholar in Radiological Research of the James Picker Foundation

MATERIALS AND METHODS

Several types of real time ultrasound imaging devices are under development, as evidenced by other articles in this volume. We have been exploring the potential of one of these devices for use in medicine. Figure 1 shows a schematic diagram of a liquid surface imager being used currently. Details of its operation can be found elsewhere.[1,2]

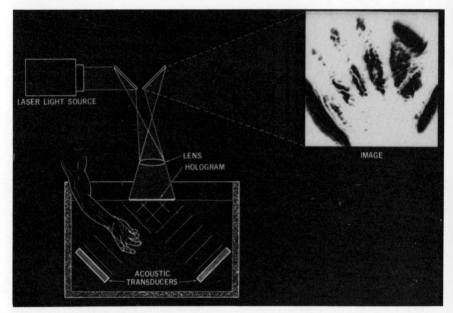

Figure 1. Schematic diagram of liquid surface "holographic" imager currently being tested.

Briefly, a "reference" ultrasonic beam and an "object" beam are made to converge at a liquid surface, where they produce a minute ripple pattern. An image of the object can be obtained by imaging a portion of the light reflected from the liquid surface. The resulting image can be viewed directly on a ground glass viewing screen, photographed, or directed into a television camera. A real time image of some of the internal structure of the object is produced. The device provides an image with a continuous grey scale, and has a line pair resolving capability quite satisfactory for medical applications.[3]

RESULTS

Figure 2 shows a composite of the x-ray and ultrasound images of the four standard roentgenographic densities. Reading from left, the plastic test tubes contain air, (liquid) fat, water, and "metal" densities (concentrated angiographic contrast material). Air is very lucent in the x-ray beam, fat is somewhat more lucent than water, and the organic iodinated contrast agent is very opaque. The ultrasound picture is entirely different -- the air is totally reflective, while the other materials are similarly lucent.

Figure 2. X-ray image of the four basic roentgenographic densities -- air, fat, water, and "metal" above, and the ultrasound image below.

Of course, in the human body the connective tissue elements in various tissues contribute greatly to the absorption and reflection of ultrasound. This point is emphasized in figure 3, which shows the x-ray and ultrasound image of two tubes of blood. The blood in the left tube has been treated to prevent clot formation, while the blood in the right tube was allowed to clot. The x-ray image shows no differentiation between the two tubes. The ultrasound picture shows a striking image of the organized clot.

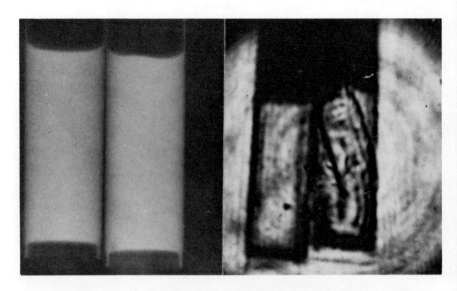

Figure 3. X-ray image on left, ultrasound image on right.
X-ray beam is indifferent to the presence of a blood clot
in the right hand tube, while a sharp image of the clot is
seen in the ultrasound image.

One of the earliest biological images made with this
device was that of the hand of the operator. Figure 4 shows
a composite image of the hand, wrist and forearm of a 4 year
old child compared to an x-ray view of the forearm. Note
that soft tissues and bone are seen in both images. The x-
ray is vastly superior for bone detail, but very little differentiation of any soft tissue structures can be made. The
ultrasound picture currently lacks the sharp definition of
the roentgenogram, but a great deal more soft tissue structure can be seen. Many of the longitudinal lines seen in
the image of the forearm represent muscles and tendons. Differentiation of these structures is much more striking when
viewed in real time, with the subject moving his hand and
wrist.

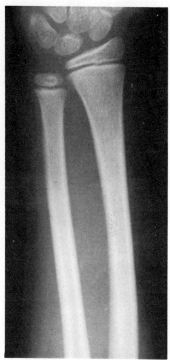

Figure 4.   Comparison of ultrasound and roentgenographic images of the forearm.   X-ray is superior for bony detail, while the ultrasound image reveals some of the soft tissue structure (muscles and tendons) of the forearm.

A practical application of this imaging capability is suggested by figure 5, which shows a composite of ultrasound and x-ray images of the hind leg of a dog.   A portion of a large tendon has been severed, and then reconstructed inside an experimental plastic tube implanted over the cut tendon. Note that neither the tendon nor the tube are visible on the x-ray, while the tube and its relationship to the tendon can be seen in the ultrasound image.

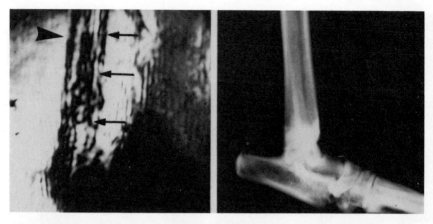

Figure 5.   Ultrasound and x-ray images of the same dog hind
limb.   Large tendon (large arrow) and plastic tube encompas-
sing a severed portion of the tendon (small arrows) are
clearly visible only in the ultrasound image.

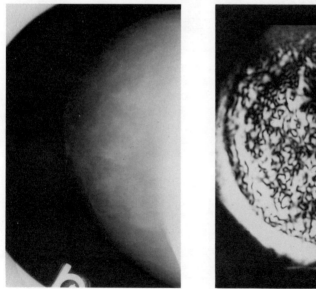
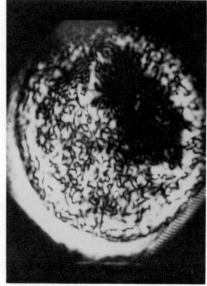

Figure 6.   X-ray "mammogram" on left, ultrasound image of a
different excised breast on the right.   Dense black area
represents a malignant tumor in the ultrasound picture.

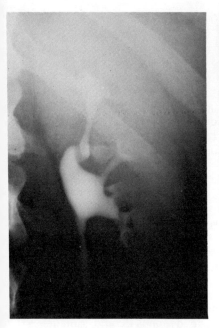 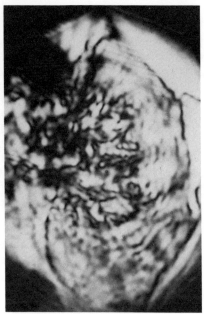

Figure 7. X-ray study of renal collecting system (left) requires the injection of contrast material, while many of
these same structures are visible in the ultrasound image of
an excised kidney (right) with no added contrast agents.

    Figure 6 compares an x-ray "mammogram" with an ultra-
sound image of a different breast. Masses in the female
breast can be diagnosed with a moderate degree of success
with the x-ray mammogram, but the method is far from ideal.
Structures within the breast obviously take on a different
appearance in the ultrasound image. The large black density
represents a breast cancer. The degree to which x-ray and
ultrasound images of the breast may complement each other is
still to be determined.

    Figure 7 compares a conventional roentgenographic image
of a kidney in vivo to the ultrasound image of an excised
kidney. The collecting system of the kidney is outlined by
injected contrast agent on the x-ray. Note that some of the
same central collecting system structures are visible in the
ultrasound image without the need for any sort of artificial
contrast material.

CONCLUSIONS

Real time ultrasonic images of the human body may provide anatomical information quite unique to this imaging modality. The real time feature of the image eliminates the problem of patient motion encountered in some B scanning applications, and makes dynamic studies possible. A continuous grey scale exists in the image which may allow differentiation of smaller changes in ultrasonic density than is possible with pulse-echo techniques. The similarities with roentgenographic images provides a familiar frame of reference for radiologists and non-radiologists alike, a point which deserves emphasis. Significant information may be extracted from an ultrasound image which might not be comprehensible without the aid of a comparison x-ray of the same diseased structure. The extent to which roentgenographic and ultrasonic images may complement one another, for example in the investigation of masses in the female breast, deserves prompt attention.

Applications of acoustical holographic imaging in medicine in the near future may include diagnosis and management of tendon trauma, examination of the non-calcified cartilage about the hip and knee in infants and children, and examination of the female breast.

Future technological improvements may make possible a wide range of other applications. Clear images of abdominal organs will aid in the diagnosis of diseases of the liver, pancreas, and kidney. The pregnant uterus may be examined without the genetic risks posed by the x-ray or radioisotope examinations now used. Blood clots within arteries or veins may be visible in some parts of the body, such as the lower leg.

The solution to the problems which will make possible these and other applications lie with the physicist and engineer. Let there be no doubt, however, that these new devices will be welcomed by physicians, and particularly by those of us who are presently obliged to perform hazardous examinations upon our patients.

BIBLIOGRAPHY

1. Brenden, Byron B.: A Comparison of Acoustical Holographic Methods. Acoustical Holography, Vol 1: 57-71.
   (Plenum Press, New York, N.Y. 1969).

2. Hildebrand, B.P., and Brenden, B.B.: An Introduction
   to Acoustical Holography. Plenum Press, New York,
   N.Y., 1972, 153-158.

3. Curtin, H.R., Anderson, R.E.: Medical Imaging Capability of Liquid Surface Ultrasonic Holography. SPIE
   Symposium Proceedings: Application of Optical Instrumentation in Medicine. Chicago, Nov. 1972.

COMPLEX ON-AXIS HOLOGRAMS AND RECONSTRUCTION WITHOUT
CONJUGATE IMAGES*

P. N. Keating, R. F. Koppelmann, R. K. Mueller
and R. F. Steinberg
Bendix Research Laboratories

Southfield, Michigan 48076

ABSTRACT

Holograms representing both real and imaginary parts
have been obtained with an on-axis reference from an experimental acoustical underwater viewing system and digitally
reconstructed to yield only the real image. The absence of
both the conjugate imagery and dc and autoconvolution terms
allows maximum efficiency in utilizing the relatively
limited amount of sampling often available for microwave and
acoustic radiation and reduces conjugate image noise. Improved imagery of underwater targets which has been obtained
in this way with relatively light sampling (20 x 20) is
presented.

1. INTRODUCTION

Conventional holography with an off-axis reference,[1]
such as that generally employed in optical holography, suffers a number of disadvantages due to the fact that only the
intensity of the optical field is measured. One disadvantage is that the sampling efficiency is only 25% because
half of the image space is occupied by the autoconvolution
of the image and the remaining 50% must hold both real and
conjugate images. With optical holography, this in itself

---

*Work sponsored in part by the Office of Naval Research
515

is not a problem since the intensity field is readily sampled at a very large number of points. However, in the case of acoustical or microwave holography, the sampling of the field is either time-consuming or expensive and a fourfold increase in sampling efficiency is highly desirable. A second disadvantage of the presence of the dc and autoconvolution terms and the conjugate image lies in the fact that their reconstruction is normally out of focus in the real image plane, causing noise in the 25% of image space assigned to the real image. Spatial filtering can reduce this, but it is still a definite problem and an undesirable complication.

While linear detectors are not normally used at optical frequencies, it is frequently convenient to use linear detectors for acoustical or microwave radiation. The full complex field can then be sampled and it is thus possible to reconstruct without the conjugate images or dc and autoconvolution terms.[2] In a previously-described[3] underwater viewing system employing acoustical holography, the processing used was such as to avoid the autoconvolution term. However, both the conjugate imagery and the dc term are present when such holograms are reconstructed optically. The present article describes an experimental investigation of underwater viewing with hologram acquisition and digital reconstruction techniques such as to allow 100% sampling efficiency to be obtained.

Section 2 of this article consists of a description of the holographic acquisition and reconstruction techniques used in the investigation while Section 3 describes the experimental results. These are compared with results previously obtained on the underwater viewing system[2,3] by means of the previous data acquisition approach and with optical reconstruction, i.e., before the system modifications described in Section 2 were implemented.

2. COMPLEX HOLOGRAM ACQUISITION AND RECONSTRUCTION

The main changes made in the system from that described previously[3] can be subdivided into two types: those dealing with data acquisition, and those dealing with processing, or reconstruction. In the first case, it is necessary to acquire twice as much information as before, since both real and imaginary parts of the acoustic field must be sampled.

In the second case, it is necessary to integrate these two measurements into a complex hologram and reconstruct it in an appropriate manner.

2.1 Complex Hologram Acquisition

In the underwater viewing system which was developed under the sponsorship of the Office of Naval Research and was described earlier,[3] the hologram was obtained as a sampled 20 x 20 array of quantities

$$A_k \cos (\phi_k - \phi_k')$$

where A_k, ϕ_k are the amplitude and phase of the acoustic field at the kth sampling point, respectively, and ϕ_k' is the phase of the (internal) reference signal at that sampling point. An off-axis reference was synthesized to separate the real and conjugate images, as is customary following Leith and Upatniek's work.[1] In the present case, it is necessary to form both

$$A_k \cos (\phi_k - \phi_k')$$

$$A_k \sin (\phi_k - \phi_k')$$

However, it is unnecessary to use an off-axis reference since there will be no conjugate image or dc term. It was found most convenient to use an on-axis internal reference (e.g., $\phi_k' = 0$, all k) and all measurements described in this article were made in this way.

Rather than double the number of channels in the existing underwater viewing system and increase the cost considerably, it was decided to modify the existing acquisition system so that it could be operated to obtain both types of data in either of two ways:

(a) To obtain both the cosine and sine data by two successive transmit/receive cycles of operation of the system.

(b) To obtain both types of data on one cycle (for moving targets with, in effect, only half the array being available).

The former mode of operation is a special form of synthetic aperture technique. The second mode reduces the sampling efficiency to 50% again, of course, but was considered useful for test purposes for moving targets and for removing the noise due to the defocused conjugate image in this case.

To allow both types of operation, the reference signals applied to all of the odd-numbered transducer array columns were in phase (e.g., 0°) and in quadrature (e.g., +90°) with the signal applied to the even-numbered columns. The two reference signals could then be switched so that the odd-numbered reference was at 90° and the even-numbered was at 0°. Thus, for one-shot operation, a 10 x 20 complex array is essentially obtained while, on two successive transmit-receive cycles, a 20 x 20 complex array can be acquired. All of the data presented here were acquired in two 'shots' to form such a 20 x 20 complex array.

The underwater viewing system, as used in the present experimental study, consists essentially of three portions: (a) the underwater, or acquisition, portion of the system, (b) the terminal and A/D interface, and (c) the CPU, or central processor, which carries out the reconstruction. The underwater portion has been described previously,[2,3] although without the foregoing modifications. This section of the system detects the linear acoustic field, mixes the resulting signal in each channel with a coherent internal electronic reference, and integrates the result over a suitable time period. The resulting dc levels in each channel are serially multiplexed and transferred to the interface. This signal is sampled and digitized into 8-bit words after being biased positively so that the dc signal is always positive and suitable for use with a CRT display. In one case, the 400 8-bit words are stored in a buffer memory, converted to ASCII characters, and transmitted by data-link to a time-shared computer for reconstruction. A second, and preferable, approach is the use of an on-line Tempo II mini-computer which accepts the 8-bit binary words themselves and carries out the reconstruction *in situ*.

The input data to the computer consist of four 400-element arrays (20 x 20) of 8-bit words. Two of these consist of base-line data obtained without an acoustic signal and these data are subtracted from the other two arrays containing the holographic data to remove the dc bias levels. The subtraction of the two array sets additionally provides

a method of removing data errors which result from dc offsets in the signal processing circuitry.

The two arrays resulting from this subtraction represent the data appropriate to a complex 20 x 20 array and
can be combined to form such an array. The latter complex
data are then multiplied by the Fresnel phase factor, as
required by Eq. (5). A two-dimensional Fast Fourier Transform subroutine is then utilized and the absolute value
of the result is then either displayed on a CRT screen
after transmission back to the terminal, or plotted.

2.2 Reconstruction

The reconstruction processing of the analog dc levels
obtained at the integrator outputs in the manner described
earlier was carried out by digitizing the serial analog
output from the multiplexer, storing these data on a buffer
memory, transmitting it by telephone data link to a time-
sharing computer system where it is processed via the FFT
routine.[4]

It can readily be shown that, in the Fresnel approximation, which is valid in the experiments described in this
article, the reconstructed complex field is, apart from unimportant factors,

$$F(\underset{\sim}{x}) = \int d\underset{\sim}{x}^{2}{}' \; f(\underset{\sim}{x}') e^{\frac{-ik}{2R} (\underset{\sim}{x} - \underset{\sim}{x}')^2} \tag{1}$$

where $\underset{\sim}{x}$, $\underset{\sim}{x}'$ are two-dimensional vectors in the image and
hologram planes, respectively; $f(\underset{\sim}{x}')$ is the complex field
at the hologram; k, R are the radiation wave-vector and
the object range, respectively.

A two-dimensional complex Fast Fourier Transform carries out the operation

$$G_{m_1 m_2} = (N_1 N_2)^{-1/2} \sum_{\ell_1=1}^{N_1} \sum_{\ell_2=1}^{N_2} g_{\ell_1 \ell_2} \; x \tag{2}$$

$$\exp \{ 2\pi i [(\ell_1-1)(m_1-1)/N_1 + (\ell_2-1)(m_2-1)/N_2] \}$$

In the present case $N_1 = N_2 = 20$. The complex hologram field is

$$f_{\underset{\sim}{\ell}} = C_{\underset{\sim}{\ell}} + iS_{\underset{\sim}{\ell}} \tag{3}$$

where $\underset{\sim}{\ell} = (\ell_1, \ell_2)$ and $C_{\underset{\sim}{\ell}}$, $S_{\underset{\sim}{\ell}}$ are the cosine and sine portions of the data, i.e.,

$$C_{\underset{\sim}{\ell}} = A_{\underset{\sim}{\ell}} \cos (\phi_{\underset{\sim}{\ell}} - \phi)$$

$$S_{\underset{\sim}{\ell}} = A_{\underset{\sim}{\ell}} \sin (\phi_{\underset{\sim}{\ell}} - \phi) \tag{4}$$

Comparing Eqs. (1) and (2), we write

$$F_{\underset{\sim}{m}} = N^{-1} e^{\frac{-i\pi R\lambda}{(Na)^2} \underset{\sim}{m}^2} \sum_{\underset{\sim}{\ell}} (C_{\underset{\sim}{\ell}} + iS_{\underset{\sim}{\ell}}) \times$$

$$e^{\frac{-i\pi a^2}{R} \underset{\sim}{\ell}^2} e^{\frac{2\pi i}{N} (\underset{\sim}{\ell}-1)\cdot(\underset{\sim}{m}-1)} \tag{5}$$

where a is the separation between adjacent hologram sampling points. In other words, the Fresnel holograms are reconstructed by forming $C_\ell + iS_\ell$, multiplying this by the Fresnel factor to form the quantity in the square brackets, and using the two-dimensional complex FFT routine on this quantity. If only the intensity field in the image plane is required, then forming the absolute value squared of the result obtained by this procedure is all that is necessary.

Image fields reconstructed in this way contain neither a conjugate image term nor any dc or autoconvolution terms.

3. EXPERIMENTAL RESULTS

Experimental data for this work were obtained by operating the underwater viewing system at a depth of 5 ft in a pool 13.5 ft wide by 45 ft long by 10 ft deep. The range to all the target objects was 5 ft. With this experimental geometry, all reflections from surfaces other than the targets were either out of the field of view or discriminated against by range gating.

The first recontructed image, shown in Fig. 1, represents the fundamental imaging quality of the system or its
point-spread function. (The dashed line in Fig. 1 is not
part of the image.) A small, 1.5 cm diameter source transducer was placed approximately 5 ft away from the array and
excited by a 250 kHz signal synchronous with the system internal reference. This approximates a point source to show
the resolution and other capabilities of the system. The
display shown in Figs. 1 and 3 was set so that anything
−14 dB or more below the maximum value in the constructed
image is represented by a blank (see Fig. 2, for example).
At this range, the image space in the reconstruction is
sampled every 5 cm. In other words, the distance between
scale markings on the border of the display represents 5 cm
at a range of 5 ft or approximately 150 cm. It will be
noted from Fig. 1 that the image display shows no zero-order
contribution, which would occur at a point two positions
above and three to the right of the image point in this display, and no conjugate image contribution. Closer examination

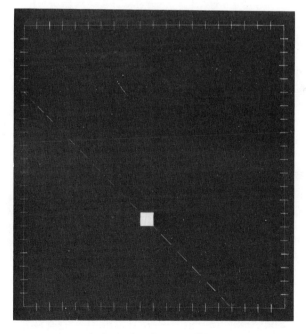

Fig. 1   Reconstructed image of a small source at 5 ft range.
         The dashed line is not part of the imagery but is
         referred to the text and in Fig. 2.

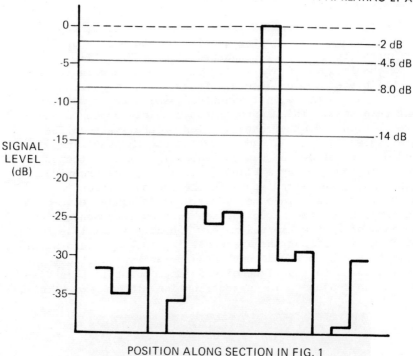

POSITION ALONG SECTION IN FIG. 1

Fig. 2   Reconstructed image intensity in a logarithmic plot
along the section represented by the dashed line in
Fig. 1.

of the reconstructed data shows that the maximum intensity
in the neighborhood of the conjugate image is 22 dB below
the image point intensity.

An impression of the noise levels in the background of
the reconstructed image can be gained from Fig. 2, where the
signal levels in the image space along the section represented by the dashed line of Fig. 1 are shown. No attempts
were made to compensate for gain and phase errors between
channels, although such compensation, which can readily be
applied, will be used in future studies.

It is also interesting to make a crude comparison between results obtained in the present study and those obtained earlier[5] using real-time optical reconstruction for
three reflecting targets illuminated by an acoustic source
near the receive array. The reconstructions obtained with

the real-time optical reconstructor tube[5] are better than those obtained by conventional optical methods because of the partial rejection of the low spatial-frequency light which can be obtained on the basis of polarization. It must be emphasized that the comparison must be made with caution since the data were taken at different times, and the characteristics of the two types of plot or display are quite different. Furthermore, the computer reconstruction is shown for the same target configuration as used for the optical reconstruction shown in Fig. 4 and presented previously in reference 5. The object field consists of three 3 in square targets, each 8 in up and 8 in across from its neighbor. Thus, the targets are 1.5 units square in the display of Fig. 3.

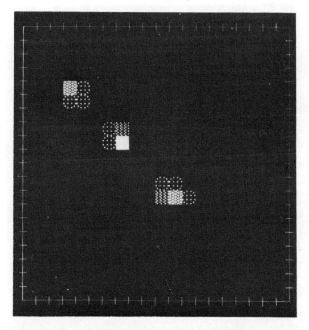

Fig. 3 Computer reconstruction of three square reflecting targets at 5 ft. Representations: ☐ : above 2 dB below I_M, the maximum intensity; ▓ : above 4.5 dB below I_M; ▒ : above 8 dB below I_M; ▒ : above 14 dB below I_M; blank, more than 8 dB below I_M.

There are a number of other advantages to computer reconstruction which are worth sketching briefly here. For
example, quantitative results may be readily obtained. More
important, a great deal of flexibility in processing is obtainable, such as various enhancement and nulling methods.[6,7]
However, on the other side of the coin, optical reconstruction can be carried out almost immediately,[5,8] whereas a
greater lag is normally encountered with computer reconstruction. Nevertheless, computer reconstruction times are
now quite short and a delay of a few seconds is frequently
acceptable, especially in view of the other advantages to
be obtained. The current Tempo II minicomputer reconstruction takes a fraction of a second, and the time limitation
in our experimental system actually lies in pushing switches
and buttons on the acoustical equipment.

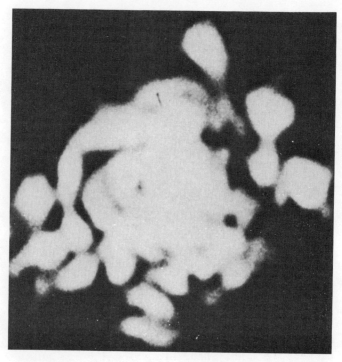

Fig. 4   Optical reconstruction for the same target distribution as in Fig. 3 (but not the same data), as in
reference 5.

Our final point worth noting is related to the question of appropriate sampling in the image space. Mueller has shown[9] that an intensity display should be sampled at a higher rate than that used in the present studies in order to make full use of the information available when both sine and cosine data are taken. This has not yet been carried out but does seem to be necessary for the most general class of objects. It can be achieved by adding zeros to the holographic arrays to quadruple their size; the image information is then interpolated by the Fourier transform reconstruction process itself. Alternatively, the complex reconstructed image field might be displayed.

ACKNOWLEDGEMENTS

The authors wish to thank Dr. G. G. Goetz for making the photograph shown in Fig. 4 available. This work was carried out with the partial support of the Office of Naval Research.

REFERENCES

1.   J. Upatnieks and E. M. Leith, J. Opt. Soc. Am., $\underline{54}$, 1295 (1964).

2.   R. K. Mueller, Proc. IEEE, $\underline{59}$, 1319 (1971).

3.   H. R. Farrah, E. Marom, and R. K. Mueller, Acoustical Holography, Vol. 2, p. 173 (Plenum Press, New York, 1970); E. Marom, R. K. Mueller, R. F. Koppelmann, and G. Zilinskas, Acoustical Holography, Vol. 3, p. 191 (Plenum Press, New York, 1971).

4.   J. W. Cooley and J. W. Tukey, Math Comput., $\underline{19}$, 297 (1965).

5.   G. G. Goetz, R. F., Koppelmann, and R. K. Mueller, Proc. Electro-Optic Systems Design Conference, New York (September 1971).

6.   R. K. Mueller, R. R. Gupta, and P. N. Keating, J. Appl. Phys., $\underline{43}$, 457 (1972); P. N. Keating, R. R. Gupta, and R. K. Mueller, J. Appl. Phys., $\underline{43}$, 1198 (1972).

7.   P. N. Keating, R. F. Koppelmann, R. K. Mueller, R. F.
     Steinberg, and G. Zilinskas, to be published.

8.   G. G. Goetz, Appl. Phys. Lett., <u>17</u>, 63 (1970).

9.   R. K. Mueller, to be published in <u>Critical Reviews</u>.

A COMPUTERIZED ACOUSTIC IMAGING TECHNIQUE INCORPORATING AUTOMATIC OBJECT RECOGNITION

J. P. Powers
Department of Electrical Engineering
Naval Postgraduate School
Monterey, California 93940

LCDR D. E. Mueller
Federal German Navy

ABSTRACT

A computerized technique combining backward wave propagation and an automatic edge detection scheme has been developed and tested. The class of objects considered is limited to those with edge boundaries since it can be shown that a universal automatic reconstruction scheme cannot be obtained for all possible objects. Using samples of the acoustic diffraction pattern as input data, this technique enables the computer to predict the most likely locations of objects and to produce graphical output of the objects. A simplified edge detection scheme conserving both memory space and computer time was used. Test results are presented for both computer generated diffraction patterns and one set of experimental data.

INTRODUCTION

The existence of linear ultrasonic transducers that can directly record the amplitude and phase of an ultrasonic field leads to the presence of many imaging techniques that have no analog in optics. One such technique is "backward wave propagation".[1,2,3] In this technique the

527

ultrasonic field is sampled in both amplitude and
phase by a scanning transducer or a transducer
array. This information is then given to a computer for processing. The diffraction equations are
programmed in the computer and the input data is
processed to produce two-dimensional cuts in a
volume defined by the program. Hence perfect reconstructions are possible only for planar objects.
The results of trying to reconstruct a three-
dimensional acoustical object are identical to
observing the real image in the reconstruction of
an optical hologram. The shape of a three-
dimensional can only be reconstructed in a series
of sections through its real image. Of course,
other difficulties[4] attendent to computerized
imaging techniques are also present (e.g. quantization errors, limited memory space, scan nonuniformities, etc.).

One of the greatest drawbacks of digital
reconstruction is that the distance from the hologram to the object must be known for use in the
diffraction equations. In optical holography this
problem never occurs in the case of visual reconstruction of an object from its virtual image because the focus is found automatically by an eye-
brain interaction. Similarly in scanning the
visual real image with a screen, it is the eye-
brain interaction that provides feedback to decide
the optimum location for the screen for maximum
object recognition and focus. The highly complex
interaction between the eye and brain is not presently amenable to computer simulation. Nevertheless it would be of great help to find a method
which brings about automatic focusing of the digital reconstruction or at least narrows down the
region in space where the image is located. To
conserve operator time and effort, it is desirable
to have the computer consider the images of all
cross-sections that it calculates to decide the
most likely position or positions of image location and present only those images for operator
consideration and object identification. In this
study it is assumed that the medium of propagation
is linear and not distrubed by turbulence or con-
vection.

In programming the propagation equations, a spatial frequency approach [5] has been used rather than a straight forward programming of the Fresnel or Fraunholfer integral propagation relations. Use of this technique has the advantages of efficient algorithms such as the Fast Fourier transform (FFT) [6] and a wide flexibility of propagation distances since no assumption on distance is made as in the Fresnel or Fraunholfer approximations.

BACKWARD WAVE PROPAGATION

To summarize the technique of backward propagation in terms of the spatial frequency approach we consider a given complex valued diffraction pattern $U_0'(x_0,y_0)$ located in plane a distance z from an object $U_1(x_1,y_1)$. The spatial Fourier transforms $U_0'(f_x,f_y)$ and $U_1'(f_x,f_y)$ of the patterns are related by [5]

$$U_0'(f_x,f_y) = H(f_x,f_y)\ U_1'(f_x,f_y) \tag{1}$$

where $H(f_x,f_y)$ is the transfer function and is given by:

$$H(f_x,f_y) = \exp\left[j\frac{2\pi z}{\lambda}\sqrt{1-(\lambda f_x)^2-(\lambda f_y)^2}\right] \tag{2}$$

For spatial frequencies such that $(\lambda f_x)^2+(\lambda f_y)^2<1$ the transfer function is a phase shift. For $(\lambda f_x)^2+(\lambda f_y)^2>1$ the transfer function behaves as a negative exponential, and the corresponding waves (called "evanescent waves") decay after propagating a few wavelengths. Hence, the propagation transfer function is often written in bandlimited form for propagation distances greater than a few wavelengths:

$$H(f_x,f_y) = \begin{cases} \exp\left[j\frac{2\pi z}{\lambda}\sqrt{1-(\lambda f_x)^2-(\lambda f_y)^2}\right] & \text{if } f_x^2+f_y^2<\frac{1}{\lambda^2} \\ \\ 0 & \text{elsewhere} \end{cases}$$

$$\tag{3}$$

The reverse propagation problem is attacked by solving the transfer relation Eq. (1) for the object given the diffraction pattern

$$U_1'(f_x,f_y)=H^{-1}(f_x,f_y)\ \underline{U_o'}(f_x,f_y) \tag{4}$$

where H^{-1} is defined by the relation $H\ H^{-1}=1$. Hence

$$H^{-1}(f_x,f_y)= \begin{cases} \exp\left\{-j\frac{2\pi z}{\lambda}\sqrt{1-(\lambda f_x)^2-(\lambda f_y)^2}\right\} \\ \qquad\qquad\qquad\qquad f_x^2+f_y^2<\frac{1}{\lambda^2} \\ \\ 0 \qquad\qquad\qquad\qquad \text{elsewhere} \end{cases} \tag{5}$$

It is important to note that the information contained in the evanescent waves has been irretrievably lost as these waves die out (assuming propagation distances of at least several wavelengths). Since this information contributes to the image, we can never expect a perfect reconstruction of the object using backward propagation. A theoretical discussion of this problem is found in Ref.7.

The computer programming of Eq.(4) to accomplish the backward propagation consists of:

(1) Taking the discrete Fourier transform of the input data with the FFT.
(2) Multiplication by $H^{-1}(f_x,f_y)$ at a given distance z.
(3) Taking the inverse Fourier transform of the result to find the acoustic field at that distance.

OBJECT RECOGNITION

Given that the diffraction pattern can be backward propagated to any plane, a technique is now required to determine if the computed diffraction pattern corresponds to an object in that plane. At this point it is necessary to make some

assumptions about the object since without them
there is no reason to favor the reversed diffraction pattern at the object plane over that
obtained at any other location.[8] In this study
we chose to search for objects that had edges,
i.e. connected regions where the acoustical amplitude underwent a sharp increase from some background level. Although this assumption provides
a limitation on the class of objects that can be
recognized (e.g. separated point sources or line
objects cannot be recognized), the class of objects is of enough interest to warrant investigation. Because of the quantization of dimensions, the smallest object that could be detected
is composed of four sampling points of high intensity next to each other, arranged in a square.
This is not a severe requirement since it implies
detection of at least all of one wavelength by
one wavelength objects if the Nyquist sampling
rate is used.

Existing detailed edge detection computer
codes were applied to the problem but were found
to take an inordinate amount of time to investigate each plane of the volume of interest. Hence
a simple intuitive scheme[8] was devised and implemented. In essence this edge detection scheme
scans the plane of interest and determines the
points associated with large changes of amplitude.
An investigation is then made to find if these
points are connected. A numerical index is kept
which increases as more connected points are
found. The planes are then ranked by the value
of this index and the corresponding diffraction
patterns are presented for operator inspection
beginning with the highest value. The number of
planes presented or the maximum difference in the
numerical index in the presented patterns can be
operator-controlled to optimize the number of
planes presented.

It should be noted that in seeking objects
with edges that it is not enough to just find the
plane that has the maximum change in amplitude
between adjacent points. For example, a lens
surrounded by an opaque aperture will produce the

maximum amplitude change at its focal point if
illuminated by a collimated beam. A technique
incorporating an edge detection scheme will have
a higher edge detection index in the lens location than at the focal point (as demonstrated in
a case considered below) because of the higher
weighting given to connected points.

The concept was tested using many computer
generated diffraction patterns and one sample
of experimental measurements. The Fourier transform and propagation algorithms were tested using
simple patterns and comparing the results with
theoretical results. Test cases of varying complexity were then applied to exercise the program.

COMPUTER GENERATED TESTS

The simplest object was a 16λ by 8λ rectangle of unit amplitude against a zero amplitude
background. All sample points in the object were in
phase thus simulating a specular object or an
illuminated slit. The diffraction pattern was
calculated over a 32λ by 32λ observation plane
(the same in all cases) at a distance of 175.3λ
in front of the object. The data was introduced
to the imaging program and the volume between
160λ and 192λ (from the diffraction data plane)
was searched. The reconstruction (Fig. 1) was
obtained at 175.25λ in 2.6 minutes.

The same object was also used with a background that had a uniformly distributed random
amplitude between zero and .2 of the object
amplitude and a uniformly distributed phase between zero and 2π radians. The diffraction pattern was calculated at a distance of 177.2λ.
Using the imaging program, the reconstruction of
Fig. 2 was found in 2.6 minutes. The position
was in error by $.05\lambda$.

A similar object was used in another test
case with a parabolic phase distribution added,
thus simulating a cylindrical lens against a
noisy background. The equivalent focal length

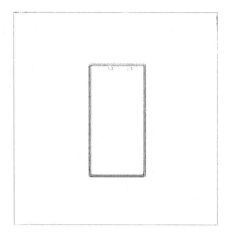

Fig. 1. Amplitude contour plot of reconstructed
16λ by 18λ rectangle.

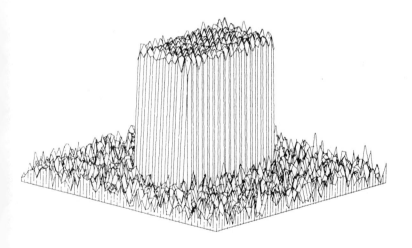

Fig. 2. Surface plot of amplitude of reconstruction of rectangular object against random
background.

of the lens was 10λ. Using a diffraction pattern
calculated at 177.1λ the correct reconstruction
was accurately found in 3.3 minutes. The focal
line was ignored (since the algorithm is insensitive to line objects).

A case was run where the object was a 16λ
by 8λ rectangle with a uniformly distributed random amplitude varying between .8 and 1.0 and a
uniformly distributed random phase between 0 and
2π. The background also was random with parameters as described in the previous cases. The
diffraction pattern was calculated at a distance
of 177.5λ. With this pattern the computer predicted two possible locations at 177.5λ and
163.75λ. The edge detection index of the first
was 1.5 that of the second. If the operator were
not confident of the differences in the index,
a cursory inspection of the images (Figs. 3a and
3b) would give the correct object location.

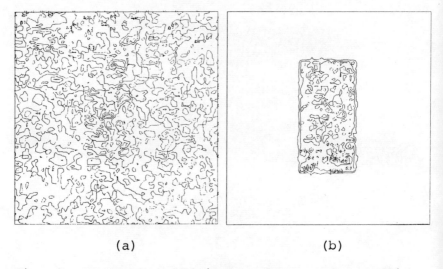

(a)   (b)

Fig. 3.   Reconstructed images from random ampli-
          tude and phase rectangle at a) erroneous
          location and b) correct location.

Planar objects of other configurations and sizes were also investigated and easily detected.

The next set of tests consisted of trying to reconstruct objects existing in two parallel planes. Since the real image is constructed in backward wave propagation, one difficulty is obvious--in the planes of the object there will be indications of the second out of focus object (Fig. 4a and 4b). The images are similar to the twin images of an in-line hologram and require more operator interpretation.

The simplest object was two 16λ by 8λ rectangles with unit amplitude and all sample points in phase. The objects were 15λ apart, and one rectangle was oriented 90° with respect to the other with a 10% overlap. The diffraction pattern was calculated 100.4λ from the front rectangle. Inspecting a volume 45λ deep, the computer took 8.5 minutes to locate the front (Fig. 4a) and back (Fig. 4b) rectangle within $.15\lambda$ of their actual location.

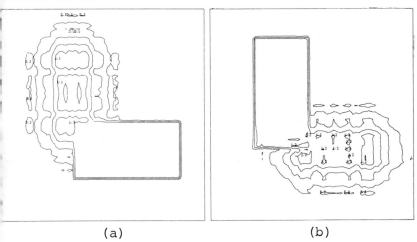

(a)   (b)

Fig. 4.   Reconstruction of double rectangle object at a) plane of front rectangle and b) plane of back rectangle.

Other tests of varying spacing and degrees of overlap were conducted.[8] In the complicated patterns it was found helpful to find and define the front object first and then subtract that object from the diffraction pattern in the front plane. Using the altered pattern as data, the second object location could then be found more quickly by the computer and understood more easily by the operator. It is noted that part of the back object was obscured by part of the front object in these tests; hence recognition of the back object was more difficult for the operator. Also the assumed perfect coherence of the sound and the subsequent interference of the waves from the two objects complicates the recognition pattern. It was, of course, observed that the processing time increased with the complexity of the object.

TEST WITH MEASURED DATA

In a final test a set of experimental data from an actual diffraction pattern was used. The pattern (which was originally used for a different purpose) was recorded at an approximate distance of 73λ from a square edge-clamped transducer with an active area of 4.5 cm by 4.5 cm at a frequency of 1 MHz. To conserve recording time approximately one-half of the pattern was recorded and folded symmetrically. This results in an unnatural symmetry in the input (Fig. 5) and output data. Another fault with this set of data is that the sampling increment was $.9\lambda$ rather than $.5\lambda$; i.e. the sampling was performed at a rate below the Nyquist rate. Aliasing effects are therefore to be expected, mostly at high spatial frequencies leading to some distortion of the edges. Despite these faults the data was used since it was readily available on tape. An improved sampling unit is presently under construction that will eliminate these problems.

A volume between 60λ (from the input data phase) and 80λ was investigated in 4.4 minutes. Two possible object locations were found at 74.66λ

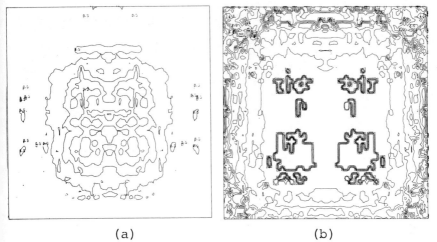

(a)   (b)

Fig. 5.   Contour plot of experimental input
           a) amplitude data and b) phase data of
           square transducer.

(edge measure index of 29.9) and at 67.68λ (edge
measure of 26.4).   It is noted that the higher
edge measure corresponds to the measured location
that was a nominal distance of 73λ. Since we are
in the very near field of the transducer, the amplitude of the diffraction field does not change
very much with distance, and the amplitude patterns at the two locations to a large degree is
the same as the input data (Fig. 5a). The phase
patterns, however, change rapidly in the near
field. It is seen in Fig. 6a that all points on
the transducer face have the same phase (or
very nearly so) while in Fig. 6b (in front of
the transducer), a distinct variation can be observed. Hence even with the flawed data, a general corroboration of the technique is supplied.
More comprehensive tests with better data is proposed for the future.

SUMMARY

    Based on these experimental and computer
results it appears that the techniques of incorporating edge detection into a backward propagation computer code can provide an automatic reconstruction from sampled data. Processing times

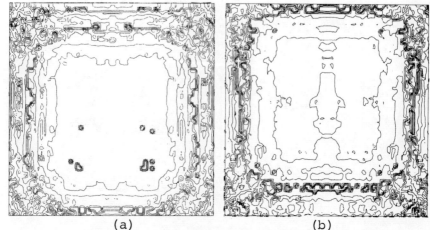

(a)   (b)

Fig. 6.   Contour plot of reconstructed object
           phase at a) object location and
           b) erroneous location.

quoted are indicative of handling 64 by 64 complex
valued arrays using the Fast Fourier transform
and our edge detection scheme.   Improvements could
be made to improve the efficiencies of both the
FFT and edge detection program as no extraordinary
effort was made to optimize these algorithms beyond achieving run times on the order of minutes.
Generally speaking however, results presented are
quite encouraging as they show that for many situations the computer can indeed make correct deci-
sions as to the object location.

REFERENCES

1. M. M. Sondhi, "Reconstruction of objects from
    their sound diffraction patterns," Journal of
    the Acoustical Society of America, 46(5):
    1158-1164, 1969.

2. A. L. Boyer et al., "Computer reconstructions
    of images from ultrasonic holograms," Acous-
    tical Holography, Vol. 2, A. F. Metherell and
    L. Larmore, Eds., Chapter 15, pp. 211-223,
    Plenum Press, New York, 1970.

3. A. L. Boyer et al., "Reconstruction of ultra-
    images by backward wave propagation," Acous-
    tical Holography, Vol. 3, A. F. Metherell,
    Ed., Chapter 18, pp. 333-348, Plenum Press,
    New York, 1971.

4. J. W. Goodman, "Digital image formation from
    detected holographic data," Acoustical
    Holography, Vol. 1, A. F. Metherell, H.M.A.
    El-Sum and L. Larmore, Eds., Chapter 12,
    pp. 173-185, Plenum Press, New York, 1969.

5. J. W. Goodman, Introduction to Fourier Optics,
    Chapter 3, McGraw-Hill, New York, 1968.

6. W. T. Cochran, et al., "What is the Fast
    Fourier Transform?" IEEE Trans. on Audio and
    Electro-acoustics, AV-15(2):45-55, 1967.

7. T. R. Shewell and E. Wolf, "Inverse diffrac-
    tion and a new reciprocity theorem," Journal
    of the Optical Society of America, 58(12):
    1596-1603, 1968.

8. D. E. Mueller, A Computerized Acoustic Imag-
    ing Technique Incorporating Automatic Object
    Recognition, Electrical Engineer's Degree
    Thesis, Naval Postgraduate School, Monterey,
    California, 1973.

Computer Enhancement of Acoustic Images

Mikio Takagi,* Nie But Tse, Glenn R. Heidbreder, Chin-Hwa Lee, Glen Wade
*University of Tokyo; University of California, Santa Barbara

ABSTRACT

Acoustic imaging may be useful in such applications as nondestructive testing, biomedical diagnosis, underwater viewing and microscopy. Unfortunately, however, the quality of images available from acoustic systems is not always satisfactory to human observers. A digital image-processing technique, employing minicomputers and relatively inexpensive input-output equipment, has been used for image enhancement and noise removal in images produced by a Bragg-diffraction system. This paper describes the technique and shows the results of applying the method in the case of a particularly simple image of a hook.

I. INTRODUCTION

This paper describes the initial techniques and results of an attempt to use the digital computer as a tool for processing acoustic images. The purpose of this research is to enhance these images, and to demonstrate the capability of digital-image processing with inexpensive equipment. Acoustic images are digitized, entered into a digital computer, processed to improve image quality, and optically displayed. Noise removal, contrast

enhancement, gamma correction, and spatial frequency filtering are achieved.

II. DIGITAL IMAGE PROCESSING SYSTEM

Digital-image processing can be divided into
three steps as follows:

(1) data acquisition from the original
     image,

(2) processing, and

(3) display.

A. Data Acquisition from the Original Image

The picture to be processed is divided into
picture elements, the density of each picture element is digitized by an analog-to-digital converter,
and the data are transferred to a computer and
stored in a large-capacity memory such as a magnetic disk or tape. Because there is no suitable
equipment for image digitization available at our
facility, we use a flying spot scanner at the
University of Southern California Image Processing
Laboratory as an input device. The digitized data
are stored in the image file of the IBM 360/44 at
USC, transferred to UCSB via the ARPA Computer
Network[1,2] and processed in the IBM 1800 of the
UCSB Department of Electrical Engineering and
Computer Science. Locally the IBM 360/75 computer
at the campus' Computer Center receives the data
from the ARPANET. It is then transferred via the
SEL 810B computer to the IBM 1800 and stored in a
magnetic disk file. The block diagram of input
data flow is shown in Fig. 1.

Our work to date involves only the processing
of hard copies of images obtained with a Bragg-
diffraction acoustic-imaging system.[3] This system
gives its output to a television camera and monitor and hard copies are obtained photographically.
The image conditioning for computer input may be

ARPA NETWORK

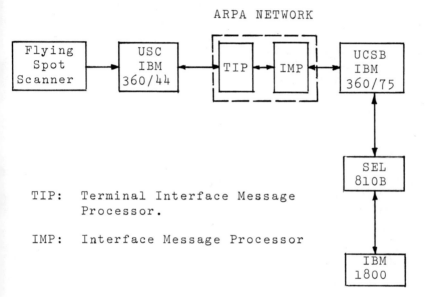

TIP: Terminal Interface Message
        Processor.

IMP: Interface Message Processor

Fig. 1   Block diagram of input data
          flow.

substantially simplified by the interposition of a
buffer memory or scan converter between the television camera of the Bragg system and the computer.
Ideally this device compresses the redundant television camera data to obtain a low data rate suitable for computer input. Two relatively low-cost
approaches to data compression are: (1) a video
compressor utilizing subframe rate sampling of the
television camera output, and (2) and analog buffer
memory in the form of a charge-storage tube with
television-rate-data write in and computer-rate
read out. It may, in fact, be satisfactory to
utilize the same storage tube for scan conversion
in both computer input and output operations.
Figure 2, which is a composite block diagram of
the proposed image-processing system, shows the input data from the Bragg-imaging system. Also shown
is an input data link from an acoustic holographic
system with an array transducer.[4] With an input
from this system, scanning may be accomplished
under computer control by time multiplexing the
electrical outputs of the various elements of the
array. (Two electrical outputs of each element
are required if both amplitude and phase information
are gathered.)

B. Computer Processing

     A variety of digital-image processing techni-
ques are potentially useful for enhancing acoustic
images and may be implemented with a minicomputer.
Among these are the following:

     (a) <u>Correction of Geometrical Distortion</u>.
A test pattern with grids or reseau marks may be
viewed through the imaging system to obtain the
characteristic geometrical distortion of the system. Measurement on the distorted image may then
be used to determine corrects to be supplied to the
X-Y coordinates of each image point in the computer.
Then if an object is imaged by the system, each
picture element is moved to compensate for the geometrical distortion inherent in the system. This
technique can be applied to change the aspect ratio
of an image, to correct the distortion of a lens,
or to obtain the image from another viewpoint.

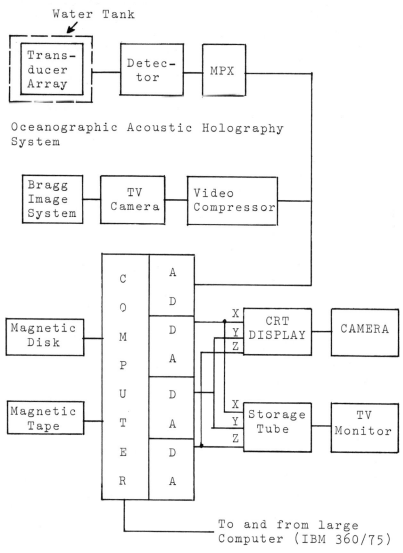

AD: Analog to digital converter.
DA: Digital to analog converter.

Fig. 2   Acoustic imaging system with interactive computer-image processing.

(b) <u>Density Correction</u>. Contrast enhancement is readily obtained in the computer. A nonlinear transformation may be applied to the datum
of each picture element so that the dynamic range
of the data most satisfactorily matches that of the
system; e.g., a small range of density in the data
may be expanded to cover the full system dynamic
range. Moreover the density correction is readily
made spatially selective so that shading and high-
lighting corrections may be accomplished.

(c) <u>Multiple Image Processing</u>. Signal-to-
noise ratio enhancement may be achieved through
multiple imaging. If the noise in successive
images is independent, averaging images will effect
an improvement in signal-to-noise ratio. Averaging
may be done in the computer using only the memory
capacity required for a single image. Some removal of systematic noise may be achieved by subtracting from the image of an object, the original background image obtained by the system in the absence
of the object.

(d) <u>Spatial Filtering</u>. Spatial filtering
may be simply accomplished in the computer in the
space domain if the point-spread function of the
filter is of small extent. Filtering in the spatial frequency domain is relatively simple if the
image data array is not large. (Fourier transformation of a 256×256 array is feasible on a minicomputer with large auxiliary memory using line-by-
line FFT and matrix partitioning techniques.)[5,6]
Spatial filtering in the computer has the advantage,
relative to optical filtering, of software flexibility and ease of combining filtering with nonlinear transformations.

C. Display of Processed Images

The devices immediately available at our facility for image display include standard laboratory
oscilloscopes and a Tektronix 611 storage CRT normally used for alphanumeric display of computer-
generated data. The former have limited persis-
tence so that direct visualization of an image

read out at computer data rates is impossible,
whereas the latter has only a two-level intensity
modulation capability. Tektronix 547 and 561A
oscilloscopes have been utilized to obtain hard
copies of processed images by time-exposure photography. The processed datum of each picture element and the corresponding coordinates are converted to analog signals to control the oscilloscope
displays. The image is raster scanned with intensity modulation of the electron beam of the oscilloscope and the film is exposed for the time required to scan the entire image. More convenient
image visualization is provided with the Tektronix
611 storage-tube display. A technique similar to
that used in photographic printing provides half-
tone displays. Varying intensity levels are represented by different sizes of two-tone pixels. A
stationary display with eight gray levels has been
obtained in this manner. A charge-storage tube and
television monitor may be used to provide a superior, but yet economical, display for interactive
processing. The charge-storage tube serves as a
scan converter with computer data rate write in and
television scan rate readout to periodically refresh
the television monitor display. This display system has the capability to display at least 10 gray
levels and is suitable for quick viewing of the
processed images pending additional processing or
a decision to make a hard copy using the oscillo-
graphic display.

III. RESULTS

Results are shown in Figs. 3 through 6.
Figure 3 shows the original image of a hook taken
by the Bragg-imaging system. Figure 4 shows the
unprocessed image after digitization and display.
The original image is divided into 160×220 picture
elements with seven-bit quantization. The digitized image is displayed with 32 levels on a CRT.
Figure 5 shows the image after correction for
shading. Figure 6 shows the image after removing
unwanted small-amplitude noise by means of nonlinear amplitude correction.

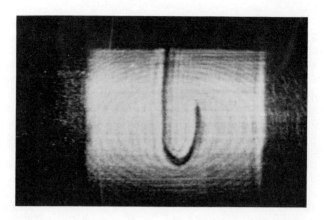

Fig. 3   The original picture of a hook
         taken by the Bragg-imaging sys-
         tem.   The upper and lower regions
         are darkened due to shading.

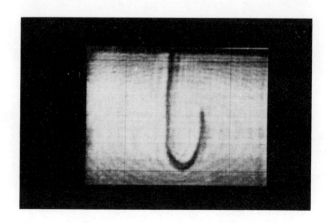

Fig. 4   The unprocessed image after
         digitization and display.

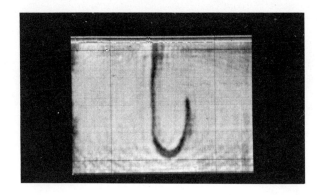

Fig. 5   The image after correction for
          shading.

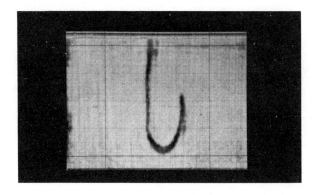

Fig. 6   The image after removal of unwanted
          spurious noise (the concentric ellip-
          tical rings of the previous pictures).

IV.   CONCLUSION

Although computer-image enhancement is sugges-
tive of computers with large working memories, and
of expensive equipment for image sampling, digitization and display, a considerable potential exists
for computer enhancement of acoustic images using
minicomputers and relatively inexpensive input-
output equipment.

V.   REFERENCES

1.   W. K. Pratt and R. P. Kruger, "Image Processing
     over the ARPA Computer Network," ARPA Report
     under Contract No. F08606-72-C-0008.

2.   Robert E. Kahn, "Resource-sharing Computer
     Communications Network," Proc. IEEE, Vol. 60,
     No. 11, Nov. 1972.

3.   G. Wade, J. Landry and J. P. Powers, "Ultra-
     sonic Imaging of Internal Structure by Bragg
     Diffraction," Appl. Phys. Letters, Sept. 15,
     1969.

4.   G. Wade, M. Wollman and K. Wang, A Holographic
     System for use in the Ocean, Acoustical Holo-
     graphy, Vol. 3:225-245.   Ed. A. F. Metherell,
     Plenum Press, New York, 1971.

5.   B. R. Hunt, "Computational Considerations in
     Digital Image Enhancement," Proc. Two Dimen-
     sional Digital Signal Processing Conference,
     Oct. 1971, Columbia, Mo.

6.   J. O. Eklundh, "A Fast Computer Method for
     Matrix Transposing," IEEE Trans. on Computers,
     Vol. C-21, No. 7, July 1972.

IMAGE RECONSTRUCTION BY COMPUTER IN ACOUSTICAL HOLOGRAPHY

Yoshinao Aoki

Department of Electronic Engineering
Hokkaido University
N 12, W 8, Sapporo (060), Japan

INTRODUCTION

In acoustical holography the reconstruction techniques of images from recorded acoustical holograms are classified mainly into three categories, that is, 1) optical reconstruction,[1] 2) numerical reconstruction,[2-4] 3) reconstruction by non-optical waves.[5] In these reconstruction techniques, the optical reconstruction is mostly investigated for its excellent capability for processing large amount of two-dimensional holographic information and its easiness in reproducing images with visible light. However the optical reconstruction has its inconvenience, that is it needs a photographic process when the acoustical holograms are recorded on films, making it impossible to reconstruct images in real time. The image reconstruction by non-optical waves, for example acoustical waves,[5] now is used for special purpose and few investigations are conducted on this technique.

The numerical reconstruction by computer is not so popular as the optical reconstruction, but it has potentiality in long-wavelength holographies such as acoustical and microwave holographies. Since the wavelengths of acoustical waves are much longer than those of optical waves, the ratio of the hologram aperture versus the wavelength is smaller in acoustical than in optical holography, resulting in recording of less hologram information in an acoustical hologram. This fact suggests that an image can be calculated with less capacity of computer memory in shorter execution time from

the acoustical hologram than from the optical hologram.
Secondly a scanned type holography, where the two-dimensional
hologram information is decomposed into one-dimensional time
signals, is often used in long-wavelength regions and it is
convenient to convert the hologram data for computer input
in this type of holography. Moreover many kinds of processing operations such as filtering, image shift, hologram
conversion, etc are easily done in numerical reconstruction.
These advantages suggest that the numerical reconstruction
by computer is a promissing technique in acoustical holography. If an acoustical holographic system is constructed
linking the hologram constructing system, data transmission
and transform system and computer, images may be reconstructed in very short time by on-line processing.

In this paper an algorithm for computer reconstruction
of images from acoustical holograms is proposed and some
discussions of the criteria concerning sampling number and
available aperture for shifted image are conducted. In
calculating the discrete Fresnel transform of hologram with
the aid of the fast Fourier transform (FFT) algorithm, periodic images appear under certain circumstances due to the
FFT algorithm. Discussion of these periodic images is also
conducted. For the experimental discussion acoustical holograms are constructed with sound-waves in audio region by
simulating the reference wave electronically. Images are
reconstructed from the photographically recorded holograms
according to the algorithm developed here and the validity
of the theoretical analysis is examined.

THEORY

Computer Reconstruction of Images from
Fresnel Transform Holograms

A hologram is an interference pattern of an object wave
diffracted by the object and a reference wave of known wavefront. The object wave on the hologram plane x_1- y_1 as shown
in Fig. 1 is expressed by the convolution integral of the
object function $o(x_0, y_0)$ on the object plane x_0 - y_0 and
the propagation function $p_1(x, y)$ under the paraxial ray approximation as follows,[6]

$$(o \circledast p_1)(x_1,y_1) = \int_{-\infty}^{\infty} \int_{-\infty}^{\infty} o(x_0,y_0)p_1(x_1-x_0,y_1-y_0)dx_0dy_0 \quad (1)$$

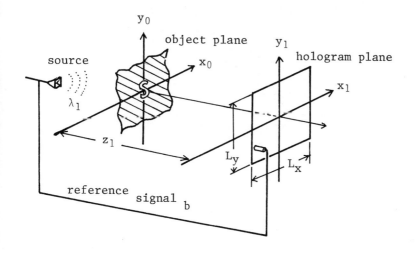

Fig. 1   Construction of acoustical holograms.

where the symbol ⊛ denotes the convolution integral.   The
propagation function is defined by Eq. (2),

$$p_{1,2}(x,y) = \frac{i}{\lambda_{1,2}z_{1,2}} \exp[-\frac{i\pi}{\lambda_{1,2}z_{1,2}} (x^2+y^2)] \qquad (2)$$

where the suffixes 1 and 2 of $p_{1,2}$ correspond to those of
$\lambda_{1,2}$ and $z_{1,2}$. λ_1 and λ_2 are the wavelengths of the waves
used in the hologram construction and image reconstruction
processes respectively. z_1 and z_2 are the distances from the
object plane to the hologram plane and from the hologram
plane to the image plane as shown in Figs. 1 and 2.   For simplicity of analysis, an acoustical hologram is considered to
be recorded with square-law-detection and is represented as
follows,

$$h(x_1,y_1)=|b+o \circledast p_1|^2=|b|^2+|o \circledast p_1|^2+b^*(o \circledast p_1)+b(o \circledast p_1)^* \quad (3)$$

where b is a plane reference wave and the symbol * denotes
complex conjugate.

Since the hologram of Eq. (3) records the Fresnel diffraction pattern of the object, this is a Fresnel transform
hologram.   Therefore images can be reconstructed by performing

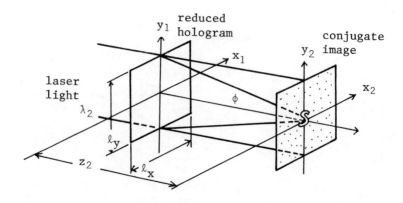

Fig. 2.   Optical reconstruction of images from reduced
          acoustical holograms.

the inverse Fresnel transform of the hologram. Mathematically speaking, the image $\phi(x_2,y_2)$ on the image plane x_2-y_2 of
Fig. 2 can be obtained by calculating the convolution integral of hologram and propagation function $P_2{}^*$ as expressed
in Eq. (4),

$$h \circledast p_2{}^* \ni (o \circledast p_1) \circledast p_2{}^* = \phi(x_2,y_2) \; . \tag{4}$$

In Eq. (4) only the true image term $o \circledast p_1$ of Eq. (3) is considered and the constant term b^* is neglected. Using the
theorem that the Fourier transform of the convolution integral is a product of Fourier transforms of each function of
integrand,[6] Eq. (4) can be written as follows,

$$\phi(x_2,y_2) = \mathcal{F}^{-1}\mathcal{F} \, [(o \circledast p_1) \circledast p_2{}^*]$$

$$= \mathcal{F}^{-1} \, [\frac{1}{R^2} \, O(\frac{\xi}{R}, \, \frac{\eta}{R}) \cdot P_1(\frac{\xi}{R}, \, \frac{\eta}{R}) \cdot P_2{}^*(\xi,\eta)] \tag{5}$$

where $\mathcal{F}$ and $\mathcal{F}^{-1}$ express Fourier and inverse Fourier transform
operations respectively. The Fourier transform $O(\xi,\eta)$ of the
object function $o(x,y)$ is defined by Eq. (6),

$$O(\xi,\eta) = \mathcal{F} \, [o(x,y)] = \int_{-\infty}^{\infty}\int_{-\infty}^{\infty} o(x,y) e^{-i2\pi(\xi x+\eta y)} dxdy \; . \tag{6}$$

The Fourier transforms $P_{1,2}(\xi,\eta)$ of the propagation functions

$p_{1,2}(x,y)$ can be obtained analytically as follows,

$$P_{1,2}(\xi,\eta) = \mathcal{F}[p_{1,2}(x,y)] = \exp[i\pi\lambda_{1,2}z_{1,2}(\xi^2+\eta^2)] \cdot \quad (7)$$

In Eq. (5) R denotes a reduction ratio and this constant appears because the original hologram is reduced to 1/R in length for optical reconstruction as shown in Fig. 2. This reduction of the hologram is inevitable in a system where holograms are constructed by long-wavelength waves such as sound-waves and images are reconstructed by short-wavelength waves such as laser lights.

Substituting the Fourier transforms P_1 and P_2^* of Eq.(7) into Eq. (5), Eq. (5) can be rewritten as follows,

$$\phi(x_2,y_2) = \mathcal{F}^{-1}[\frac{1}{R^2}0(\frac{\xi}{R},\frac{\eta}{R}) \cdot \exp[(i\pi)(\frac{\lambda_1 z_1}{R^2} - \lambda_2 z_2)(\xi^2+\eta^2)]] \cdot$$
$$(8)$$

Where the following condition is assumed to be satisfied,

$$R^2 = \frac{\lambda_1 z_1}{\lambda_2 z_2} \quad (9)$$

the original object function is obtained as follows,

$$\phi(x_2,y_2) = \mathcal{F}^{-1}[\frac{1}{R^2}0(\frac{\xi}{R},\frac{\eta}{R})] = \mathcal{F}^{-1}\mathcal{F}[o(Rx_2,Ry_2)] = o(Rx_2,Ry_2) \cdot$$
$$(10)$$

In Eq. (10) the variables of the reconstructed object function are multiplied by R. This means the image is reconstructed in reduced size of 1/R of original object in length because of the hologram reduction.

Image Reconstruction from Sampled Hologram Data

A procedure to calculate the discrete Fresnel transform is shown in Fig. 3. First the hologram h is sampled and the Fourier transform H is calculated by the discrete Fourier transform (DFT) algorithm. The Fourier transform P_2 of the propagation function p_2 is also sampled and a product $H \cdot P_2$ is calculated. A conjugate image is obtained by calculating the inverse Fourier transform of $H \cdot P_2$, whereas a true image is calculated with the complex Fourier transform P_2^*.

Assuming a reduced hologram of aperture $\ell_x \times \ell_y$, where the original aperture is $L_x \times L_y$, the hologram is sampled with sampling numbers M and N with respect to x_1, y_1 coordinates of the hologram plane. Then the sampling intervals with respect to these coordinates are as follows,

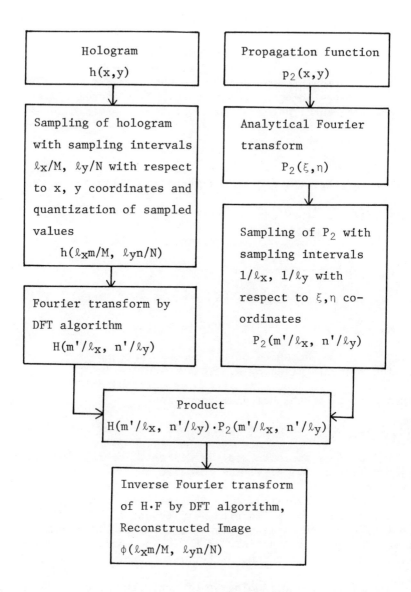

Fig. 3. Procedure for calculating images from Fresnel
transform acoustical holograms by computer.

hologram plane   Fourier spectrum plane

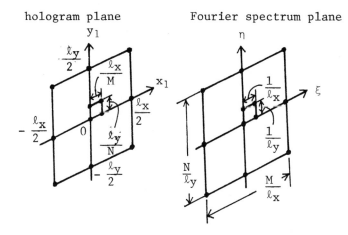

Fig. 4. Relation between the hologram aperture $\ell_x \times \ell_y$ and
the sampling intervals $1/\ell_x$, $1/\ell_y$ on the Fourier
spectrum plane, where M and N are the sampling
numbers.

$$\frac{\ell_x}{M} = \frac{L_x}{RM} \quad , \quad \frac{\ell_y}{N} = \frac{L_y}{RN} \quad . \tag{11}$$

According to the DFT algorithm, the Fourier spectra of the
hologram sampled with the intervals of Eq. (11) are obtained
at discrete points separated by $1/\ell_x$ and $1/\ell_y$ with respect
to ξ and η coordinates on the spectrum plane as shown in
Fig. 4. Therefore the discrete Fourier transform of the
hologram can be written as follows,

$$H(m',n') = \frac{1}{MN} \sum_{m=0}^{M-1} \sum_{n=0}^{N-1} h(\frac{L_x}{M}m, \frac{L_y}{N}n) e^{-i2\pi(\frac{m'm}{M} + \frac{n'n}{N})} \quad . \tag{12}$$

In Eq. (12) non-negative integers (m,n) represent the sampling coordinates on the hologram plane and (m',n') represent
the sampling coordinates on the spectrum plane.

The Fourier transform P_2 of Eq. (7) is also sampled
with the sampling intervals $1/\ell_x$, $1/\ell_y$ with respect to ξ
and η coordinates,

$$P_2(m',n') = \exp[\frac{i\pi\lambda_1 z_1}{L_x^2} (m'^2 + (\frac{M}{N})^2 \cdot n'^2)] \quad . \tag{13}$$

In deriving Eq. (13), the sampling intervals of Eq. (11) are assumed to be equal ($\ell_x/M = \ell_y/N$) and the condition of Eq. (9) is used. The image can be reconstructed by calculating the inverse Fourier transform of $H(m', n') \cdot P(m', n')$ after calculating $H(m', n')$ of Eq. (12), both by the FFT.

Next a criterion on the sampling number is discussed. Considering a point object $o(x_0, y_0) = \delta(x_0, y_0)$ placed at the origin of the image plane, the hologram of this point object consists of the propagation function p_1 of Eq. (2) and its complex conjugate p_1^*. Since the hologram of a general object can be considered as the superposition of the holograms of point objects, the criterion on the sampling number is discussed with the hologram of a point object. The phase of the propagation function of Eq. (2) changes rapidly as the edges of the hologram aperture are approached and the phase change over one sampling interval L_x/M at the hologram edge $x_1 = L_x/2$ is obtained as follows,

$$\frac{\pi}{\lambda_1 z_1} \left(\frac{L_x}{M} \right)^2 \left[\left(\frac{M}{2} \right)^2 - \left(\frac{M}{2} - 1 \right)^2 \right] \simeq \frac{\pi L_x^2}{\lambda_1 z_1 M} \qquad (14)$$

where the sampling number M is assumed to be a large even number. The phase change of Eq. (14) should be samller than π and from this condition a criterion on the sampling number M is determined. The same discussion applies to the sampling number N with respect to y_1 coordinate with the result,

$$M \gtrsim \frac{L_x}{\lambda_1 z_1} \qquad , \qquad N \gtrsim \frac{L_y}{\lambda_1 z_1} \qquad . \qquad (15)$$

Image Shift and Periodic Image due to the DFT Algorithm

In the numerical reconstruction discussed in this paper, the aperture of the reconstructed image plane is limitted to the hologram aperture $\ell_x \times \ell_y$. Therefore an image shift is necessary when the image is reconstructed on the edge or out of the aperture of the image plane. For simplicity of the analysis a point object $o(x_0, y_0) = \delta(x_0-a, y_0-b)$ placed at the point (a, b) on the object plane is considered. As the Fourier transform of the object function is $\mathcal{F}[o] = \exp[-i2\pi(a\xi+b\eta)]$, the Fourier transform of $o \circledast p$ can be written using Eqs. (5) and (7) as follows,

$$\mathcal{F}[o \otimes p_1] = \frac{1}{R^2} e^{-i2\pi(\frac{a}{R}\xi + \frac{b}{R}\eta)} P_1(\frac{\xi}{R}, \frac{\eta}{R})$$

$$= \frac{1}{R^2} e^{-\frac{i\pi(a^2+b^2)}{\lambda_1 z_1}} P_1(\frac{\xi - \frac{aR}{\lambda_1 z_1}}{R}, \frac{\eta - \frac{bR}{\lambda_1 z_1}}{R}) \cdot \quad (16)$$

Changing the variables $\xi' = \xi - aR/\lambda_1 z_1$ and $\eta' = \eta - bR/\lambda_1 z_1$ in Eq. (16), the result of $P_1(\xi'/R, \eta'/R) \cdot P_2^*(\xi', \eta') = 1$ is obtained under the condition of Eq. (9). Therefore the reconstructed image can be expressed as follows,

$$\phi(x_2, y_2) = \mathcal{F}^{-1}[\frac{1}{R^2} e^{-\frac{i\pi(a^2+b^2)}{\lambda_1 z_1}}]$$

$$= e^{-\frac{i\pi(a^2+b^2)}{\lambda_1 z_1}} \cdot \delta(Rx_2, Ry_2) \cdot \quad (17)$$

Equation (17) shows that a point image is reconstructed at the origin of the reconstructed image plane. This fact means that the reconstructed image can be shifted by replacing the Fourier spectrum of the object. In sampled coordinates this spectrum replacement is expressed as follows,

$$m' = m' - \frac{aL_x}{\lambda_1 z_1} \qquad n' = n' - \frac{bL_y}{\lambda_1 z_1} \cdot \quad (18)$$

Now the permissible region for an image shift, that is, the permissible object displacement from the center of the object plane, is discussecd. When a point object at the origin $(0,0)$ is displaced to the point (a,b), a phase term $\exp[-i2\pi(a\xi/R + b\eta/R)]$ is added to the Fourier transform of the hologram of the point object as shown in Eq. (16). The phase change of this phase term over one sampling interval $1/\ell_x$ with respect to ξ coordinate is $(2\pi a/R)(1/\ell_x)$. Whereas the phase change of P_1 in Eq. (16) over one sampling interval has the maximum value at the edge of the spectrum plane $\xi = M/2\ell_x$ and this maximum phase change can be obtained using Eq. (7) as follows,

$$\frac{\pi\lambda_1 z_1}{R^2}(\frac{1}{\ell_x})^2[(\frac{M}{2})^2 - (\frac{M}{2} - 1)^2] \simeq \frac{\pi\lambda_1 z_1 M}{L_x^2} \cdot \quad (19)$$

The phase change of the phase term due to the object displacement should be smaller than that of Eq. (19). The same discussion applies to η coordinate, resulting in a criterion on the permissible object displacements as follows,

$$a \leq \frac{\lambda_1 z_1 M}{2L_x} \qquad\qquad b \leq \frac{\lambda_1 z_1 M}{2L_y} \quad . \tag{20}$$

The relations of Eq. (20) can be derived also from the periodic nature of the calculated Fourier spectrum. Since the Fourier spectrum of the hologram by the DFT algorithm has a periodic structure with periods M/ℓ_x and N/ℓ_y, the spectrum shifts $aL_x/\lambda_1 z_1$ and $bL_y/\lambda_1 z_1$ in Eq. (18) should be smaller than halves of the sampling numbers M and N respectively. From this condition Eq. (20) can be derived again.

In the case where the object is placed far from the center of the object plane and the shadow cast by the object does not come into the hologram aperture, the conditions of Eq. (20) cannot be satisfied and no image is expected to appear reconstructed within the image aperture. However even in this case a periodic image due to the DFT algorithm appears. If the object is a simple one, this periodic image can be used as the reconstructed image. In the DFT algorithm one period of the image aperture is displayed. But the calculated image is intrinsically a periodic one. Therefore the reconstructed image $\phi(x_2, y_2) = o(Rx_2-a, Ry_2-b)$ of a point object is expressed by Eq. (21), considering the periodicity with the periods ℓ_x and ℓ_y with respect to x_2 and y_2 coordinates,

$$\phi(x_2, y_2) = \sum_{p=-\infty}^{\infty} \sum_{q=-\infty}^{\infty} o(Rx_2-a+pL_x, Ry_2-b+qL_y) \tag{21}$$

where p, q are integers. Even when the constants a, b of Eq. (21) do not satisfy the conditions of Eq. (20), suitable integers p, q can be chosen to satisfy Eq. (22),

$$\left| a-pL_x \right| \leq \frac{\lambda_1 z_1 M}{2L_x} \quad , \quad \left| b-qL_y \right| \leq \frac{\lambda_1 z_1 N}{2L_y} \quad . \tag{22}$$

Equation (22) means the (p,q) order image of periodic images appears reconstructed within the image aperture. Figure 5 explains the appearance of this higher order periodic image in the optical reconstruction system, where the same holograms are arranged periodically and these holograms are illuminated by a collimated coherent light. Then one of the images reconstructed from these holograms falls into the image aperture as shown in Fig. 5. This image corresponds

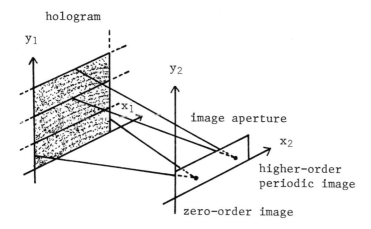

Fig. 5.   Reconstruction of periodic images from a hologram
           of periodic structure.

to the periodic image reconstructed by computer as satisfying the condition of Eq. (22).

    The difficulty of this periodic image occurs when images
from other holograms overlap this image. For example when
the diffracted light from the $(0,0)$ order hologram reconstructs an image within the same image aperture as shown in
Fig. 5, it may overlap the image reconstructed from the (p,q)
order hologram if these images are complex ones. In this
case the periodic image cannot be used as the reconstructed
image.

EXPERIMENT

Construction of Acoustical Holograms
with Electronic Reference

    The acoustical holograms are constructed in the experi-
mental arrangement of Fig. 6 [1]. An object is illuminated by
a sound-wave radiated from a tweeter and the sound-wave diffracted by the object is received by a microphone. The
received object signal is mixed with the reference signal
from the oscillator and detected, resulting in the simulation

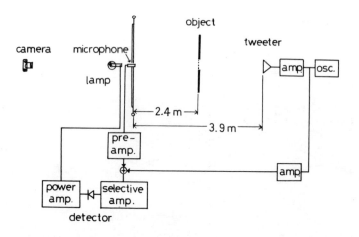

Fig. 6.   Experimental arrangement for constructing
          acoustical holograms by the electronic
          reference method.

of an electronic reference. The detected signals are amplified to lighten a lamp fixed to the microphone and a camera
records the hologram pattern. Since the electric length is
constant independently of the microphone scanning, a plane
wave is simulated in this configuration.

If the tweeter itself is considered as an object, a
hologram of a wave source, that is a zone plate, is recorded.
For example Fig. 7 shows one of such zone plates, when the
frequency of the sound-wave is chosen as 15 kHz.

Next a slit cut in a screen board forming a letter S
is chosen as an object and is placed as shown in Fig. 6.
Figure 8 show the hologram in this experimental arrangement
with the sound-wave of 15 kHz.

Optical and Numerical Reconstructions

The optical reconstruction is done by illuminating the
reduced hologram, about 4mm × 4mm in size, with laser light
(632.8nm). The optically reconstructed image is small due
to the reduction of the hologram, the image is observed by
magnifying the reconstructed image. The true image reconstructed from the hologram of Fig. 8 is shown in Fig. 9,

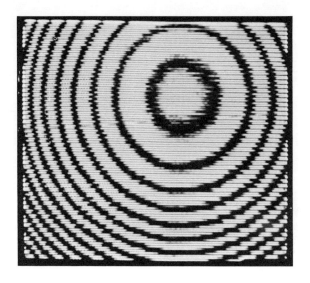

Fig. 7.   Acoustical hologram of a wave source (tweeter).

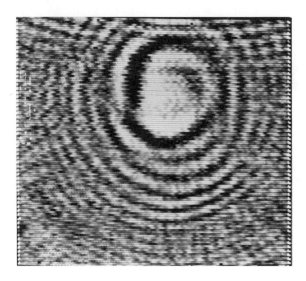

Fig. 8.   Acoustical hologram constructed in the experimental
arrangement of Fig. 6.

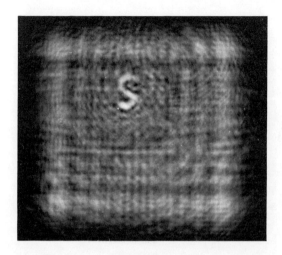

Fig. 9.   Optically reconstructed true image from the
          hologram of Fig. 8.

where the image S of the original object is observed.

For numerical reconstruction the hologram of Fig. 8 is
sampled using the photograph of Fig. 8, which is divided
into $110 \times 150 = 16,500$ cells and each cell is considered as
one sampling point. The sampled values are quantized into
5 levels according to the emulusion brightness of the photograph. Figure 10 shows the sampled and quantized hologram
printed by a line printer.

In calculating the reconstructed image from the hologram of Fig. 10 according to the procedure discussed in the
previous section, it is necessary to determine the parameter
of Eq. (23) included in the exponential function of Eq. (13),

$$D = \frac{\pi \lambda_1 z_1}{L_x^2} \quad . \tag{23}$$

This parameter determines the position of the reconstructed
image plane. The experimental values are $\lambda_1 = 2.27$cm, $z_1 = 2.4$m
and the available aperture for the computer reconstruction
$L_x = 1.5$m. These values are substituted into Eq. (23), resulting in $D = 0.076$. Therefore many images are calculated

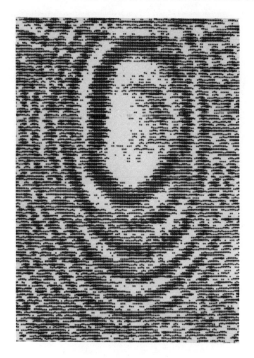

Fig. 10.   Quantized hologram of Fig. 8, where
the sampling points are $110 \times 150 = 16,500$.

changing the value of D around this calculated value.

Figure 11 shows the calculated images for D=0.08, 0.09
and 0.10. The display of the images is done by quantizing
the absolute values of calculated values into 6 levels and
adopting a multiple printing technique to display gray tone.
In Fig. 11 the best image is reconstructed for D=0.09 and
the images at other Fresnel transform planes are out of
focus. Comparing the image of Fig. 11-b with the optically
reconstructed image of Fig. 9, it may be said that the computer reconstructed image has a quality comparable to the
optically reconstructed image in this example. This experimental result demonstrates that if the image is not complex
one, a satisfactory image can be reconstructed by computer
according to the algorithm developed here. The execution
time for calculating one image of Fig. 11 is about 4 minites
using FACOM-230-60 system.

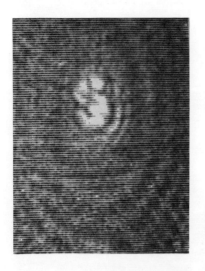

(a) D=0.08   (c) D=0.10

(b) D=0.09

Fig. 11.   Numerically reconstructed images from the
          hologram of Fig. 10 by computer, where
          the parameter D indicates different
          Fresnel transform planes.

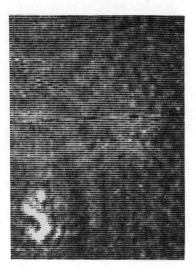

Fig. 12.   Shifted image.

Image Shift and Periodic Image

In the optical reconstruction, the reconstructed image can be shifted within the image plane by inclining the illuminating light beam incident upon the hologram. In the numerical reconstruction, this inclination of illuminating light beam is simulated equivalently by replacing the spectrum coordinates as expressed by Eq. (18). For example Fig. 12 shows the reconstructed image by replacing the spectrum coordinates according to Eq. (18) under the conditions of $a=L_x/2$ and $b=L_y/2$. This reconstructed image shows clearly the image shift within the image aperture.

For the experimental discussion of periodic images due to the DFT algorithm, the hologram of Fig. 7 is used for the computer reconstruction. Sampling points are chosen as $110 \times 112 = 11,200$ and the sampled values are quantized into 2 levels, that is 0 or 1. Figure 13-a shows this binary hologram. The computer reconstruction is done according to the same procedure mentioned in the preceding sections. Thus the reconstructed image is shown in Fig. 13-b, where the white part shows the reconstructed image of a tweeter.

Next an image is reconstructed from the hologram of

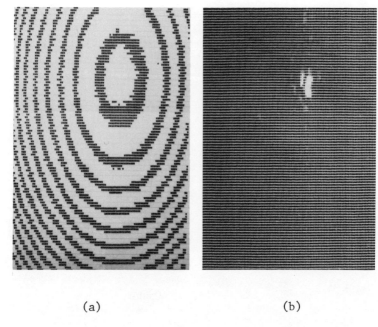

(a)   (b)

Fig. 13.   Binary hologram (a) produced from the hologram
           of Fig. 7 and numerically reconstructed image
           (b).

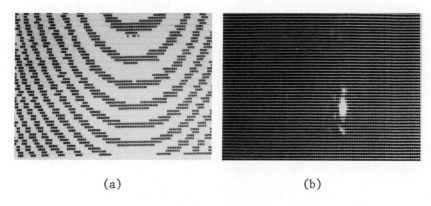

(a)   (b)

Fig. 14.   A part of the hologram of Fig. 13-a,
           (a) and numerically reconstructed
           periodic image (b).

Fig. 14-a which is half of the hologram of Fig. 13-a (therefore the sampling points are 50✕112 in Fig. 14-a). Since
in Fig. 14-a the center of the zone plate is out of the
hologram aperture, the image is not expected to be reconstructed within the aperture of the image plane. However
an image appears in the calculated image plane as shown in
Fig. 14-b, where the image position does not coincide with
the center of the zone plate. This is the periodic image
due to the DFT algorithm as discussed before. This experimental result suggests the possibility that the periodic
image can be used as the reconstructed image when the normal
image is reconstructed out of the aperture of the image plane.

Image Reconstruction with Data of Various
Sampling Numbers

Since the two-dimensional FFT program developed for the
research in this paper can process hologram data of arbitrary sampling number, the reconstructed images can be observed
by changing the sampling numbers. Though reducing the sampling number corresponds to reducing the hologram aperture
in this experiment, the diffraction effect due to the reduced hologram aperture itself does not appear in the numerical
reconstruction.

Considering only the hologram aperture represented by
$\text{rect}(x/\ell_x) \cdot \text{rect}(y/\ell_y)$, where the rectangular function is de-
fined as $\text{rect}(x)=1$ for $|x| \leq \frac{1}{2}$ and $\text{rect}(x)=0$ for else. Then
the Fourier transform of the aperture is as follows,

$$[\text{rect}(\frac{x}{\ell_x})\text{rect}(\frac{y}{\ell_y})] = \ell_x \ell_y \ \text{sinc}(\ell_x \xi)\text{sinc}(\ell_y \eta) \qquad (24)$$

where $\text{sinc}(x) = \sin\pi x/\pi x$. Since the coordinates ξ, η on the
spectrum plane are sampled with sampling intervals $1/\ell_x$,
$1/\ell_y$ due to the DFT algorithm, the right hand side of Eq.(24)
has non-zero value only at $\xi=\eta=0$ and zeros at other sampling
points. This means that the sinc function resulted from the
diffraction effect of the limitted aperture becomes an impulse function and the effect of the limitted aperture
vanishes in the computer reconstruction. Whereas in the
optical reconstruction, the effect of the limitted aperture
always appears.

The image are reconstructed by changing the sampling
number N of the horizontal direction of the hologram of

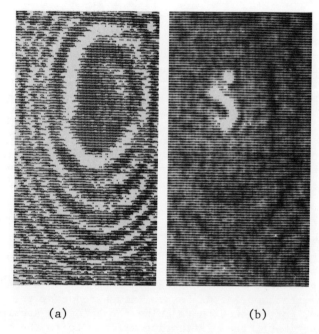

(a)   (b)

Fig. 15.   Hologram (a) and numerically reconstructed
           image (b), where the sampling points are
           110 × 100=11,000.

Fig. 10 (in Fig. 10 the sampling number is N=150, but the
hologram is displayed with the portion of N=134 because of
the limitted space of the line printer).   The sampling
number M of the vertical direction is unchanged.

        Figure 15-a and Fig. 16-a are the holograms and Fig.
15-b and Fig. 16-b are their reconstructed images for the
sampling numbers M× N=110 × 100 and M × N=110 × 60 respectively.
Comparing the image of Figs. 11-b, 15-b and 16-b, the appearance of the reconstructed images is observed for different sampling numbers.   Naturally the image cannot be reconstructed when the sampling number is reduced too much.
In the example of Fig. 16-b the image quality does not deteriorated significantly, where the sampling number is reduced to 2/5 of that of Fig. 11-b.

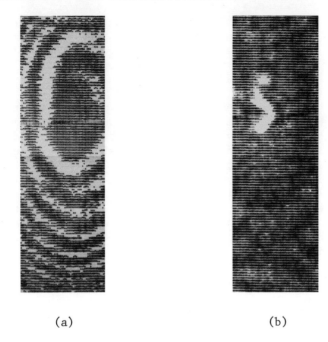

(a)   (b)

Fig. 16.   Hologram (a) and numerically reconstructed
           image (b), where the sampling points are
           110 × 60=6,600.

CONCLUSION

Theoretical discussions of the computer reconstruction
of images from acoustical holograms are conducted and the
experiment is followed by reconstructing images from sound-
wave holograms. From the experimental results, it can be
said that the numerically reconstructed images have a quality comparable to the optically reconstructed images in the
case examined here and better images are expected to be reconstructed when the computer reconstruction system, especially the hologram data converting system, is improved.
If the suitable A-D converter or flying spot scanner can be
utilized, the sampling and quantization of the hologram data
can be done with much accuracy in much shorter time, whereas
the sampling and quantization of the hologram data are done
manually in this experiment.

The display of the calculated image is another problem

to be studied. The display technique of the calculated
image by a line printer as studied in this paper has its
limitation because the line printer is not too well suited
to display the image output. The CRT or plotter display
will be useful to display the better images.

Though the algorithm to reconstruct images from Fresnel
transform holograms developed in this paper is principally
a satisfactory one, many problems are left for future discussions. For example, since the aperture size of the image
plane is fixed to that of the hologram in this algorithm,
it is difficult to reconstruct images out of this aperture.
The image shift and periodic images are discussed to solve
this problem in this paper, but it is not a satisfactory
solution. Further investigation should be done on these
problems for the development of the computer reconstruction
technique in acoustical holography.

ACKNOWLEDGEMENTS

The present work is supported by Matsunaga Science
Foundation, to which the author is very grateful.

REFERENCES

(1) Y. Aoki, Acoustical Holography Vol. 2 (Plenum Press,
    1970), chap. 23.
(2) Y. Aoki, IEEE Trans. on Audio and Electroacoustics,
    AU-18(1970), p.258.
(3) J. Powers, J. Landry and G. Wade, Acoustical Holography
    Vol. 2 (Plenum Press, 1970), chap. 13.
(4) A. L. Boyer, J. A. Jordan, Jr., D. L. Van Rooy,
    P. M. Hirsch and L. B. Lesem, Acoustical Holography
    Vol. 2 (Plenum Press, 1970), chap. 15.
(5) J. L. Pfeifer, Acoustical Holography Vol. 4 (Plenum
    Press, 1972), p.317.
(6) J. W. Goodman, Introduction to Fouries Optics
    (McGraw-Hill Book Co., 1968), p.9, p.57.
(7) J. W. Cooley, P. A. W. Lewis and P. D. Welch,
    IEEE Trans. on Autio and Electroacoustics, AU-15(1967),
    p.79.
(8) J. A. Glassman, IEEE Trans. on Computers, C-19(1970),
    p.105.

THE EFFECTS OF CIRCUIT PARAMETERS ON IMAGE QUALITY IN A HOLOGRAPHIC ACOUSTIC IMAGING SYSTEM

J. L. Sutton, J. V. Thorn, J. N. Price

Naval Undersea Center

San Diego, California 92132

INTRODUCTION

The design of holographic acoustic imaging systems using hydrophone arrays at the Naval Undersea Center has led to the development of special signal processing electronics which will be described in detail. The tolerances on electronic component values in this circuit, the major noise sources in the circuit, and other properties of an imaging system will be used to calculate the expected imaging performance of a holographic imaging system, as measured by its image dynamic range. An alternative version of the circuit will also be discussed, analyzed, and compared with the original circuit. Finally, an analogy between the signal processing circuit and a cross-correlator will be made, indicating possible future developments.

MOTIVATION

The Naval Undersea Center is involved in the development of acoustic imaging systems for underwater viewing. One such system being developed as an experimental tool is a small holographic acoustic imaging system called the System for Evaluation and Simulation (SES) (Ref. 1). The 20 by 20 element hydrophone array portion of this system is shown in Fig. 1. In the process of developing circuitry for the conversion of SES to Automatic SES (ASES), with which real-time data may be gathered, an attempt has been made to

Fig. 1. 20 by 20 Hydrophone Array

relate ASES circuit parameters to the overall performance of ASES as an imaging system.

Specular reflections in an acoustic image tend to be much brighter than the lower-level, nonspecular returns. In order to preserve both kinds of image information, it is necessary that the imaging system be capable both of processing the high-level specular returns and of maintaining the noise level in the image below the level of the nonspecular returns. The measure of how well an imaging system preserves the entire image in this way is called Image Dynamic Range (IDR). In turn, the goal for an acoustic imaging system is to have as large an IDR as practical, given the usual economic constraints of price, size, weight, etc. IDR will be discussed in more detail later in the paper.

THE IMAGING SYSTEM

The type of imaging system of interest here is a holographic acoustic imaging system (Fig. 2). It consists of a discrete array of hydrophones as the acoustic detector, followed by special electronics to perform the signal processing, and a separate unit, the holographic reconstructor, which derives the acoustic image. We will assume that the reconstructor introduces negligible noise to the system and hence does not affect the IDR. The use of a separate channel of signal processing electronics behind each

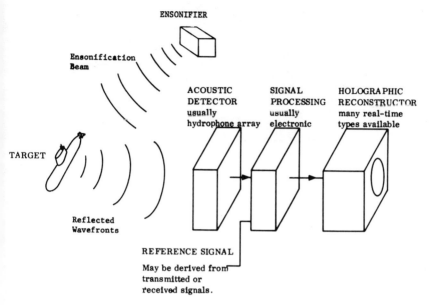

Fig. 2. ASES Simplified System Block Diagram

hydrophone in the array, called a channel processor, enables the system to gather the complete hologram at one time, similar to the manner in which a photographic camera records a snapshot.

THE CHANNEL PROCESSOR

A block diagram of the channel processor is shown in Fig. 3. The hydrophone converts the sound-pressure signal to an electrical signal, $S \sin(\omega t + \varphi)$, which the preamplifier amplifies to give $AS \sin(\omega t + \varphi)$. The signal then enters two essentially identical paths consisting of mixer, integrator, and multiplex switch. The mixer multiplies the signal during the receive gate, T, by a reference wave, $\cos \omega t$, or by its quadrature signal, $\sin \omega t$. The use of the quadrature reference has been explained by Mueller (Ref. 2). The result is a dc term proportional to either $AS \sin \varphi$ or $AS \cos \varphi$, plus harmonics of the frequency ω. The integrator filters out all the harmonics, exactly, as long as the receive gate remains on for an integral number of cycles of the frequency ω. The integrator also samples the dc voltage, produced by the mixer, or more accurately, integrates the dc term over the interval T. This

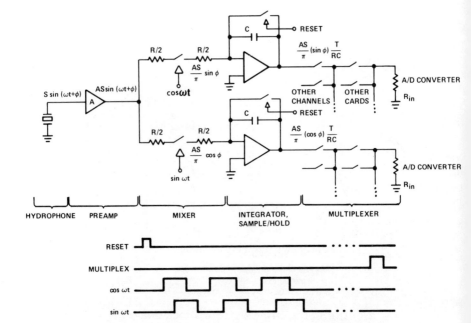

Fig. 3.   ASES Channel Processor Block Diagram

distinction between sampling and integrating will prove important
later.   The integrator also serves as a sample-and-hold circuit,
so that at the end of the receive gate, dc voltages proportional to
$(AS/\pi)\sin\varphi$ and $(AS/\pi)\cos\varphi$ are present at the outputs of the integrator.   The multiplexer supplies these voltages one at a time to an
analog-to-digital converter, the assumed input of our reconstructor.
A switch in each integrator resets that integrator to zero following
the multiplex period.   The timing diagram, Fig. 3, indicates the
operation of reset, multiplex, and mixer analog switches.

Any errors introduced by the circuitry in performing the above
signal processing functions will enter the reconstructed image as
noise, as discussed by Thorn (Ref. 3).   Thus amplitude and phase
nonuniformities between channels, thermal noise, and any dc voltage offsets should be kept to a minimum.   It should be noted, however, that the actual gain and the actual phase shift (as measured
with respect to some arbitrary standard phase) of a channel are not
crucial as long as they are uniform from channel to channel.

A Detailed Channel Processor Circuit

Figure 4 presents the full schematic of the active integrator channel processor designed for use in ASES. The hydrophone is a PZT ceramic disk, having a resonance frequency slightly above the system operating frequency of 250 kHz, and may introduce both amplitude (sensitivity) and phase (position) errors to the signal.

The next stage is the preamplifier. It consists of a low-noise, high-input impedance amplifier for matching to the hydrophone; a band pass filter for protection from unwanted signals, such as those from other sonars, which can saturate the electronics; and a high-gain amplifier to provide most of the signal gain. The preamplifier may introduce noise to the system in the form of thermal noise or gain and phase variations. The analog switch in the feedback loop of the high-gain amplifier is for gain-switching. When the switch is closed, the high-gain stage has a gain of only one. This is to prevent saturation of the integrator at longer receive gate times.

The mixer consists of a COS/MOS analog switch, which multiplies by either 1 or 0. When driven at the frequency ω, the switch effectively performs multiplication by a sine wave plus its higher harmonics. If there is no distortion in the amplified acoustic signal, and if the receive gate is an integral number of cycles long, the harmonics contribute nothing to the dc output of the channel. The mixer may contribute gain errors due to varying "on" resistance in the analog switch, phase errors due to varying propagation delays through the switch, dc offset due to capacitive feedthrough of the driving reference signal, and current leakages during the hold period.

The integrator integrates the signal from the mixer, S_M, as given by

$$S_{\text{Integrator}} = \frac{T}{RC} \int_T S_M \, dt,$$

where T is the receive gate length, R is the mixer resistance, and C is the integrator capacitor value. In combination with the COS/MOS analog switch of the mixer, the integrator also performs a sample-and-hold function. Low offset voltage, offset current, and

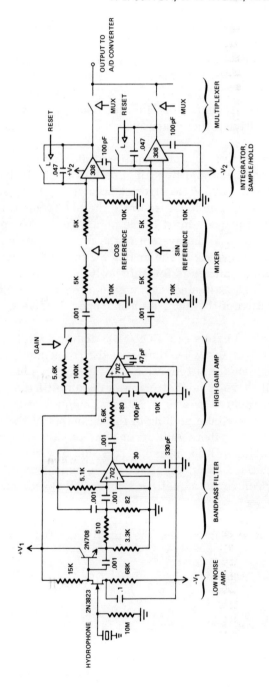

Fig. 4. Channel Processor for ASES, Schematic

drift characteristics are important parameters in these applications. Major error sources in this stage are the operational amplifier's input offset voltage and input offset current, both of which lead to voltage offset, and the tolerances on the R and C values which lead to gain variations.

Finally, the multiplexer is a two-level arrangement of COS/MOS analog switches, the first level of which selects the particular output on the column to be sampled, while the second level selects the column. As with the mixer, the multiplexer also suffers varying "on" resistance of the switches and current leakage during the hold period, which can, in addition, reduce the IDR of the system.

Noise Sources

We have mentioned four types of errors or noise that the electronics can introduce: gain errors, phase errors, dc offsets, and thermal noise. The first two occur most often and are expected to be the most severe. Hence, to eliminate these two errors, two resistors in the preamplifier are adjusted during a calibration procedure. The gain is standardized by trimming the 100k resistor in the high-gain amplifier, and the phase is calibrated by trimming the 82-ohm resistor in the band pass filter. These two trims compensate for all of the gain and phase error sources within the channel processor, because only the output is observed and the input is known. Of course, the calibration can be accurate at only one temperature; gain and phase errors will certainly still appear as the temperature varies and as other conditions change. This cannot be helped, but only those variations that fall unpredictably or randomly on either side of the average will lead to noise in the image. If all channels start from the same initial gain and all of them drift in the same direction at the same rate, then the gains will always be uniform, and these drifts will not give rise to image noise.

As for the dc offset term, two more trimming resistors could be inserted in the integrator circuits to cancel out all the offsets in the channel. The cost of the extra trims, however, would probably be more than the expected gains in performance. A second option, the use of a circuit with lower offsets, will be considered later.

Thermal noise can be minimized by conventional techniques, such as a low-noise preamplifier and narrow bandwidths in the band pass filter. It can also be shown that the amount of error due to thermal or stochastic noise is inversely proportional to $\sqrt{T}$, the square root of the receive gate length. Thus, longer receive gates tend to reduce thermal noise.

IMAGE DYNAMIC RANGE

The measure of image quality we have chosen is Image Dynamic Range (IDR) as developed by Thorn (Ref. 3). IDR is one type of signal-to-noise level, being the level of peak image intensity minus the level of average background noise. In a medium where target specularity can cause wide differences in echo intensity, which will be observed in the image, a large IDR is necessary for seeing many of the details of the object being viewed.

IDR for an array of L by N hydrophones is defined as follows:

$$\text{IDR} \;=\; \text{AG} + \text{CDR} = 10 \, \log\!\left(\frac{\text{Intensity of image highlight}}{\text{Intensity of background noise}}\right)$$

$$=\; 10 \log L \cdot N - 20 \log \left[\left(\frac{\Delta A}{A} \right)^2 + \Delta\varphi^2 + \left(\frac{1}{M} \right)^2 \right]^{1/2} ,$$

where

AG $\;=\;$ Array Gain = 10 log L·N

CDR $\;=\;$ Composite Dynamic Range =

$$-20 \log \left[\left(\frac{\Delta A}{A} \right)^2 + (\Delta\varphi)^2 + \left(\frac{1}{M} \right)^2 \right]^{1/2}$$

$\dfrac{\Delta A}{A} \;\equiv\;$ the rms fractional error in voltage gain over all the channels

$\Delta\varphi \;\equiv\;$ the rms phase error over all the channels

$M \;\equiv\;$ the ratio of signal amplitude to average noise voltage of the channel, including both offset and thermal noise contributions.

Composite Dynamic Range (CDR) is simply the level of the root-sum-square of all the error sources of a channel. In a sense, it is an expected value of channel processor error over the whole array. Therefore, CDR is, in fact, a measure of the performance of a single, "typical" channel processor. In broad terms, CDR is a measure of signal processing performance.

The IDR, then, is composed of two very different terms, one of which is related solely to the number of receivers in an array (array gain), and the other which describes the quality of those receivers (CDR). Since this paper is concerned primarily with the relationship between receiver quality and image quality, and less with array size, the CDR will be taken as the measure of imaging system performance. IDR differs from CDR only by a constant equal to 10 log L·N, so that IDR can be derived from CDR by simply adding the appropriate AG. For example, the ASES 20 by 20 array has an array gain of 26 dB. Therefore, a graph of CDR versus signal level may be converted to one of IDR versus signal level by merely adding 26 dB to all vertical axis values.

CHANNEL PROCESSOR PROPERTIES AND PERFORMANCE

Table 1 is a list of the parameters and their tolerances, where appropriate, which apply to the ASES active integrator channel processor of Fig. 4. Some of the numbers have been measured in the laboratory, such as preamplifier gain and mixer offset voltage. Others were deduced from component specification sheet data. Where necessary, values pertaining to a specific imaging system design are based on the 20 by 20 element ASES array.

Table 2 is a list of the error contributions from various sources in the ASES active integrator channel processor which were calculated using the values from Table 1. In many cases, the error terms are straightforward and simple to calculate. In others, Taylor series expansions and simplifications are used to reduce the expression and retain only the first order terms.

The CDR of the active integrator channel processor is plotted as a function of input signal level from the hydrophone in Fig. 5. The second curve, that for the RC filter channel processor, will be derived shortly. The first curve indicates that we can expect a CDR

Table 1. ASES Channel Processor Parameters and Tolerances

Parameter	Symbol	Value	Tolerance
1. Hydrophone - preamplifier voltage gain	G	1400	$\frac{\Delta G}{G} = 1\%$
2. Mixer resistance	R	$10k\Omega$	$\frac{\Delta R}{R} = 0.5\%$
3. Mixer offset voltage	V_{Moff}	5mV	
4. Mixer switch "on" resistance	R_{on}	200Ω	$\Delta R_{on} = 50\Omega$
5. Integrator Capacitance	C	$0.047\mu F$	$\frac{\Delta C}{C} = 0.3\%$
6. Integrator offset voltage	V_{Ioff}	1.4mV	
7. Integrator offset current	I_{Ioff}	0.4nA	
8. Multiplex switch "off" resistance	R_{off}	$5 \times 10^{10}\Omega$	$\Delta R_{off} = 2 \times 10^{10}\Omega$
9. Multiplex switch offset current	i_{off}	80pA	
10. Multiplex switch input or output capacitance	C_S	4pF	$\Delta C_S = 0.3pF$
11. A/D converter input impedance	R_L	10k	$\frac{\Delta R_1}{R_1} = 0.5\%$
12. Number of analog switch capacitances in parallel in multiplexer	n	61	
13. Number of multiplex switches in series	q	2	
14. Noise equivalent input	$\frac{N}{\sqrt{\Delta f}}$	$10nV/\sqrt{Hz}$	
15. Receive gate length (minimum)	T	$100\mu sec$	
16. Average hold time before multiplex	t	16msec	
17. Signal amplitude (out of hydrophone)	S	$\leq 700\mu V$	

Table 2. ASES Channel Processor Error Terms

Error Source	Equation	Error Ratio
GAIN VARIATIONS		
Hydrophone - preamplifier gain	$\dfrac{\Delta G}{G}$	0.01
Integrator capacitance tolerance	$\dfrac{\Delta C}{C}$	0.003
Mixer resistance tolerance	$\dfrac{\sqrt{\Delta R^2 + \Delta R^2_{on}}}{R + R_{on}}$	0.007
Multiplexer gain variation	$\dfrac{\Delta R_{on}}{R_L}\sqrt{q}$	0.007
TOTAL GAIN VARIATIONS (fractional error)	$\dfrac{\Delta A}{A}$	0.014
PHASE VARIATIONS		
Preamp phase shift variations	$\Delta \phi_p$	0.017
Mixer phase shift variations	$\Delta \phi_M$	0.001
TOTAL PHASE SHIFT VARIATIONS (radians)	$\Delta \phi$	0.017
NOISE-TO-SIGNAL RATIO		
Mixer - integrator offset voltage	$\dfrac{\pi (V_{Ioff} + V_{Moff})}{GS}$	$\dfrac{1.4 \times 10^{-5}}{S}$
Integrator offset current	$\dfrac{\pi I_{Ioff}\, t\, R}{GTS}$	$\dfrac{1.4 \times 10^{-6}}{S}$
Thermal and acoustic noise	$\dfrac{N}{\sqrt{\Delta f}}\sqrt{\dfrac{2}{T}}\,\dfrac{1}{S}$	$\dfrac{1.4 \times 10^{-6}}{S}$
TOTAL NOISE-TO-SIGNAL RATIO	$\dfrac{1}{M}$	$\dfrac{1.4 \times 10^{-5}}{S}$

$$IDR = AG - 20 \log \sqrt{(2.2 \times 10^{-2})^2 + \frac{(1.4 \times 10^{-5})^2}{S^2}}$$

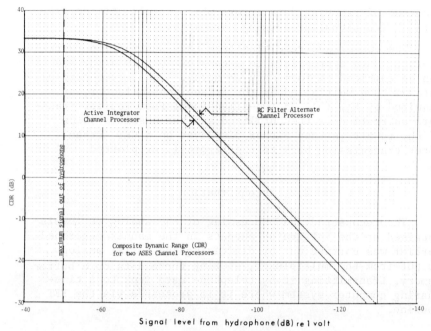

Fig. 5.   Composite Dynamic Range (CDR) for Two ASES Channel
          Processors

up to 32 dB from the channel processor and an IDR up to 58 dB from
our 20 by 20 array.   That is, for a relatively simple, specular target, we should be able to distinguish (at 0 dB signal-to-noise level)
image detail from background noise as low as 58 dB below the
brightest specular return.   The dotted line, labeled "maximum
signal out of hydrophone", shows the signal from the hydrophone
that will just saturate the channel processor's preamplifier, an
undesirable situation which would lead to signal distortions and to
further noise in the image.

There are two basic regions in the graph of CDR.   The plateau
region represents the signal level above which further increases in
signal cannot raise CDR and IDR.   In this region performance is
limited by the gain and phase errors of the channel processor.   In
the sloped region, CDR and IDR are limited by thermal and dc offset voltage noise.   The channel processor is, thus, noise-limited
over most of its useful range, i.e., at input signal levels less than

maximum. This appears to provide adequate performance, however, since no further increase in IDR or CDR can be obtained by further increasing the input signal level.

AN ALTERNATE CHANNEL PROCESSOR AND ITS PERFORMANCE

In an attempt to find a simpler and cheaper channel processor, which might also possess better performance properties, the alternate channel processor of Fig. 6 was analyzed in a manner similar to that employed for the active integrator channel processor. The basic difference between this new circuit and the channel processor of Fig. 3 and 4 is the substitution of a simple RC filter in place of the active integrator. It is true that an RC filter can be considered an integrator if the receive gate, T, is much shorter than the filter time constant, $T \ll RC$. The noise performance of the channel will be best, however, if the RC filter is operated as an averager or sampler, i.e., where $T \gtrsim RC$.

The elimination of the two operational amplifiers certainly makes the channel processor simpler and less expensive. In addition, the simple RC filter makes the use of a single reset switch for the whole array possible, in place of the pair of reset switches required by the active integrators. In this circuit all multiplex switches are closed at the same time as the reset switch, as shown in the timing diagram of Fig. 6, to insure that all RC filters start at the same dc value. Since a major source of limiting noise in the active integrator channel processor, from Table 2, is integrator voltage offset—an error the passive RC filter does not have—it is reasonable to expect that the performance of the RC filter channel processor may, in fact, be better than that of the active integrator channel processor.

The results of the noise source analysis of the RC filter channel processor are listed in Table 3. Here it can be seen that several new error sources have been introduced over those in Table 2, because the RC filter lacks the buffering properties of the operational amplifier. Many dc voltages are subject to exponential rolloffs in the RC filter that the operational amplifier active integrator prevented, but these rolloffs are very predictable and uniform from channel to channel because of the passive nature of the components. As was discussed previously, identical (or proportional)

Table 3. Alternate ASES Channel Processor Error Terms

Error Source	Equation	Error Ratio
GAIN VARIATIONS		
Hydrophone – preamplifier gain variations	$\dfrac{\Delta G}{G}$	0.01
RC time constant variation	$\dfrac{T\sqrt{\Delta R_{on}^2 + R^2}}{(R + R_{on})^2 C(e^{T/RC} - 1)}$	0.006
Multiplex switch "on" resistance variations	$\dfrac{\sqrt{q}\,\Delta R_{on}}{R_L}$	0.007
Multiplex switch capacitance variations	$\dfrac{\sqrt{\Delta C_s}}{C}$	0.00005
Multiplex $R_L C$ and C variations	$\dfrac{\Delta C}{C}\left(7 \times 10^{-6} - \dfrac{T}{RC(e^{T/RC}-1)}\right)$	0.0026
A/D converter input impedance variations	$\dfrac{7 \times 10^{-6}\,\Delta R_L}{R_L^2}$	0.00007
TOTAL GAIN VARIATIONS (fractional error)	$\dfrac{\Delta A}{A}$	0.014
PHASE VARIATIONS		
Preamp phase shift variations	$\Delta\phi_p$	0.017
Mixer phase shift variations	$\Delta\phi_M$	0.001
TOTAL PHASE SHIFT VARIATIONS (radians)	$\Delta\phi$	0.017
NOISE-TO-SIGNAL RATIO		
Mixer offset voltage	$\dfrac{\pi V_{Moff}}{GS}$	$\dfrac{1.1 \times 10^{-5}}{S}$
Mixer offset current	$\dfrac{\pi(.283)I_{Moff}}{CGS}$	$\dfrac{1.1 \times 10^{-6}}{S}$
Leakage current variations during "hold"	$\dfrac{\pi(.283)R_{off}}{CR_{off}^2}$	1.5×10^{-4}
Leakage current variations during "hold"	$\dfrac{\pi 7n \times 10^{-6}}{R_{off}C}$	5.7×10^{-7}
Thermal and acoustic noise	$\dfrac{\pi N}{\sqrt{\Delta f}}\sqrt{\dfrac{1 - e^{-2T/RC}}{2RC}}\left(\dfrac{1}{S}\right)$	$\dfrac{4.2 \times 10^{-7}}{S}$
TOTAL NOISE-TO-SIGNAL RATIO	$\dfrac{1}{M}$	$\left[\dfrac{(1.1 \times 10^{-5})^2}{S^2} + (1.8 \times 10^{-3})^2\right]^{1/2}$

$$IDR = AG - 20\log\sqrt{(2.21 \times 10^{-2})^2 + \frac{1}{S^2}(1.1 \times 10^{-5})2}$$

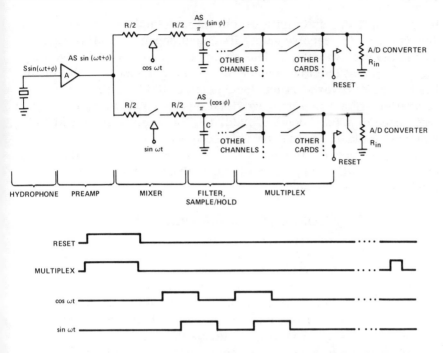

Fig. 6. Alternate Channel Processor Block Diagram

changes in all channel processors have no affect on CDR and IDR. The graph of CDR versus input signal level, shown in Fig. 5, indicates that the RC filter can, in fact, perform slightly better than the active integrator channel processor. The more important result, however, is that the simpler and cheaper RC filter circuit performs at least as well as the active integrator circuit over the range of useful input signal levels.

    In spite of the very real advantages of an RC filter over an active integrator, there is a cogent reason for selecting the original channel processor with its active integrator. The RC filter has, in effect, a very short memory—on the order of a time constant or two. Its output depends only on the voltages present during its memory span, and all earlier data is lost. In essence, therefore, the RC filter has a fixed receive gate length of T = RC. The active integrator, on the other hand, performs a true integration (i.e., an infinite sum of infinitesimal increments) during the entire receive gate. Its output depends on all the data during the receive gate, which implies a depth of field for the imaging system that is truly variable and dependent upon only the sonar limitations of reverberation, focal depth of field, and integrator saturation. For some imaging situations, such as a single target in free space or a fixed depth of field, the RC filter would certainly be a prime candidate for the channel processor. However, on the basis of the superior flexibility of the integrator over the fixed nature of the RC filter, ASES will be built using the active integrator channel processor.

A CORRELATION PERSPECTIVE

Considering the active integrator channel processor block diagram in Fig. 7, we see that under certain conditions the channel processor is really a cross-correlator, that is, a mixer followed by an ideal integrator. Indeed, the correlation function of two sinusoids with variable phase is either $\cos\varphi$ or $\sin\varphi$, the output of the channel processor. It is in these terms that the real flexibility of the active integrator channel processor makes itself known. The acoustic signal no longer needs to be a simple, single-frequency, coherent tone burst; more complex wave forms can be used to give better noise performance for comparable depths of field in the image, or shorter depths of field for comparable noise performance, or comparable performance at reduced transmitted power levels. Special waveforms may also be employed for special purposes, such as the ability to tolerate Doppler shifts, or to collect target information besides reflected intensity, such as shape or composition of the target. In short, the full formalism of cross-correlation processing developed for sonar over the past 30 years is available for improving system performance in acoustic imaging systems at a minimal cost in materials. Further investigation in this area is planned at the Naval Undersea Center.

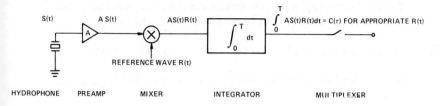

Fig. 7. A Correlation View of the ASES Channel Processor

CONCLUSIONS

The active integrator channel processor has been shown to be a flexible and even a powerful signal processor for acoustic imaging applications. Using readily obtainable circuit components, a channel processor capable of 58 dB image dynamic range in a 20 by 20 element hydrophone array is indeed feasible. The fabrication and testing of 400 such active integrator channel processors for the ASES holographic acoustic imaging system at the Naval Undersea Center will verify these predictions.

The passive RC filter channel processor has been shown to be an equally effective alternative to the active integrator in applications where a fixed receive gate is satisfactory.

Further work will be directed at making more effective use of the active integrator's capability as a cross-correlator circuit, both in improving system performance and in exploring new ways of employing acoustic imaging techniques.

REFERENCES

1. Booth, N. O. and Saltzer, B. A., "An Experimental Holographic Acoustic Imaging System," Acoustical Holography, Vol. 4, Plenum Press, New York, 1972, Page 371.

2. Mueller, R. K., Proceedings of IEEE, Vol. 59, 1971, Page 1319.

3.   Thorn, J. V., "Gain and Phase Variations in Holographic
      Acoustic Imaging Systems," Acoustical Holography, Vol. 4,
      Page 569.

ALGEBRAIC RECONSTRUCTION OF SPATIAL DISTRIBUTIONS OF
ACOUSTIC ABSORPTION WITHIN TISSUE FROM THEIR TWO-
DIMENSIONAL ACOUSTIC PROJECTIONS

J. F. Greenleaf*, S. A. Johnson*, S. L. Lee*,
G. T. Herman†, and E. H. Wood*

Department of Physiology, Mayo Foundation,
Rochester, Minnesota*, and SUNY, Buffalo, New York†

Introduction

It has been known for many years that three-dimensional
information concerning the spatial distribution of energy
absorbers within an object could be obtained from two-
dimensional shadow projections of the energy absorption of
the object (1). Two-dimensional projections or shadows of
the absorption of an object can be obtained using many forms
of energy such as light, x-radiation, electrons, or sound.
The problem was treated in abstract mathematics as early as
1917 by Radon (2). The first practical solution of this
problem was obtained by Bracewell in 1954 (3) who applied the
technique to radioastronomy. The first application of these
kinds of techniques to biology were probably done by DeRosier
and Klug who obtained the cross-sectional structure of the
tail of a bacteria phage from one-dimensional projections
obtained with electrons in an electron microscope (4).

The attainment of information concerning the three-
dimensional distributions of energy absorbers within a body
from its two-dimensional projections requires special analytic techniques because it is equivalent to the problem
of solving large systems of algebraic equations or integral
equations. In the past few years, several new techniques
have been developed for solving the large number of algebraic
equations which are necessary to obtain this information. A
technique which has been especially suitable for noisy biologic data obtained in particular from projections of x-radiation, has recently been developed by G. T. Herman of SUNY

and has been called algebraic reconstruction technique or ART (5).

The algebraic reconstruction technique of solving large simultaneous systems of linear equations is essentially a relaxation technique which is basically immune to noise and which can be shown to be stable (6).

The reconstruction technique requires that the intensity at each point on the shadow of an object be related directly to the absorption within the object projected to that point. For this reason, the use of ultrasonic energy to obtain two-dimensional shadow projections (C scans) for the algebraic reconstruction of acoustic absorption within a three-dimensional body presents several special problems. One of these problems is that sound reflects from interfaces having different acoustic impedances and this loss of energy from the transmission shadow is not easily calculated. Another problem is that sound refracts within a tissue due to gradients in acoustic impedance. This causes energy to be measured on the projection shadow at points unrelated to the regions of absorption encountered by the refracted ray. The combination of these two effects adds some degree of uncertainty to the accuracy of two-dimensional sound projections of a structure since the energy measured at various points in a transmission C scan will not have a one-to-one relationship with the absorption within the structure.

The purpose of this paper is to describe initial attempts at reconstructing the three-dimensional distribution of acoustic impedances within a structure from its two-dimensional projection of ultrasonic energy obtained in the form of a C scan. The distributions of acoustic impedances will be obtained using an algebraic reconstruction technique, under the assumption that reflection and refraction problems are negligible.

METHOD

C scans of objects were obtained using the scanning arm of a modified computer-controlled Picker Magna IV scintiscanner onto which were mounted transmitting and receiving ultrasonic transducers. The length of the scan was 15 cm with a duration of approximately four seconds per scan. Pulsed sound energy was used having a center frequency of approximately 5 MHz and a bandwidth of 2 MHz. The transmitting

transducer was a Panometrics VIP-5[*] and the receiving transducer was a Panometrics VIP-2.25[*] The transmitting transducer emitted sound through a lucite lens which had a focal
length of 10 cm. The receiving transducer was connected to
a PR 50/50 Panometrics receiver which in turn was connected
to a Biomation Transient Recorder(BTR)[**] model 8100 analog-
to-digital converter.

The BTR is a high-speed analog-to-digital converter
which has a 2,048 word-buffer memory and which can convert
analog signals to 8-bit digital words at rates of up to one
hundred million words per second. The BTR was used as a
buffer input to a Control Data Corporation 3500 general purpose computer which, through digital output lines, controlled
all of the digitizing variables of the Biomation Transient
Recorder. These variables included delay from trigger to
sampling, sample interval, voltage gain, voltage offset, and
coupling (7).

The computer was used to trigger the transmitter and to
move the arm of the scintiscanner on which was mounted the
transmitting and receiving transducers. Using a low-speed
A/D converter (1000 samples/second), the computer continually
measured the voltage on a linear potentiometer attached to the
arm of the scanner and, at equal intervals along the scan length,
sent transmitter synchronization signals to both the transmitter and the analog-to-digital converter (BTR). A schematic
of the setup is shown in Figure 1. After an appropriate delay
allowing traversal time of the pulse across the transmitter/
receiver gap, the BTR began to digitize the received signal
at a rate of twenty million samples per second for a period
of 2,048 samples. After waiting one millisecond, which was
ample time to digitize the signal, the computer sent a strobe
signal to the BTR, at which time the computer read into its
memory the 2,048 samples within the buffer memory of the BTR,
which represented the digitized ultrasonic signals.

The 2,048 computer words representing the digitized
ultrasonic pulse from the receiver were analyzed for each of
the 200 positions along the scan length at which the transmitter was pulsed. Generally, only 200 digitized samples
representing 10 μseconds of received signal were required to
encompass the received pulse. The remainder of the 2,048 words

---

* Panometrics, Waltham, Massachusetts.
** Biomation Transient Recorder, Inc., Cupertino, California.

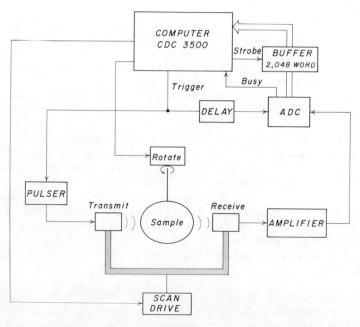

Figure 1.   Schematic of computer-controlled data acquisition
            system which obtains digitized ultrasonic pulses
            from B and C scans. Pulser and A/D converter are
            triggered by computer at 200 equispaced points
            along the scan length of 15 cm. Received signal
            is converted into 2,048 eight-bit words at a rate
            of 20 words per microsecond. Non-busy signal from
            ADC causes buffer memory of digitized data to be
            strobed into computer. At completion of scan the
            sample is rotated and scan is repeated. Position
            of scanner arm is measured by computer as voltage
            across linear potentiometer connected to arm (not
            depicted in schematic).

represented multiple reflections and were deleted resulting
in the equivalent of a range-gated receiver. Total energy
within the pulse was calculated as the log of the sum of the
squares of the values representing the received pulse. For
each line scanned, this resulted in an array of 200 values
each of which were proportional to the log of the energy of
the received signal at a specific point on the one-dimensional
C-scan line representing the shadow projection of the acoustic absorptions in a plane within the sample.

ANALYSIS: BACKGROUND

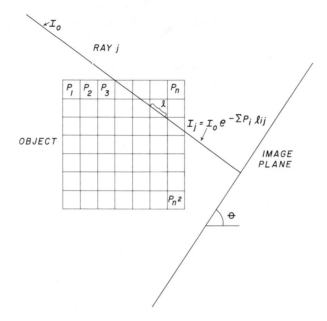

Figure 2.   Geometry of ultrasound transmission for algebraic
reconstruction. Grid is fixed to coordinates of
the object. Absorption is assumed to be constant
within each cell of the grid. Initial intensity
of ray I_{0j} is absorbed within cell i according to
$e^{-P_i l_{ij}}$ where P_i is the absorption coefficient
and l_{ij} is the length of traversal of ray j through
cell i. Total intensity absorption for ray j is
given by

$$\frac{I_j}{I_0} = e^{-\sum_{i=1}^{n^2} P_i l_{ij}} \quad \cdot$$

Acoustic density of object along ray j is given by

$$D_j = \ln (I_0/I_j) = \sum_{i=1}^{n^2} P_i l_{ij} \quad \cdot$$

In order to derive the equations necessary to obtain the
three-dimensional distributions of acoustic absorption from
two-dimensional projections, we construct a geometry as shown
in Figure 2. The grid of squares is fixed to the coordinate
system of the object. Acoustic absorption within each square
of the grid is assumed to be constant. Line P represents the
projection plane or the locus of the receiver during a one-
line C scan onto which is projected a ray which represents
the path taken by one of the pulses of ultrasonic energy.
This projection line is considered to be at a general angle
(θ) to the grid coordinates of the body. The density encountered by a ray after it has passed through the object can
be expressed by the following Beer-Lambert equation:

$$D_{ik} = \ln\left(\frac{I_o}{I_i}\right) = \sum_{j=1}^{n^2} L_{ijk}P_{jk} \qquad \text{(Eq. 1)}$$

where the L_{ij} represent the length of ray i as it passes
through individual square j within the grid of the body, the
P_j represent the absorption coefficients within each square
on the grid and the D_{ik} represent the density of the object
measured for ray$_i$ along the scan line taken at angle θ_k to the
object. There is one of these equations for each point i on
each scan line k.

For the reconstructions to be reported here, solutions
were obtained in which the number n on the grid on the body
was 64 resulting in a total of 4,096 individual picture cells
or unknowns within the desired picture. It is clear that
more than one projection is required in order to obtain enough
measurements to solve this set of equations. For this reason,
the projections or C scans of the object must be obtained at
several angles of rotation of the scanning plane relative to
the grid coordinates of the body.

In order to obtain these multiple ultrasonic one-dimensional C scans through the sample, the object was rotated
about an axis perpendicular to the scanning line with a computer-controlled stepper motor. At the end of each scan, the
object was rotated 5.4° and the scan repeated. This procedure
was repeated through 35 steps (approx. 180°). Since there are
35 views, each with 200 samples along the line, this procedure
resulted in 7,000 data values, each representing the density
or absorption of a sound ray projected through the body. This
results in a set of simultaneous algebraic equations which can

be expressed as:

$$D = LP: \qquad\qquad\qquad (Eq.2)$$

where D is a $M \times 1$ matrix of density measurements, L is a $M \times n^2$ matrix of picture cell intersection lengths which are known from geometric considerations and P is a $n^2 \times 1$ matrix of unknown picture cell densities. M must be greater than n^2.

Equation 2 contains values which have been measured and thus contain noise; therefore, we have chosen a technique for

Figure 3.   Photograph of phantom used to evaluate ultrasonic reconstruction techniques. Phantom consists of two 1.6-cm o.d., 0.9-cm i.d. silastic tubes which can be filled with fluid. Between the tubes is a 2-mm diameter rod. Next to one tube is a 5-mm rod and next to the other tube are three 1-mm diameter wires separated by 1 cm. The phantom was immersed in water and ultrasonically scanned using the system described in Figure 1.

(35 x 5.4° Steps)

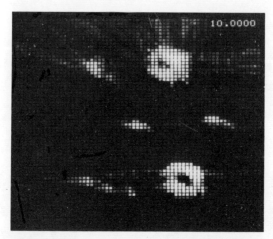

Figure 4.   Algebraic reconstruction of acoustic absorption
within cross section through submerged phantom.
Picture is photograph of computer-driven CRT and
is made up of 64 x 64 picture elements. Water within silastic tubes is resolved as are three 1-mm
wires and larger rods. The silastic tubes and rods
appear thicker than they should, although overall
geometric fidelity is qualitatively good.

obtaining its solution which has recently been developed by
G. T. Herman of the State University of New York, Buffalo (6).
This technique will only be referenced here, but has been
shown to be stable and relatively immune to noise.

RESULTS

In order to test the entire data acquisition and analysis system, a phantom was constructed (Figure 3) which consisted of two silastic tubes capable of being filled with various
fluids and several wires of various sizes and separations all
mounted in parallel. This phantom was immersed in water and
scanned with ultrasound at each of 35 rotations separated by
5.4°. The silastic tubes were filled with water. The resulting algebraic reconstruction was plotted by computer on an
oscilloscope and photographed and is shown in Figure 4. The
whiteness of each picture cell represents the calculated
acoustic absorption for that region of the cross section

through the phantom. The picture resolves three of the 1-mm
diameter wires, but some geometric distortions of the silastic tubes have occurred although the density of the water
within the water-filled silastic tubes was found to be the
same as the density of the surrounding water. It is clear
that the walls of the silastic tubes are depicted as being
thicker than their true dimension. This was apparently
caused by refraction of the sound away from the receiver-
transducer at the edges of the tubes resulting in a loss of
energy which appeared as greater absorption. In addition,
the finite line spread function of the lense system added to
the measured thickness of the objects.

(35 x 5.4° Steps)

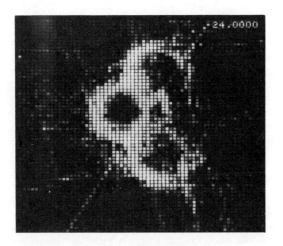

Figure 5.   Algebraic reconstruction within cross section through
            three fingers of a water-filled rubber glove. Cross
            section of water-filled rubber glove was scanned
            with system depicted in Figure 1. 64-by-64 element
            reconstruction depicts rubber membrane as being
            thicker than actual and in some areas as missing
            completely. Errors are apparently due to refrac-
            tion of sound energy away from edges of glove.

   The effect of refraction on the reconstruction of acoustic absorptions within a structure is more clearly demonstrated with a phantom made from a rubber glove. Three fin-

Figure 6. Photograph of transverse slice through canine heart.
The narrow curved opening is the right ventricle,
within which can be seen irregular surface muscle
called trabeculae. The large cavity is the left
ventricle and includes some papillary muscle cut
in cross section. An intact canine heart was
mounted on a shaft and scanned (at a level equivalent to this cross section) with the system
shown in Figure 1.

gers of a rubber glove filled with water were ultrasonically
scanned in the manner described above. A picture of the resulting reconstruction of absorptions within a plane perpendicular to the axis of the fingers is shown in Figure 5. The
rubber membrane of the fingers are depicted much thicker than
they should be. Once again, this may be due to refraction of
sound from the edges of the fingers and subsequent measurement of erroneously low transmittance of energy through the
glove resulting in a calculated high absorption coefficient.

We chose the heart as the biologic organ with which to
obtain reconstructions of acoustic absorption because the
interior of the heart contains chambers of complex geometry
filled with homogeneous material. A photograph of the cross
section through a canine heart is shown in Figure 6 and illustrates the complexity of the right and left ventricular chambers. An unfixed heart was mounted on the stepper motor

(35 x 5.4° Steps , Apex + 2 cm)

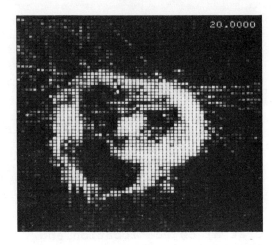

Figure 7.   Algebraic reconstruction of acoustic absorptions
within transverse cross section through canine
heart. Heart was rotated through 35 steps of 5.4°
and a C scan obtained at each step in a manner
described in Figure 1 and text. This 64-by-64
element reconstruction clearly depicts the left
and right ventricles. Myocardium is depicted as
having varying degrees of absorption. The geometric accuracy of the reconstruction can be compared with typical cross section in Figure 6.

assembly and ultrasonically scanned and rotated in the manner
described previously. Reconstruction of the acoustic absorptions within a cross section through the heart at approximately the same level as the cross section shown in the photograph is shown in Figure 7. One can observe several effects
in this reconstruction, 1) the edges seem to be enhanced,
apparently because of reflection of sound away from edges,
2) various regions of the heart are depicted as having acoustic absorption very near that of water.

DISCUSSION

The determination of the regional distribution of acoustic absorption within a structure using algebraic reconstruc-

tion resulted in qualitatively accurate reconstructions.
Quantitative evaluations of the geometric and densitometric
accuracies of the reconstructions were not obtained since the
effect of refraction was seen to be less than negligible.   It
would appear, therefore, that in order to use the technique
for quantitative evaluation of acoustic absorptions, the effects of refraction and reflection must be taken into account
in the reconstruction algorithm.

Nevertheless, the application of the technique to measurement of acoustic absorption within tissues, especially
isolated organs, seems promising.   The complete determination
of the structure within an organ would require the reconstruction of many planes through the organ.   At the present time,
our algorithm requires 2.5 to 3.5 minutes of computer time to
reconstruct each plane.   Therefore, reconstruction of 10 to 20
planes through an entire organ is not an impossible task.

In addition to correcting for refraction, one might consider correcting for the beam pattern of the transducers.
This would be done by giving a thickness and pattern to each
ray through the object rather than assuming the ray to have
zero thickness, as shown in Figure 2.   This would essentially
"deconvolve" the beam pattern from the data.

ACKNOWLEDGEMENT

The authors wish to thank Dr. Rich Robb for assistance
in using the BTR and Mr. Donald Erdman for constructing the
phantom and rotation device.

This investigation was supported in part by Research
Grants HL4664, HL3532, RR-7, and HE52076 from the National
Institutes of Health, U. S. Public Health Service; NGR 24-003-0
from the National Aeronautics and Space Administration;
AHA CI 10 from the American Heart Association; and GJ998 from
the National Science Foundation.

Dr. Wood is a Career Investigator of the American Heart
Association.   Dr. Greenleaf is a Postdoctoral Research Fellow
of the National Heart and Lung Institute, National Institutes
of Health.   Dr. Herman is Professor of Computer Sciences,
State University of New York, Buffalo.

REFERENCES

1. Gordon, R., G. T. Herman (1971): Reconstruction of
    pictures from their projections. Comm ACM 14(12):
    759-768.

2. Radon, J. (1917): Ueber die Bestimmung von Functionen
    durch ihre integralwerte Laengs gewisser
    Manningfoltigkeiten (on the determination of functions
    from their integrals along certain manifolds) Berichte
    Saechsische Acadamie der Wissenschaften (Leipzig)
    Mathematische-Physische Klasse 69, 262-277.

3. Bracewell, R. N., and J. A. Roberts (1954): Aerial
    smoothing in radio astronomy. Aust J Phys 7(4):
    615-640.

4. DeRosier, D. J., and A. Klug (1968): Reconstructions of
    three-dimensional structures from electron micrographs.
    Nature 217:130-134.

5. Gordon, R., R. Bender, and G. T. Herman (1970): Algebraic
    reconstruction techniques (ART) for three-dimensional
    electron microscopy and x-ray photography. J Theor
    Biol 29:471-481.

6. Herman, G. T., and S. Rowland (1971): Resolution in ART:
    An experimental investigation of the resolving power
    of an algebraic picture reconstruction technique.
    J Theor Biol 33:213-223.

7. Robb, R. A., S. A. Johnson, J. F. Greenleaf, M. A. Wondrow,
    and E. H. Wood: An operator-interactive computer-con-
    trolled system for high fidelity digitization and anal-
    ysis of biomedical images. SPIE Proc Quantitative
    Imagery in Biomed Sci II, San Diego, August 1973
    (in press).

OPTICAL PROCESSING OF ANAMORPHIC HOLOGRAMS CONSTRUCTED IN
AN ULTRASONIC HOLOGRAPHY SYSTEM WITH A MOVING SOURCE AND AN
ELECTRONIC REFERENCE

T. Iwasaki and Y. Aoki

Department of Electronic Engineering
Hokkaido University
N 12, W 8, Sapporo, 060, Japan

ABSTRACT

An ultrasonic holography system with a moving source
and an electronic reference is studied, where anamorphic
holograms are recorded due to the relative movements of the
source, object and receiver. In this paper a technique to
reconstruct images from such anamorphic holograms using a
cylindrical lens is proposed and a theoretical analysis is
conducted. An experiment using a one-dimensional receiver
array is done to examine the theoretical discussion.

INTRODUCTION

A scanned-type holography is a well studied technique
to construct acoustical holograms, and various kinds of scanning methods have been proposed[1-3]. The authors also proposed a system where the ultrasonic holograms were constructed
combining one-dimensional scannings of a receiver and a
linear motion of a source (or object)[4]. Since the receiver
scannings were done mechanically in this system, it took
much time to construct an ultrasonic hologram. Therefore
the authors suggested a system where the mechanical scannings
were replaced by the electronic scannings using a one-dimensional array of receivers to construct a hologram in much
shorter time.

In the present paper, an ultrasonic holography system

with a one-dimensional array of receivers is constructed and a technique to record holograms using an electronic reference and a linear motion of the source is studied. Since the ultrasonic holograms constructed in this system are intrinsically anamorphic ones, an optical processing is proposed to reconstruct correct images from these holograms.

EXPERIMENTAL APPARATUS

The ultrasonic holograms are constructed in the experimental arrangement of Fig.1, where some modifications are done to the experimental arrangement described in the reference 4.

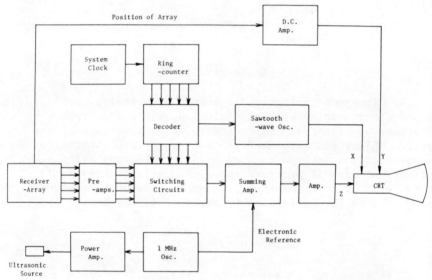

Fig.1   Block diagram of the experimental equipments

The receiver array is constituted of 30 elements of PZT transducers with each diameter of 5mm. The signals from each transducer are amplified by the preamplifiers and then fed to the switching circuits. The ring counter and decoder made of the IC components generate the pulses for electronic switching. The reference signal, which corresponds to the coherent background wave, is simulated superposing the electric signal from the oscillator and the object signals from the switching circuits. The resulted signals carrying

the hologram information are amplified appropriately and
these signals modulate the brightness of the CRT spot.   The
sawtooth wave sweeps the spot horizontally, resulting in a
one-dimensional and real-time hologram, where one electronic
scanning, that is the electronic switching of 30 transducers,
is done with the period of one cycle of the sawtooth wave.
The time required for one electronic scanning, that is one
sweep on the CRT scope, is about 15 msec.   The source is
allowed to move perpendicular to the receiver array.   The
position of the source is detected and transformed to the
dc voltage by a potentiometer to deflect the one-dimensional hologram.   Thus the position of the one-dimensional
hologram is synchronized with that of the source and the
one-dimensional hologram moves vertically on the CRT scope
as the source moves, resulting in the display of the two-
dimensional hologram.   The others of the experimental apparatus are much the same as those in the reference 4.

CONSTRUCTION OF ULTRASONIC HOLOGRAMS

Ultrasonic holograms are constructed in the experimental configuration of Fig.2, where $(0, s_i, z_i)$, (x_o, y_o, z_o)
and $(x, 0, 0)$ express the coordinates of a source, a point
on the object and an array element respectively. r_{io} and
r_o are the distances from the source to the object point and
from the object point to the array element respectively.   It
should be noted that s_i represents the y_i coordinate of the
moving source, that is s_i changes with respect to time.

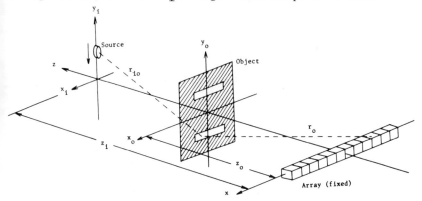

Fig.2   Experimental arrangement for constructing
        ultrasonic holograms with a one-dimensional
        receiver array

The object wave U_o at the array can be represented, neglecting the amplitude term and the constant phase factor,

$$U_o(x, 0, 0 ; t) = \exp[j\{k_1(r_{io} + r_o) - \omega_1 t\}] \qquad (1)$$

where k_1 and ω_1 are the wavenumber and the angular frequency of the employed ultrasonic wave respectively. r_{io} and r_o are expressed under the assumption of paraxial-ray approximation as follows,

$$r_{io} = z_i - z_o + \frac{x_o^2 + (s_i - y_o)^2}{2(z_i - z_o)} \qquad (2)$$

$$r_o = z_o + \frac{(x - x_o)^2 + y_o^2}{2z_o} \qquad . \qquad (3)$$

The sinusoidal time signal from the source forms an electronic reference U_r,

$$U_r(t) = \exp[-j\omega_1 t] \qquad . \qquad (4)$$

Assuming the square-law detection of the ultrasonic wave fields, the one-dimensional hologram $H(x ; s_i)$ can be expressed as follows,

$$H(x ; s_i) = (U_o + U_r)(U_o + U_r)^*$$
$$= |U_o|^2 + |U_r|^2 + U_o U_r^* + (U_o U_r^*)^* \qquad (5)$$

where the asterisk * denotes complex conjugate. The term $U_o U_r^*$ is expressed by Eqs.(1)-(4) as follows,

$$U_o U_r^* = \exp[j\frac{k_1}{2}\{ \frac{x_o^2 + (s_i - y_o)^2}{z_i - z_o}$$
$$+ \frac{(x - x_o)^2 + y_o^2}{z_o} \}] \qquad (6)$$

where the constant phase factor is neglected.

The holograms of Eq.(5) are displayed at the different vertical position on the CRT scope synchronizing to the source position s_i as mentioned in the preceding section. Thus a two-dimensional hologram displayed on the scope is recorded on a film by a camera. In this system the variable x and the parameter s_i of Eq.(5) are transformed to the coordinates (x_1, y_1) of the final hologram according to the following equations,

$$x = mx_1$$
$$s_i = ny_1 \tag{7}$$

where m and n are the reduction ratios with respect to x and y directions respectively. Substituting Eq.(7) into Eq.(5), the final hologram $H(x_1, y_1)$ is obtained as follows,

$$H(x_1, y_1) = |U_o|^2 + |U_r|^2$$
$$+ 2R_e \exp[j\frac{k_1}{2} \{ \frac{(mx_1 - x_0)^2 + y_0^2}{z_o}$$
$$+ \frac{(ny_1 - y_0)^2 + x_0^2}{z_i - z_o} \}] \tag{8}$$

where R_e denotes the real part.

RECONSTRUCTION OF IMAGES

Conditions of Image Reconstruction

In the reconstruction of the image, first it is required that the image is focused correctly. Secondly, the image is to be reconstructed as the similar figure of the original object. Here the planar object is considered for simplicity of discussion. If the object is a three-dimensional one, the discussion becomes complicated taking account of the longitudinal similarity besides the lateral similarity in the situation of different wavelengths of ultrasonic wave and laser light. Assuming an image point (x_2, y_2, z_2), the following three conditions must be satisfied to reconstruct correct images;
(1) Focusing condition; that is an image plane must be uniquely determined. To satisfy this condition the coefficients of the quadratic terms of x_1 and y_1 in the phase term of Eq.(8) must be equal.
(2) Distortion-free condition; that is the image should not be distorted in the image plane satisfying the condition (1). To satisfy this condition the coordinates x_2 and y_2 must be linear functions of x_0 and y_0 respectively.
(3) Similarity condition; that is the reconstructed image satisfying conditions (1) and (2) should be similar to the original object. To satisfy this condition the

magnifications of the reconstructed image with respect to x_2 and y_2 coordinates must be equal, that is, $\partial x_2/\partial x_0 = \partial y_2/\partial y_0$.

Anamorphic Hologram and Optical Reconstruction

The phase ψ_f recorded in the hologram of Eq.(8) is written as follows,

$$\psi_f(x_1, y_1) = \pm \frac{k_1}{2}\{A_x(x_1 - \frac{D_x}{A_x})^2 + A_y(y_1 - \frac{D_y}{A_y})^2\} \qquad (9)$$

where

$$A_x = \frac{m^2}{z_0} \quad , \qquad D_x = \frac{\dot{m}}{z_0} x_0$$

$$\qquad\qquad\qquad\qquad\qquad\qquad\qquad\qquad (10)$$

$$A_y = \frac{n^2}{z_i - z_0} \quad , \qquad D_y = \frac{n}{z_i - z_0} y_0 \quad .$$

The upper and lower signs correspond to the true and conjugate images respectively, and these signs are to be taken throughout. The hologram which records the phase of Eq.(9) is in general anamorphic one. Processing this kind of anamorphic hologram optically under the conditions of (1)-(3), a correct image can be reconstructed. It should be noted that in the experiment of the reference 4, these three conditions were simultaneously satisfied under the condition of Eq.(6) of the reference 4.

When the collimated coherent light is applied to the final hologram, the wavefronts reproduced by the terms $U_0U_r^*$ and $(U_0U_r^*)^*$ in Eq.(8) reconstruct images on the image planes. The spherical wave Φ which produces a point image at the image point (x_2, y_2, z_2) is expressed as $\Phi = \exp[(-jk_2/2z_2)\{(x_1 - x_2)^2 + (y_1 - y_2)^2\}]$. Comparing the wavefronts of Φ and the hologram of Eq.(8), the position of image plane z_2 and the coordinates (x_2, y_2) of the reconstructed image are obtained under the condition (1) as follows,

$$z_2 = \mp \frac{\mu}{A_x} = \mp \frac{\mu}{A_y} \qquad\qquad\qquad (11)$$

$$(x_2, y_2) = (\frac{D_x}{A_x} , \frac{D_y}{A_y}) \qquad\qquad\qquad (12)$$

where $\mu = k_2/k_1$ is the ratio of the wavenumbers of light wave k_2 and ultrasonic wave k_1. Under the condition (1), that is $A_x = A_y$, the following relation is obtained from Eq. (10),

$$\frac{m}{n} = [\frac{z_0}{z_i - z_0}]^{\frac{1}{2}} .$$ (13)

Note that this relation differs from Eq.(6) of the reference 4. Here the condition (2) is automatically satisfied. The magnifications of the reconstructed image is obtained from Eqs.(10) and (12) as follows,

$$\frac{\partial x_2}{\partial x_0} = \frac{1}{m}$$ (14)

$$\frac{\partial y_2}{\partial y_0} = \frac{1}{n} .$$ (15)

Since the magnifications of the image with respect to x_2 and y_2 coordinates are different, the condition (3) is not satisfied and the similarity of the image is lost. There-fore, a non-spherical lens is used to obtain a correct image. For example, the convex cylindrical lens with the focal length f_x is placed in front of the final hologram as shown in Fig.3. Then the reconstruction of the image can be re-presented by the following matrix equation[5],[6],

$$
\begin{bmatrix} x_2 \\ x_2' \\ y_2 \\ y_2' \end{bmatrix} = \begin{bmatrix} 1 & z_2 & 0 & 0 \\ 0 & 1 & 0 & 0 \\ 0 & 0 & 1 & z_2 \\ 0 & 0 & 0 & 1 \end{bmatrix} \left\{ \begin{bmatrix} 1 & 0 & 0 & 0 \\ \pm\frac{Ax}{\mu} & 1 & 0 & 0 \\ 0 & 0 & 1 & 0 \\ 0 & 0 & \pm\frac{Ay}{\mu} & 1 \end{bmatrix} \begin{bmatrix} 1 & d & 0 & 0 \\ 0 & 1 & 0 & 0 \\ 0 & 0 & 1 & d \\ 0 & 0 & 0 & 1 \end{bmatrix} \right.
$$

$$
\left. \cdot \begin{bmatrix} 1 & 0 & 0 & 0 \\ \frac{-1}{fx} & 1 & 0 & 0 \\ 0 & 0 & 1 & 0 \\ 0 & 0 & 0 & 1 \end{bmatrix} \begin{bmatrix} x_1 \\ 0 \\ y_1 \\ 0 \end{bmatrix} \mp\frac{1}{\mu} \begin{bmatrix} 0 \\ D_x \\ 0 \\ D_y \end{bmatrix} \right\} \quad (16)
$$

$$
= \begin{bmatrix} 1\pm\frac{Ax}{\mu}z_2 & z_2 & 0 & 0 \\ \pm\frac{Ax}{\mu} & 1 & 0 & 0 \\ 0 & 0 & 1\pm\frac{Ay}{\mu}z_2 & z_2 \\ 0 & 0 & \pm\frac{Ay}{\mu} & 1 \end{bmatrix} \begin{bmatrix} (1-\frac{d}{fx})x_1 \\ \frac{-1}{fx}\cdot x_1 \\ y_1 \\ 0 \end{bmatrix} \mp\frac{1}{\mu} \begin{bmatrix} z_2 D_x \\ D_x \\ z_2 D_y \\ D_y \end{bmatrix}
$$

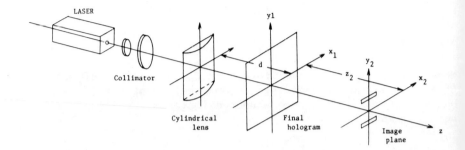

Fig.3   Arrangement for the optical processing

where d is the distance from the cylindrical lens to the hologram plane.

For simplicity of discussion only the conjugate image is considered. Since in the image plane the diffracted rays from the hologram come to one point on the image, the coordinates x_2 and y_2 of the image are determined independently of the hologram coordinates x_1 and y_1. Therefore the position z_2 of image plane is determined from Eq.(16) as follows,

$$z_2 = z_{x_2} = (\frac{A_x}{\mu} + \frac{1}{f_x - d})^{-1} \qquad (17)$$

$$z_2 = z_{y_2} = (\frac{A_y}{\mu})^{-1} \qquad (18)$$

where z_{x2} and z_{y2} are two solutions of z_2 with respect to x_2 and y_2 coordinates. When z_{x2} is not equal to z_{y2}, the condition (1) is not satisfied, resulting in the reconstruction of an astigmatic image. In Fig.4-(a) the appearance of this astigmatic image is explained schematically, where the image of a point object is reconstructed as lines at different image planes. To focus an image point at the same image plane, the cylindrical lens is used in the optical system of Fig.4-(b) and the distance d is adjusted to satisfy the condition $z_{x2} = z_{y2}$. The distance satisfying this condition is obtained from Eqs.(17) and (18) as follows,

$$d = f_x + \frac{\mu}{A_x - A_y} \qquad . \qquad (19)$$

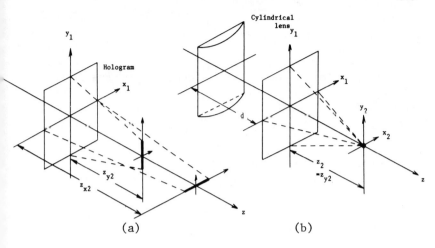

Fig.4   Astigmatic images of a point object (a)
and the correction of them by a cylindrical
lens (b).

Then the position z_2 of the image plane is uniquely determined as follows,

$$z_2 = \frac{\mu}{n^2} (z_i - z_o) \qquad . \qquad (20)$$

In the image plane determined by Eq.(20), the image coordinates x_2, y_2 are obtained from Eqs.(10) and (16) as follows,

$$x_2 = \frac{z_2}{\mu} D_x = \frac{m}{n^2} \frac{z_i - z_o}{z_o} x_o \qquad (21)$$

$$y_2 = \frac{z_2}{\mu} D_y = \frac{1}{n} y_o \qquad . \qquad (22)$$

To satisfy the condition (3), let the magnifications of the
image $\partial x_2/\partial x_o$ and $\partial y_2/\partial y_o$ be equal in Eqs.(21) and (22),
resulting in the following equation,

$$\frac{m}{n} = \frac{z_o}{z_i - z_o} \qquad . \qquad (23)$$

Therefore the hologram should be reduced according to the
relation of Eq.(23). Then the resulted magnifications are

$$\frac{\partial x_2}{\partial x_0} = \frac{\partial y_2}{\partial y_0} = \frac{1}{n} \qquad . \tag{24}$$

Thus a correct image can be reconstructed from the anamorphic hologram.

EXPERIMENTAL RESULTS

The holograms are constructed with $1MH_z$ ultrasonic wave using the one-dimensional receiver array system mentioned before. Figure 5-(a) shows one of the examples of the constructed hologram, where the hologram reduction ratios are chosen to satisfy Eq.(23). In this case the experimental values are z_i = 63 cm, z_0 = 41 cm and m/n = 41/22.

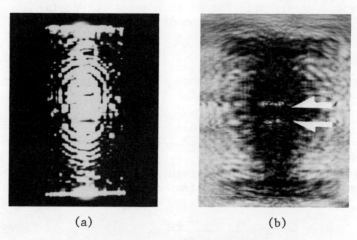

(a)   (b)

Fig. 5   Ultrasonic hologram (a) and
            optically reconstructed image (b).

Figure 5-(b) is the reconstructed image from the hologram of Fig.5-(a) using a cylindrical lens as shown in the optical system of Fig. 2. The distance d as shown in Fig.2 is adjusted to satisfy the following equation,

$$d = f_x + \frac{\mu}{m^2} \ \frac{z_0^2}{2z_0 - z_i} \tag{25}$$

where this equation is obtained from Eqs.(10), (19) and (23). The focal length of the cylindrical lens is 6 cm. The object is two-slits cut in a air-contained spongy screen as shown in Fig.6, which is placed in water.

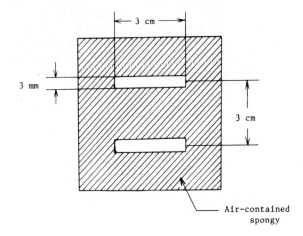

Fig. 6 Dimension of the object

In Fig.5-(b) the image of these slits appears reconstructed correctly. When the cylindrical lens is not used, the image plane cannot be determined and the image is out of focus. Moreover when Eq.(23) is not satisfied in the holo-gram reduction process, images can be hardly observed. These experimental results confirm the validity of the theoretical discussions conducted in the previous sections.

CONCLUSION

The investigation of this paper is the extension of the previous research by the authors.[4] In principle the holographic system with a one-dimensional receiver array discussed in this paper is same as that of the scanned-type system in the reference 4. However ultrasonic holograms are constructed with the electronic reference here and this causes the construction of anamorphic holograms. It is shown that correct images can be reconstructed by adjusting the hologram reduction ratios and using a cylindrical lens in the optical reconstruction process. Computer reconstruction of the correct image from those anamorphic holograms is another possibility, where the complicated processing can be done numerically.

The system where ultrasonic holograms are constructed
using the linear motion of the object instead of the source
is also possible as discussed in the reference 4. The anamorphic holograms are also constructed in this object movable system with the one-dimensional receiver array and an
electronic reference. In appendix only the analytical
results for the object movable system are shown refering to
the results obtained in the text.

ACKNOWLEDGEMENTS

The authors wish to thank Dr. M. Suzuki, Dept. of
Electronics, Hokkaido University for supporting the present
work. They also thank Mr. M. Fukushima, Hitachi Co. Ltd.
and Mr. H. Hayashizaki, Nippon Telegraph and Telephone
Public Co. for their helps in the experiment.

REFERENCES

(1) A. F. Metherell and S. Spinak, " Acoustical holography
     of nonexistent wavefronts detected at a single point
     in space ", Appl. Phys. Lett., Vol. 13, p.22(1968).

(2) B. P. Hildebrand and K. A. Haines, " Holography by
     scanning ", J. Opt. Soc. Am., Vol. 59, p.1(1969).

(3) N. H. Farhat, W. R. Guard and A. H. Farhat, " Spiral
     scanning in longwave holography ", Acoustical Holography,
     Vol. 4, Plenum Press, New York., p.267(1972).

(4) T. Iwasaki and Y. Aoki, " Ultrasonic holography in a
     source- and object-movable system ", Acoustical
     Holography, Vol. 4, Plenum Press, New York, p.653(1972).

(5) Y. Aoki, " Acoustical holograms and optical reconstruc-
     tion ", Acoustical Holography, Vol. 1, Plenum Press,
     New York, p.223(1969).

(6) Y. Aoki, " Higher-order images from grating-like
     acoustical holograms and their multiplexing and
     multicolor applications ", Acoustical Holography,
     Vol. 2, Plenum Press, New York, p.305(1970).

APPENDIX

For the object movable system,

$$A_x = \frac{m^2}{z_o} \quad , \quad D_x = \frac{m}{z_o} x_o$$

$$A_y = \frac{z_i}{z_o(z_i - z_o)} n^2 \quad , \quad D_y = \frac{z_i n}{z_o(z_i - z_o)} y_o \tag{A-1}$$

$$\frac{m}{n} = \left[\frac{z_i}{z_i - z_o} \right]^{\frac{1}{2}} \tag{A-2}$$

$$z_2 = \frac{\mu}{m^2} \frac{z_o z_i}{z_i - z_o} \tag{A-3}$$

$$\frac{m}{n} = \frac{z_i}{z_i - z_o} \tag{A-4}$$

$$d = f_x + \frac{\mu}{m^2} z_i \tag{A-5}$$

Equations (A-1), (A-2), (A-3), (A-4) and (A-5) correspond to Eqs. (10), (13), (20), (23) and (25), respectively. It should be noted that the distance d of Eq. (A-5) does not depend on the position z_o of the object. This means that the optical processing is more convenient in the object movable system than in the source movable system.

AN ACOUSTIC IMAGE SENSOR USING A TRANSMIT-RECEIVE ARRAY

M.G. Maginness, J.D. Plummer, J.D. Meindl

Electrical Engineering Department
Stanford University

Stanford, California 94305

ABSTRACT

A preliminary model of a novel acoustic image sensor which utilizes a multiplexed area array of transducers operating in a transmit-receive mode has been developed. The major components of the system include: 1) an acoustic lens for focusing the transmitted energy of an array element on the corresponding object element; 2) an area or two-dimensional array of acoustic transducers formed within a monolithic wafer of piezoelectric ceramic material; 3) an array of silicon monolithic integrated circuits located immediately adjacent to the piezoelectric array; each integrated circuit performs the transmit/receive multiplexing for several transducer array elements; 4) an analogue signal processor consisting of high power transmit circuitry as well as sensitive receiver circuitry including preamplifiers, swept gain amplifiers and function generators, filters, gain compressors and detectors; 5) a digital controller which permits a) any single array element or any row or column of array elements to operate in the A-scan mode, b) any row or column of array elements to operate in the B-scan mode or c) the full array of elements to operate in the C-scan mode; and 6) a display subsystem consisting of a cathode ray display and interface circuitry which provides an M-scan mode and interpolation in the C-scan mode. The image sensor operates at an acoustic frequency of 3.5 MHz and is intended for use in medical diagnostics.

1. INTRODUCTION

The Integrated Circuits Laboratory at Stanford University has recently begun work on an ultrasonic imaging system specifically directed to medical applications. In cooperation with the Stanford University Medical Center work has particular concentrated on systems suited to non-invasive cardiac study with emphasis on real time, high resolution performance.

In keeping with the projected application, importance has been attached to a compact, self-contained design and a facilit to operate and present information in modes that are already familiar from the use of present ultrasonic and X-ray instruments. The total requirements impose some very stringent demands[1] on such a device as a whole and in particular on the transducer employed. Since this item appears to be the most critical in the whole system a significant portion of our work to date has concentrated on applying techniques developed in our laboratory for integrated circuit fabrication to the proble involved in realizing two dimensional ultrasonic imaging arrays

At the same time, some preliminary complete system prototypes have been made to better evaluate the operational possibilities of the transducer concept and permit initial clinical evaluation.

In this paper we first discuss more specifically the syst design requirements. We then describe the initial form of our system and the transducer development details.

2. SYSTEM REQUIREMENTS

Since information is to be gathered by non-invasive means and with negligible patient risk both a low level of transmitted power and very high receiving sensitivity are mandated.

Vilkomerson's[2] calculations indicate that at a range of 10 cm. a 1 mm. target insonified at 3.5 MHz (wavelength = 0.43 mm.) will have an echo about 90dB down on the incident power, even when a large fraction of the reflected energy is gathered. If spreading of the transmitted power before reaching the target is also considered total loss may exceed 100db. At the other extreme, a fat-muscle interface in a superficial location may return the signal only 30db down. Thus not only is high limiting sensitivity needed but also wide dynamic range and the most efficient use of transmitted power. This excludes all but piezoelectric transducers from practical consideration.

Our second requirement for real time imaging of an extended volume prohibits mechanical transducer scanning and taken with the above sensitivity and dynamic range specificati

demands an electronically accessed piezoelectric array construction.

It was initially apparent that past attempts to meet this
goal[3],[4] suffered from serious limitations in ease of fabrication and performance largely as a result of trying to adapt
existing semiconductor devices of unsuitable physical form or
characteristics to the task. Attack on these disabilities
has been our main task.

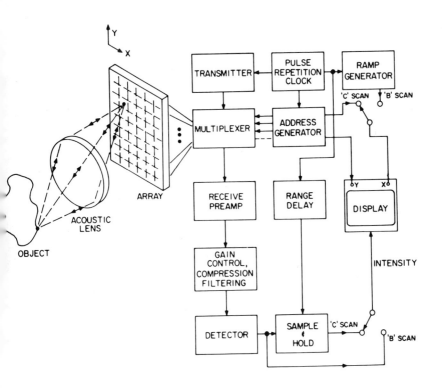

Figure 1. Block Diagram of System

3. SYSTEM CONFIGURATION

Early in our considerations of potential transducer array addressing means it became apparent that it would be possible to fabricate switching devices capable of sufficient large signal capacity to transfer transmit levels to individual transducers without compromising the low loss and noise require ments for receiving microvolt signals. Thus the system designs are based on a capability to both transmit and receive on any selected element of a full two dimensional array. A basic block diagram is shown in Figure 1.

Operating in C-scan mode, an element of the array is selected and used to transmit a burst of sound. After passage through the lens energy is focussed onto a small area of the target region. Any target scatter is collected by the lens and refocussed on the same element as originally transmitted. With this element now acting as a receiver a signal is obtained and passed through a chain of amplifying, gain control, compres sion and detection stages much as in conventional instruments. Time gating defines the precise surface within the lens depth of field and separates a value for one display spot. Repeti tion of this process for each element in turn at a sufficiently fast rate builds up a continually refreshed image of the target area.

One advantage of the transmit capability is immediately apparent. Since the signal received on any element has emmanated from the same position, the total delay from trans mission instant to receive instant for any point on the focal surface and corresponding array element is constant for all elements. Thus very accurate and narrow time gating is easily achieved with one fixed setting for all points.

With an $f/4.5$ lens such as we are using, acceptable focus is maintained over a 3-4 cm. depth of field. Thus if we use just one column or row of array elements in the general fashion described above but display the echo history for each element over the depth of field region on one axis against element location on the other a 'B-scan' presentation of a slice per pendicular to the array can be formed. Distinct adjacent slices can be viewed by using differing rows or columns of transducers without physical movement of the probe.

Selection of any single element can provide a simple 'A-scan' display and with further slow movement of an intensity modulated trace on the screen in a storage mode the dynamic motion of a given reflecting object can be presented in the time-motion (M) format commonly employed by cardio logists. It is important to note that the versatility of this

system in terms of its display modes is a direct result of
the ability to transmit and receive with the same array elements.

Our first implementation of such a system used a 5 x 5
element array formed from a 4 cm. square wafer of lead metaniobate and operated at a center frequency of 3.5 MHz with a 1 MHz
bandwidth. Retaining this frequency and total array size
progression has now been made to a 10 x 10 array with 3.5 mm.
square elements. As discussed later, development to a 32 x 32
element array within this size appears practical and at this
point the element spacing (1 mm.) will match the lens resolution. Use of the particular piezoelectric material chosen
enables very wide bandwidths to be achieved without resort to
elaborate backing and matching proceudres and it has advantages
compared to the more usual lead zirconate-titanates in minimizing electroacoustic coupling between elements.

With our design requirement of real time display, a
complication arises with large numbers of elements. If a
maximum range of 20 cm. in the body is needed and a lens of
approximately unity magnification is used, the total acoustic
path is about 80 cm. At 1500m/s this takes 530μS. Thus if
a strictly sequential operation is used, a 100 point image
requires at least 53mS to generate, giving about 20 frames
per second. This is marginally sufficient for real time
viewing and thus for larger arrays it becomes necessary to
simultaneously access a number of elements or to time share
by transmitting from one, then after a short delay for address
switching (about 5μS) from the next, and so on, returning to
the first element in time to receive the echo and then
readdressing the successive elements for the correspondingly
delayed echos from the same depth. Either of these techniques
can be used with the array multiplexing technique we have
devised.

A very distinctive capability of the transmit-receive
array is the effective doubling of resolution for a given
lens aperture compared to a receive only device viewing a
uniformly insonified target. This action of a focussed transmit-receive system is well known in synthetic aperture radar[5],
but has not been previously practical in high frequency real-
time ultrasonic systems. This effect is of considerable
practical significance in cardiac examination since access is
possible only through the relatively limited intercostal spaces.

One further result of the transmit-receive capability
is the improvement in uniformity of target insonification.
With physically distinct transmitters it is often difficult
to control the energy distribution and the resultant fluctuations have a marked effect on the final images[6]. Here each

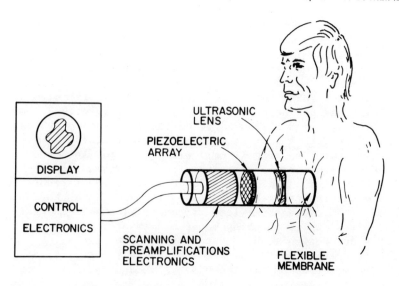

Figure 2.   Application Concept

target point is insonified by a specific transducer point and only when information is desired about it. This has the further effect of reducing artifacts from equirange as well as other scatterers since they are only weakly insonified even if very close to the object. Further, the control over energy distribution reduces the average power density for a given penetratio compared to a receive only array and "floodlight" insonificatio

     In physical embodiment and application our device takes the form of fig. 2. A tube filled with water contains the lens and array structure. The front end is sealed by a flexible membrane for direct application to the patient and by moving the lens-array assembly relative to the front various depth regions may be brought into focus. The hardware is shown in figs. 3 and 4.

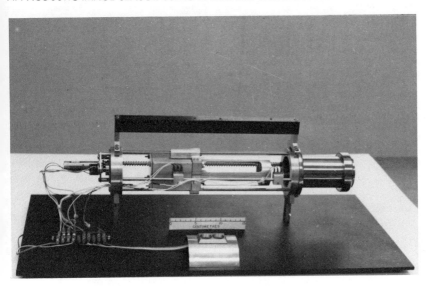

Figure 3.   Camera Interior Components

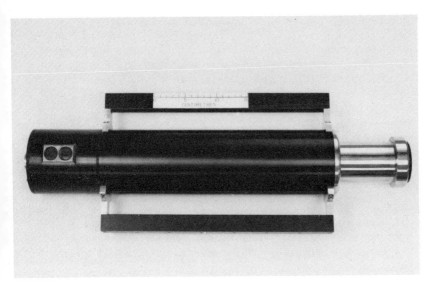

Figure 4.   Camera Assembly

4. A BI-DIRECTIONAL TRANSDUCER ARRAY

The general transducer and addressing configuration adopted for a two dimensional array is shown schematically in fig. 5. Rather stringent requirements are placed on the switching devices M. During receive a low 'on' resistance compared to the transducer element impedance is needed to preserve signal amplitude and minimize injected noise. With lead metaniobate the impedance level of a 3.5 MHz water loaded element 2 mm. square is approximately 2000 ohms. Thus the switch resistance cannot exceed about 200 ohms. If the transducer $T_{1,1}$ is addressed with all other switches to $T_{2,1}$, $T_{3,1}$, $T_{N,1}$, off, then (N-1) times the stray capacitance of a single switch loads the signal line S_1. At a 2000 ohm impedance level only 15 pF is tolerable for operation up to 5 MHz. Since we contemplate N values up to 32 this requirement is severe but can be eased by broad tuning with a small inductor L. With this inductor, a 1 MHz bandwidth can be maintained with up to 75 pF of total parasitic capacitance. Field effect transistors (FET's) chosen for this application have off resistances in the megohm region and this is more than adequate.

In the figure separate signal lines S_1, S_2, switchable to any member of a column by the orthogonal address lines are shown. This is suited to the large array C-scan requirement of multiple access. Where a single access port suffices a further row of switches P attached to lines S_1, S_2, S_3 etc. may be provided. In either case access to any one of the N^2 elements of a N x N array is available with 2N leads. Within the array N^2 connections are still required. These can be provided in mass produced form by adaption of the masking and conductor deposition methods of integrated circuit fabrication. This adaptation has formed one section of the work.

Consider now the requirements if we are to transmit from any selected element. To achieve the requisite power levels with a short pulse, a signal of 100 volts or more is needed. Peak current flow may reach 0.25-0.5 amperes in the 'on' switch depending on the element size while the off switches must withstand the voltage without breakdown, and retain sufficiently low leakage that nominally off transducers are not appreciably excited.

The best current commercial devices can meet the electrica requirement for a receive only system well but are in a physically inconvenient form for assembly in close proximity to the array as needed for minimal crosstalk. Such devices are limite to 15 to 20 volts when off although they can handle the 25-50 m

transmit current required at this voltage level. Thus the
second section of our transducer work has been concerned
with developing a new version of the field effect transistor
structure suited to the transmit specification without degredation of low signal performance. Devices fabricated to date
have shown 25-50 ohms on resistance, with 250 volts blocking
capacity and 0.5 amperes current limit, while exhibiting
less than 2 pF leakage capacitance.

The general physical form taken by an array using such
devices is shown in figure 6. The piezoelectric wafer is
scribed to about one half the total thickness for element
isolation. Mechanical integrity is provided by an intermediate support layer of insulating material, typically
oxidized silicon. Conductive paths established through this
layer at each element position join the transducer elements
electrically to the top surface of the layer on a cell by

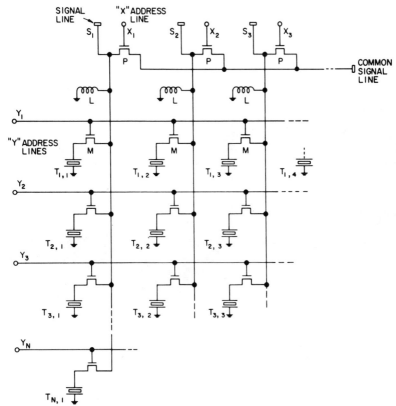

Figure 5. Schematic of Array Multiplexing

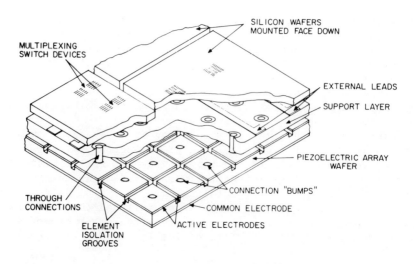

Figure 6.   Array Construction

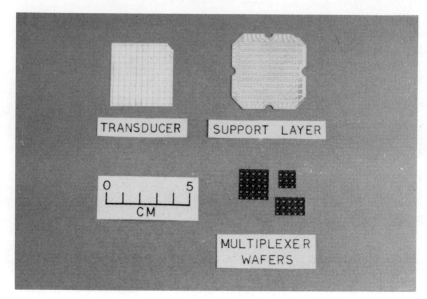

Figure 7.   Transducer Array Components

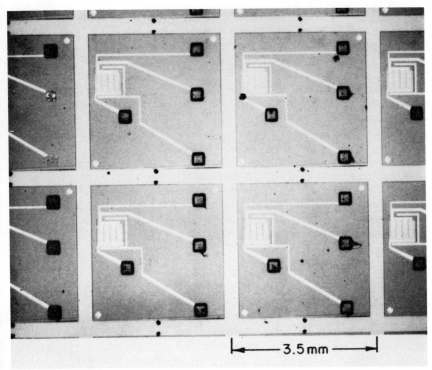

Figure 8.   Enlargement of Multiplexing Devices

cell basis. Further deposited conductors provide for the
address and signal transfer functions, with silicon chips
containing the multiplexing transistors finally bonded face
down on top to complete the assembly.

Examples of these items, in this case for a 10 x 10
array with 3.5 mm. square elements are shown in figure 7.
Figure 8 shows a processed silicon wafer with the multiplexer devices on 3.5 mm. centers. It is apparent that even
with the lower resistance and thus larger transistors required
for this element size only a small portion of the total area
is actively used. At present the lower size limit appears
to be about 1 mm. square, including both the transistor and
additional space required to make external connections.

5. INITIAL RESULTS

In the period before the specialized transistors became available an early system was constructed with commercially available devices.

Although limited in transmit capacity echos could be clearly perceived from the grosser details of the normal human heart. More recently a 50 element system using higher voltages has been made. The increase in transmitter power enables very strong echos to be obtained and early clinical trials are planned. At the present time a 100 element system is under construction. This system will be the first to use fully the modular array construction described above. It is this construction technique and the use of new FET devices for multiplexers that we feel will make possible the construction of much larger arrays in the near future.

6. CONCLUSIONS

The most conspicuously deficient item of ultrasonic imaging systems to date has been in transducer operation. In particular this is true for realtime inspection of moving structures. Our work on this aspect has exploited techniques especially adapted to the core problem of connection to a large number of elements at close spacing and further to achieving this without compromising basic transducer performance. Additionally it has been possible to obtain transmit as well as receive capability giving a number of unique system advantages. Currently we consider it feasible to construct an array of at least 32 x 32 elements on 1 mm. centers for any frequency in the lower megahertz range. Larger sizes are possible through the use of a number of such array modules. Further, the very small dimensions in which complex circuitry may be accomodated gives the longer term possibility of incorporating signal preamplifiers and a higher level of address decoding immediately adjacent to the module and thus maximizing signal to noise ratio and reducing still further the external leads.

We feel that this modular array device provides a basic building block suited to not only the type of system described above but to holographic arrangements as well. Such application is a long term interest.

ACKNOWLEDGMENTS

The work reported here was performed under grant number 5 P01 GM17940-03 from the National Institute of General Medical Sciences. Contributions to the project have been made by J. Beaudouin, R. Knapp, J. Long, H. Mussman, S. Norton, M. Pocha, T.J. Rodgers, J. Shott, K. Stafford, A. Susal M.D., T. Walker and S. Wetterling.

REFERENCES

1. P.S. Green, L.F. Schaefer and A. Macovski. "Considerations for diagnostic ultrasonic imaging." Proc. 4th Intl. Symp. on Acoust. Holography. Plenum, New York, 1972. pp. 97-112.

2. D. Vilkomerson. 'Analysis of various ultrasonic holographic imaging methods for medical diagnosis.' Ibid. pp. 401-430.

3. N. Takagi, T. Kawashima, T. Ogura and T. Yamada. 'Solid state acoustic image sensor.' Ibid. pp. 215-236.

4. S.O. Harrold. 'Solid state acoustic camera. Ultrasonics, 7 pp. 95-101, April 1969.

5. L.J. Cutrona et. al. 'A high resolution radar combat surveillance system.' IEEE Trans. on Military Electronics. MIL-5 pp. 127-131, April 1961.

6. M.G. Maginness, G.B. Cook and L.G. Higgens. 'A small scale model for seismic imaging systems. Proc. 4th Intl. Symp. on Acoustic Holography. Plenum, New York, 1972. pp. 195-213.

ADVANCES IN THE SOKOLOFF TUBE

John E. Jacobs
Donald A. Peterson

Northwestern University
Evanston, Illinois 60201

INTRODUCTION

The Sokoloff Tube, embracing modern electron tube technology, has emerged as one of the most sensitive yet simple high speed image converters for two dimensional display of ultrasound fields.[1] A number of studies have identified the limitations of the tube as regards sensitivity and image forming capabilities.[2] The principal limitations are 1)a limited effective aperture when achieving maximum detail resolution, and 2)a constraint on the maximum size of the imaging surface that permits maximum detail resolution. The work reported here describes two technological advances that have, for all practical purposes, eliminated these factors.

TECHNIQUE USED TO INCREASE THE EFFECTIVE APERTURE

The effective aperture of the modern version of the Sokoloff Tube (shown as Figure 1) when operated to achieve maximum detail and sensitivity is such that the sound incident to the converting surface will only be detected within several degrees of normal to the surface. Experimental work as well as theoretical considerations regarding the detail resolution of the Sokoloff Tube has indicated that the maximum resolution obtainable is equal to the thickness of the transducer plate at its first resonant point.[3,4,5] This is the thickness that corresponds to one-half wavelength of the incident sound in the conversion plate material, usually X-cut quartz. This

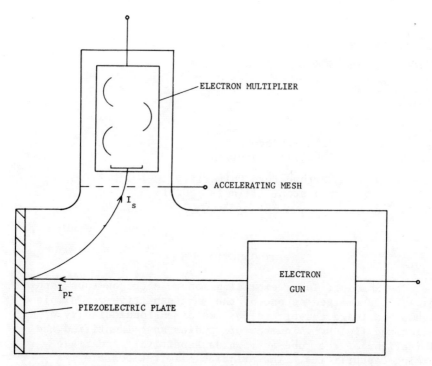

FIGURE 1   SCHEMATIC REPRESENTATION OF MODIFIED SOKOLOFF TUBE
The operation of the modified Sokoloff image tube is as
follows.   A high velocity electron scanning beam stabilizes
the scanned surface of the quartz imaging plate at a potential approximately three-volts positive relative to the second anode of the electron gun.   The secondary emission
electrons produced by the electron scanning beam are modulated by the piezo electrically induced potential resulting
from the impinging sound field.   These secondary electrons
form a virtual cathode whose ultrasonically modulated electrons are directed into the electron multiplier by the accelerating mesh.   The use of the electron multiplier eliminates,
for all practical purposes, stray signal inputs to the system.
At the input frequency of 3.58 Mhz used for these studies,
an input level of 10^{-7} watts/cm^2 produces a discernable output
from the tube.

resolution is obtained when the conversion plate is operated
at the frequency that corresponds to the first resonance.
Even though the tube is operated at odd harmonics of this
frequency, the resolution is not improved. Experimentally
obtained resolvable detail studies have verified that sound
intensity patterns separated by a distance equal to the
thickness of the plate are discernable. Use is made of this
operational characteristic in the technique to be described
for increasing the spatial bandwidth of the converter surface. An added improvement in performance is that sensitivity is increased due to the fact that the matching units
couple the incident sound to the converting surface without
a major reflection loss.

The majority of the two-dimensional imaging systems utilizing the Sokoloff Tube make use of a liquid, normally water,
to couple the ultrasound energy to the image converting surface. With the availability of the sensitivity to acoustic
impedance changes inherent in the color display,[6] it becomes
mandatory that a liquid coupling media be used, since any
media other than a free-flowing liquid produces an impedance
discontinuity between the object being viewed and the image
converter that exceeds the impedance discontinuity sensitivity of the system.

When water is used as the coupling media, the characteristic acoustic impedance of the material needed to provide
an impedance match between water and the quartz plate is
equal to 4.7×10^{6} rayls. An examination of characteristic
acoustic impedances of readily available materials indicates
the desired value can be obtained by the use of metallic pow-
ders embedded in a suitable casting plastic. It was determined experimentally that the use of a fine copper powder in
a casting plastic would produce a composite material having
the desired acoustic impedance yet exhibit low absorption of
the incident sound energy.

The characteristic acoustic impedances as well as the absorption of the fabricated composite materials were evaluated
experimentally by pulse measurements using bulk samples and
by means of layers deposited on the quartz plate. These layers were of a variable thickness such that several odd multiple quarter wavelength resonant points were obtained in the
layer. Figure 2, which illustrates the image obtainable with
such layers, verifies that in those regions where the acoustic
impedance is being matched by the proper thickness, the output

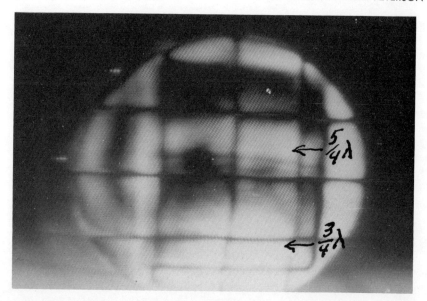

FIGURE 2   IMAGE DISPLAYED FROM SIGNAL PLATE HAVING AN APPLIED
VARIABLE THICKNESS MATCHING LAYER OF CORRECT ACOUSTIC IMPED-
ANCE
The layer thickness at the points of maximum displayed intensity corresponds to 3/4 and 5/4 wavelengths of the incident sound. Frequency is 3.58 Mhz. Input surface of tube
used for these studies has diameter of 5cm.

of the system increases. It may be seen that not only is
the impedance matched properly, but also the sensitivity of
the system is enhanced.

    The use of an extended layer of the proper thickness on
the surface of the Sokoloff Tube was ruled out due to the
fact that it was determined experimentally that such a layer
in intimate contact with the quartz conversion plate had the
net effect of mechanically coupling adjacent elements with
a resultant loss of resolution. This coupling of adjacent
elements by the applied matching layer prevents the electromechanical decoupling that is responsible for achieving maximum resolution capability of the Sokoloff Tube. This be-

havior dictates that not only must the layer be of such a
material that the impedance is properly matched, but also it
must be of a discrete element form to avoid mechanical coupling of adjacent imaging elements of the conversion surface
by the layer.

CONSIDERATIONS FOR THE DESIGN OF THE DISCRETE ELEMENT ARRAY

Design of the extended array of coupling elements is
based on the fact that the directivity of a linear array of
finite elements is the product of the directional pattern of
each separate element and the directional pattern of the linear point array. Since a hemisphere is "half-omnidirectional"
the linear array will have an extended geometry and have the
"half-omnidirectional" response.

A rectangular array of elements has the response of a
line array but in two perpendicular directions. The radiation pattern of a plane surface may be considered as due to
two perpendicular line sources. The net result of these considerations is that the rectangular array of hemispheres acts
as one large hemisphere having a half-omnidirectional response.

The requirement for discrete elements of a particular
shape places rather stringent requirements upon the fluid
flow as well as the long-term acoustic impedance characteristics of the composite material chosen for the impedance
coupling elements.

It was determined that the above mentioned material could
be formed into half-spheres by appropriate silk-screening
techniques, provided that attention was paid to the time the
material was mixed and screened. By careful control of the
time between the addition of the copper powder to the casting plastic, it is possible to screen the material and achieve
the correct ratio of surface tension to setting time such
that reasonably good spherical shapes are formed (Figure 3).

The design of the discrete matching elements is such that
they are hemispheres whose diameter is ideally equal to one-
half wavelength of the sound in the material. These elements
are deposited on centers ideally one-half wavelength apart
on the converting surface. They serve, therefore, to assure
coupling onto the imaging surface of sound arriving at angles

FIGURE 3   EQUIPMENT USED FOR FABRICATING SHAPED MATCHING
ELEMENTS
The discrete matching elements are formed by squeezing the
copper powder-casting plastic mixture through the metal
screen shown. After the material has been squeezed through
the plate onto the piezoelectric converting surface, the
screen is removed, following which the material flows to
assume a half-sphere shape prior to hardening.

other than 2°-3° to normal. By virtue of the intimate coupling between the sound transmitted through the coupling elements and the conversion plate, as well as the fact that the
coupling elements are discrete, no loss of resolution is encountered with this system. A typical input surface is
shown as Figure 4.

CONSTRUCTION OF LARGE AREA TUBES

The maximum resolvable detail of the imaging system is related to the wavelength of sound utilized for visualization. At the frequencies used here, i.e. 3.6Mhz, this is in the order of twenty-five elements per inch. To obtain a high quality image of rather large objects, a need exists for a large area conversion surface. The Sokoloff Tube's operation is such that the scanned side of the input surface is stabilized by secondary electrons produced by the scanning beam when operating in the high velocity mode. This mode of operation is particularly advantageous in the Sokoloff Tube, since it permits the extraction of the signal through a relatively small port placed in the side of the tube. With the incorporation of the electron multiplier and the associated Faraday screen in the Sokoloff Tube, the design and operation of large area tubes became practical.

When the tube is operating, the scanned surface of the piezoelectric plate is completely in equilibrium with a cloud of secondary electrons reasonably uniform throughout the volume of the tube. This permits signals to be extracted from the conversion surface even though the surface is visually shielded by the input surface supporting structure from the signal extraction port. As nearly as can be ascertained, no signal loss as far as the reading of the piezoelectric plate by the scanning beam occurs as a function of the effective diameter of the conversion surface, even though the signal extraction port through the electron multiplier remains the same. This is obviously a distinct advantage in the operation of the tubes.

Investigations previously reported[4] regarding the design of a supporting structure for the piezoelectric converting surface indicated that if supporting structures were such as to limit the unsupported portion of the converting surface to 1 cm,^{2} then quartz plates whose half-wave resonant frequency was in the order of 20Mhz could be readily utilized. Use has been made of this design in the fabrication of the supporting structure to permit tubes whose diameter of the converting surface is in the order of 14cm. The current supporting structure is such that the previous limitation to the maximum sensitive area of the tubes no longer is of any consequence. Figure 5 shows the 22.5cm diameter imaging tube with a 12.5cm diameter quartz conversion plate. Plans

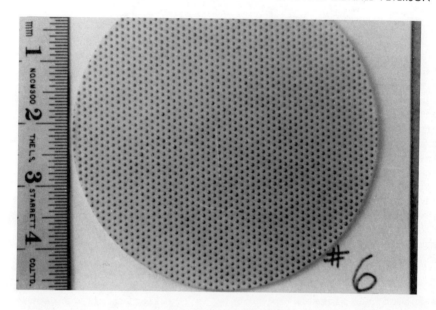

FIGURE 4   REPRESENTATIVE QUARTZ PLATE WITH DISCRETE MATCHING
ELEMENTS APPLIED
Completed plate illustrating the size and spacing of the
matching elements used for these studies. All studies reported here are at 3.58Mhz.

are to increase the input sensitive area to the design limit
for the tube, i.e. 22.5cm.

PERFORMANCE OF THE IMPROVED TUBES

A detailed examination of the sound-propagation characteristics of the half-sphere as deposited on the quartz surface discloses that the sound incident along the axis normal
to the surface of the half-sphere is directed so as to enter
the quartz surface and produce longitudinal waves which are

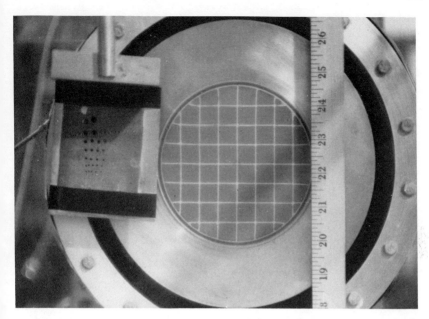

FIGURE 5   THE IMAGING TEST PLATE USED FOR RESOLUTION STUDIES
This test plate, shown in front of the newly developed 12.5cm
diameter pick-up tube, has a series of resolution holes
varying from 3mm to 0.3mm diameter.

responsible for the imaging process. All tests were run
using the resolution test plate shown as Figure 5. Figures
6 and 7 illustrate imaging characteristics of the tubes
having the "discrete element" matching surfaces.

The addition of the spheres, although they are not perfect in geometry as evidenced from Figures 8 and 9, does not
seriously degrade the overall performance of the tubes. In
fact, the measured sensitivity of the tubes has been increased
by some two to three times that which would be anticipated
from the acoustic matching theory. When the tubes are used
in A and B comparison studies with uncoated tubes to image
finite objects, it is readily apparent that additional information in terms of fine detail in the objects is in fact
being presented by the tubes.

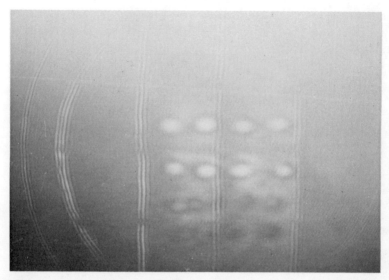

FIGURE 6   RESOLVABLE DETAIL OF IMAGE TUBE WITH SHAPED ACOUS-
TIC MATCHING ELEMENTS
Ultrasound incident along an axis normal to the quartz con-
version surface.   Smallest diameter resolved - 1.2mm.

FIGURE 7   IMAGE OBTAINED WITH ULTRASOUND INCIDENT ALONG AN
AXIS TWENTY DEGREES FROM THE NORMAL TO THE QUARTZ CONVERSION
SURFACE

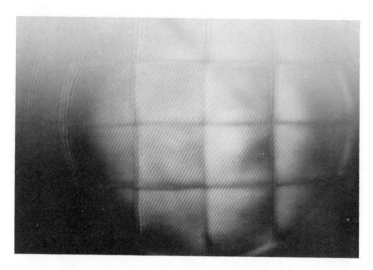

FIGURE 8   VARIATION OF OUTPUT SIGNAL FROM CONVERSION SURFACE
HAVING DISCRETE ACOUSTIC MATCHING ELEMENTS UNDER UNIFORM
ILLUMINATION
Sound incident along axis normal to the quartz.

FIGURE 9   VARIATION OF OUTPUT SIGNAL FROM CONVERSION SURFACE
HAVING DISCRETE ACOUSTIC MATCHING ELEMENTS UNDER UNIFORM
ILLUMINATION
Sound incident along axis ten degrees from normal to the
quartz.

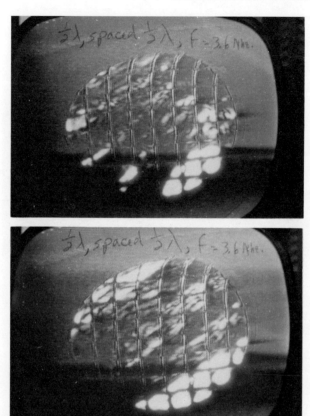

FIGURE 10   IMAGES OBTAINED ON ADULT
Images of a) adult hand b) adult forearm. Frequency 3.58Mhz.
Incident power to object 1×10^{-3} watts/cm^2.

SUMMARY

    Now that Sokoloff Tubes of large area and increased ef-
fective apertures are available, this technique of imaging
should permit detailed clinical studies in those areas of
interest where high speed two dimensional visualization
studies are indicated.

    This research supported in part by PHS Grant GM08522-11

BIBLIOGRAPHY

1. Smyth, C.N., Paynton, F.Y., and Sayers, U.F., (1963)
   "The Ultrasound Image Camera", Proc. IEE 110(1), pp.16-28.

2. Jacobs, J.E., (1968a) "Present Status of Ultrasound Image
   Converter Systems", Trans. N.Y. Acad. Sci. Series II 30
   (3), pp.444-456.

3. Hartwig, K., (1959) "Uber das Raeumliche Aufloesungsuer
   mogen von Barium-Titanat-Und Quarzplatten bei der
   Ultraschallabbildung", Acustica 9, pp.109-117.

4. Jacobs, J.E. (1967) "Performance of the Ultrasound Microscope", Materials Evaluation 25(3), pp.41-46.

5. Ahmed, M., Whitman, R.L., and Korpel, A., "Exact Response
   of Isotropic and Piezoelectric Acoustic Imaging Faceplates", to be published in IEE Trans. Sonics and Ultrasonics, 1973.

6. Jacobs, J.E., Reimann, K., and Buss, L. (1968b) "Use of
   Colour Display Techniques to Enhance Sensitivity of the
   Ultrasound Camera", Materials Evaluation 26(8), pp.155-
   159.

LINEAR RECEIVING ARRAY FOR ACOUSTIC IMAGING AND HOLOGRAPHY

G.C.Knollman, J.L.Weaver, and J.J.Hartog

Lockheed Research Laboratory

Palo Alto, California 94304

ABSTRACT

A hydroacoustic receiver intended for use in ultrasonic imaging and holography is described. An experimental model contains a 16-element rectilinear array of PZT transducers. Voltage output of each element at 2.1 MHz is amplified separately to a level at which acoustic sensitivity is limited by amplifier noise rather than by switching transients of the multiplex circuit. Amplifier outputs are sampled sequentially by the multiplexer at a rate of 8000 elements/second to produce one line of video information. Acoustic images in nondestructive test and medical applications are obtained by translating or rotating either the array or the test specimen. Utilization of such array for ultrasonic visualization in nondestructive testing is discussed, and methods for receiver miniaturization by use of hybrid-circuit techniques are proposed.

INTRODUCTION

For acoustic imaging and holography, the linear array of transducers has proven to be a utilitarian compromise between a mechanically scanned single transducer and an electronically scanned rectangular array. By translating or rotating a linear piezoelectric array across the image field of an acoustic lens (or the plane of an acoustic hologram), one can readily view ultrasonic images (or holograms). In nondestructive-test (NDT) applications, translation and/or rotation of the test specimen is frequently convenient, with the linear array remaining fixed in the image plane of a lens. Essentially real-time acoustic imaging or holography can be realized by means of a stationary linear transducer array used in conjunction with an auxiliary scanning system such as one or more rapidly revolving mirrors or a pair of counter-rotating acoustic wedges.

Applications of linear ultrasonic receiving arrays previously have been demonstrated with an underwater array for hydroacoustic imaging[1,2] and in acoustical holography with two types of mechanical scan.[3] A scanned linear array has widespread appeal for real-time acoustic imaging and holography systems owing to its versatility, simplicity, and economy.

The linear receiving array described here was designed for ultrasonic nondestructive viewing of imperfections in moving samples, with the array held stationary. It is a developmental model intended to serve as a prototype for a larger, more sophisticated array, either rectilinear or curvilinear. Following a description of the device and its performance, we will briefly cite proposed modifications for future development.

DESCRIPTION OF THE ARRAY

The prototype acoustic receiving array devised for NDT purposes contains a rectilinear arrangement of 16 transducers connected to individual high-gain amplifiers. The transducer array was fabricated by cutting deep slots into a narrow strip of PZT-5 lead-zirconate-titanate ceramic which had a thickness-mode resonance frequency at 2.1 MHz. Transducer elements are 1.6 x 1.6 mm and have a common ground electrode on one face. The array was cemented into the faceplate of the receiver housing and is protected by a Mylar sheath. Figure 1 is a photograph of the faceplate assembly and the welded aluminum housing which contains the electronic components. Although not restricted in application to submerged objects, the receiver is intended predominantly for hydroacoustic use.

Electrical leads (aluminum) were bonded ultrasonically to each transducer element. These bonds are protected by means of a plastic cover. Each transducer in the array is connected to a separate integrated-circuit amplifier, RCA

Figure 1. Exterior view of the 16-element
hydroacoustic receiver

Type CA 3022. Amplifier gains are individually adjusted to
approximately 42 dB by means of a miniature variable resistor in the negative feedback path of each amplifier. Gain
is essentially constant over the transducer bandwidth. The
amplifiers are also designed for automatic gain control,
which is a desirable feature in view of the wide dynamic
range of acoustic input. However, the requirement of equal
gain for all amplifiers at all signal levels precluded the
use of automatic gain control (AGC) in the preamplifiers.

    The 2.1-MHz output of the amplifiers is sampled sequen-
tially by a digital sampling circuit and delivered to an external video amplifier. A block diagram of the electronic
system is shown in Fig. 2. Field-effect transistors are
used as switching devices to sample the radio-frequency (RF)
output of the amplifier. Sampling is accomplished in two
steps by dividing the 16 amplifier channels into two groups
of eight. Transistors are switched simultaneously in
pairs, one from each group. The group outputs are then
sampled alternately by two additional switching transistors.
These transistors had to be selected for low drain-source
saturation voltage with 4.7 volts supplied to the gate.
Diode switching of the amplifier outputs was tried initially
since a diode is available in the CA 3022. However, cross-

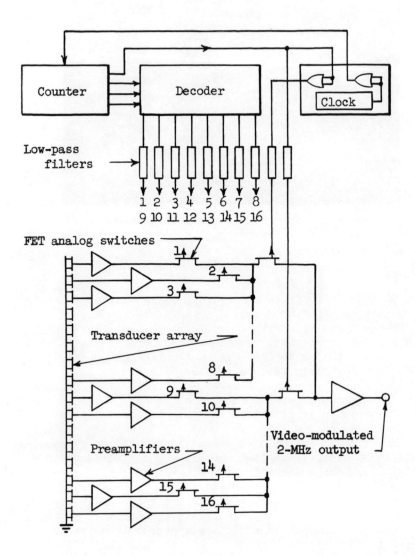

Figure 2.   Electronic scanning system

talk between   channels and transients resulting from shifts
in DC levels discouraged further use of that simple scheme.

Switching pulses (4.7 volts) are derived from a clock,
a 4-bit counter, and a decoder, as indicated in Fig. 2.
Two NOR gates generate the clock pulse at an 8-kHz rate.
The pulses occur in sequence at the outputs of an octal decoder and sequentially turn on pairs of switching transducers. The fourth bit of the counter and its complement,
which is obtained from a NOR gate, are used to select alternate groups of 8 from the 16 amplifier channels. With minor
modifications the same circuit can be used to sample 20
channels.

The 2.1-MHz sampled output of the receiver is amplified
an additional 48 dB in an external video amplifier. The RF
amplitude above an adjustable threshold level is then demodulated and used to intensity-modulate a cathode-ray oscilloscope. The video amplifier can be gated asynchronously
with the receiver clock to permit capacitive coupling to
the Z-axis of the oscilloscope. To view a sample that is
being scanned slowly, we employ a storage oscilloscope such
as the Hewlett-Packard 1201A or Tektronix 241 for display
purposes.

PERFORMANCE DATA

Although the performance of the aforementioned linear
receiving array in nondestructive testing has not yet been
fully evaluated, several parameters of significance can be
quoted. We will discuss the receiver directivity pattern,
sensitivity, and uniformity of response across the array.

The directivity pattern, or graph of relative sensitivity versus angle of sound incidence, is displayed in Fig. 3
for a single element of the array. Directivity measurements were performed with the array and an insonifier submerged in water. The orientation of the receiving array is
indicated at the center of the figure. In a plane perpendicular to the axis of the array, the total acceptance angle
designating the "6-dB down" points is approximately 26°,
as one might expect for an element width of 1.6 mm. However, in the normal plane through the axis of the array the
acceptance angle is only 14°. This angle suggests and effective element length of 2.9 mm and results from acoustical

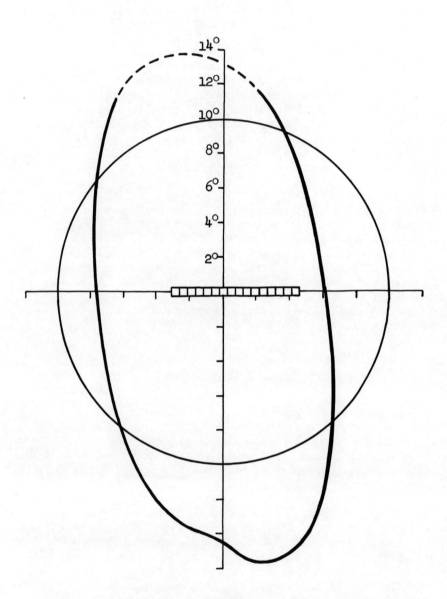

Figure 3. Directivity pattern of one transducer element

coupling to neighboring elements. A conical acceptance
angle of 26° could be obtained by completely dicing and
acoustically isolating the various elements. The array
could then be used with an F/2 acoustic lens for high-
resolution imaging.

Sensitivity of the receiver was measured by use of a
calibrated acoustic source. An approximate calibration
at 2.1 MHz was obtained by measuring the streaming force in
water for a given power input.[4] Sensitivity determined in
this manner was expressed as noise-equivalent power in the
full bandwidth of the receiver. For a bandwidth of 300 kHz,
the noise-equivalent power is 4.7×10^{-11} W/cm^2, which is
typical of this type of device. Somewhat lower noise-
equivalent power $(2.6 \times 10^{-11}$ W/cm$^2)$ is realized at the
output of the video amplifier by virtue of its smaller bandwidth.

Noise at the receiver output is chiefly white noise
originating in the amplifier. By providing sufficient
voltage gain prior to sampling, the white-noise level is
raised above the noise level of switching transients from
the sampling circuit. Amplitude of these transients is
considerably reduced by filtering out unnecessary high-
frequency components from the switching pulses. However,
at the high sample rates required in a large receiving array,
smoothing of the sampling pulses could cause electronic
crosstalk between elements.

Sensitivity of each receiving channel was adjusted to
fall within 5% of a constant level across the array. The
gain controls were separately adjusted while the faceplate
of the receiver was partially immersed in water and exposed
to acoustic power from a distant source. Figure 4 shows
the RF output of the receiver when exposed to this sound
field, with the receiver partially obscured by a straight
edge. Undulations in the envelope are due to Fresnel diffraction around the straight edge, which was placed 7 mm
from the receiver. Theoretical maxima and minima of the
diffraction pattern occur at distances $\sqrt{(2n+1)\lambda b}$ and $\sqrt{2n\lambda b}$,
respectively, from the projection of the edge on the receiver,[5]
where b is the distance between the straight edge and the
receiver and n is zero or an integer. Thus, the first maximum should occur at about 2.2 mm, or one and a half array
elements from the edge projection. In reality, we determine the position of this maximum at about two element widths.

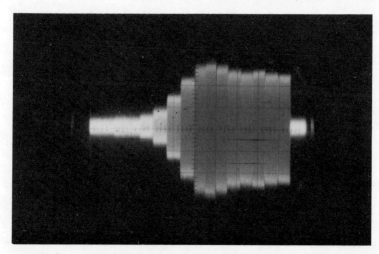

Figure 4.   Output of the 16-element array at 2.1 MHz

Although resolving capability of the array has not been measured, maximum resolution appears to be about 2.5 mm rather than one element width (1.6 mm) as one would expect. This broadening is attributed to acoustic coupling between adjacent transducer elements. We note that the minimum object diameter resolvable by the piezoelectric transducer (set by the thickness of the material at its first resonant point) is about 1.1 mm at 2.1 MHz. Thus, if required, smaller element size could be employed for better resolution at the given frequency.

PROPOSED IMPROVEMENTS

To meet the exacting requirements of acoustic non-destructive testing and medical applications, the design of our acoustic receiving array can be improved in several ways. The need for a wider acceptance angle and better resolution has already been mentioned and can be met by means of acoustically independent transducer elements. Uniformity of response across the array is another crucial requirement in the aforementioned applications, since the contrast in some test samples and in most biological tissues is inherently small. Response variations less than a few percent may be required of a linear array for it to be competitive with a mechanically scanned single element.

Although the nonuniformity in response of the 16-element array could be brought within an acceptable limit by careful adjustment of the individual gain controls, the method is tedious for large arrays. If a rough degree of uniformity is first obtained by selection of fixed resistors and amplifiers in each channel, final adjustment of uniformity might be achieved by other means. For example, an optical mask (transparency) could be prepared by photographing a cathode-ray tube display of a uniform acoustic background. A full-scale negative transparency of the nonuniform display, overlaid in registration with the oscilloscope display, would compensate for most of the residual nonuniformity. Similarly, a metal acoustic mask could be prepared by photoresist techniques to attenuate the acoustic input to high-gain channels. These methods are equally applicable to large rectangular arrays, in which the uniformity problem is more formidable. The entire problem can be avoided, however, by resorting to phase-only holography.[6]

To our knowledge, several available and applicable solid-state devices have not been exploited in the development of acoustic receivers. For example, integrated-circuit amplifiers similar to those used in the present 16-channel receiver are available as silicon chips which are comparable in size to a 2-MHz transducer element. A hybrid circuit containing one dual amplifier per channel can be mounted in close proximity to a linear array and can provide at least 80 dB of voltage gain. Thus, an acoustic signal just above the noise level can be amplified to a rectifiable level and stored on a capacitor prior to sampling.

The stored video signal can then be read out of the array by means of a charge-coupled electronic scanner.[7] This device transfers the stored charge from each channel sequentially to the video output terminal without introducing switching noise. Charge-coupled devices currently used in optical scanners are typically much smaller than a linear transducer array.

A proposed linear receiving array incorporating these new devices is shown in Fig. 5. The amplifier chips and charge-coupled scanner are assembled on a ceramic substrate which is mounted above the transducer array. Rectangular arrays can be built by stacking these linear arrays and sampling their outputs in sequence.

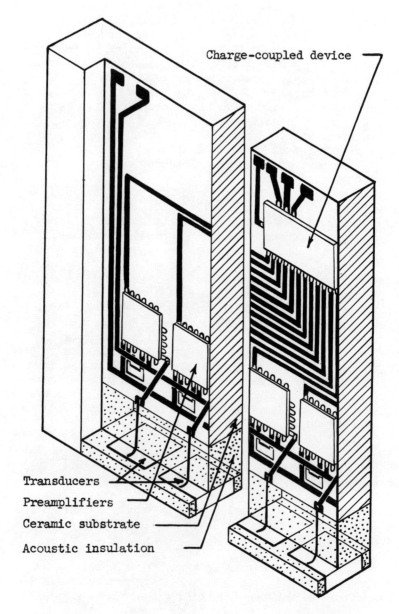

Charge-coupled device

Transducers
Preamplifiers
Ceramic substrate
Acoustic insulation

Figure 5. Linear receiving array incorporating
preamplifier chips and a charge-coupled scanner

Storage and averaging of acoustic signals in each channel during the long intervals between sampling pulses does not enhance the signal-to-noise ratio since noise output is also restored. But significant improvement could be realized by reducing the amplifier bandwidth of each channel. Active filters for this purpose are commensurate in size with the devices indicated in Fig. 5. With these innovations the linear acoustic receiving array could become a closer acoustical analog to the solid-state optical scanner.

CONCLUSIONS

A developmental linear acoustic receiving array has been described, and its performance has been partially evaluated. It has met basic requirements of sensitivity and uniformity of response, and it should prove useful as a small-aperture scanner for experimental nondestructive test systems. Necessary improvements have been mentioned and several innovations for future large-scale designs have been suggested.

REFERENCES

1. G. C. Knollman, A. E. Brown, J. L. Weaver, and J. L. S. Bellin, "Experimental Hydroacoustic Imaging System," J. Appl. Phys. 42, 2168 (1971).

2. G. C. Knollman and A. E. Brown, "Hydroacoustic Image Transducer," Rev. Sci. Instr. 42, 1202 (1971).

3. G. C. Knollman and J. L. Weaver, "Acoustical Holography with a Scanned Linear Array," J. Appl. Phys. 43, 3906 (1972).

4. T. F. Heuter and R. H. Bolt, Sonics (John Wiley, New York, 1955), Chap. 2.

5. G. S. Monk, Light, Principles and Experiments (McGraw-Hill, New York, 1937), p. 171.

6. See, for example, A. F. Metherell, "The Relative Importance of Phase and Amplitude in Acoustical Holography," Acoustical Holography (Plenum Press, New York, 1969), Vol. 1, Chap. 14.

7. M. F. Tompssett, G. F. Amalio, and G. E. Smith, "Charge-Coupled Eight-Bit Shift Register," Appl. Phys. Letts. 17, 111 (1970).

A PROGRESS REPORT ON THE SOKOLOV TUBE UTILIZING A METAL FIBER FACEPLATE

Dr. Robert C. Addison

American Optical Corp.
Research Laboratory
P.O. Box 2267
Framingham, Mass. 01701

INTRODUCTION

The conventional Sokolov tube utilizes a piezoelectric plate to convert the incident acoustic signal into an electric signal and also to serve as the interface between the vacuum chamber and the water. This paper describes experiments which use a metal fiber faceplate to serve as the interface between the vacuum chamber and the water. There are a number of advantages that result from this arrangement. The mechanical strength requirement that is imposed on the piezoelectric plate in a conventional Sokolov tube is eliminated and the size of the acousto-electric transducer can be increased in order to provide a large field of view. Furthermore, the type of acousto-electric transducer can be chosen to provide optimum image quality. The metal fiber faceplate also makes it possible to obtain a hard vacuum seal between the faceplate and the tube envelope and offers the possibility of altering the secondary electron emission characteristics of the vacuum surface of the faceplate in order to obtain a higher transconductance resulting in improved sensitivity.

Tubes utilizing metal fiber faceplates have been suggested previously [1] and the construc- of a tube of this type was reported in 1965. [2] We are not aware of any work involving sealed off tubes with metal fiber faceplates as large as those we are currently using.

In the conventional Sokolov tube, an electron beam scans a rectangular raster over the back of the piezoelectric plate so that the entire plate is covered in about 1/30 of a second. When the electron beam hits a given point on the plate, secondary electrons are generated from that re- gion. A detailed description of the secondary electron emission process is given by Kazan and Knoll. [3] For our purposes, it is sufficient to say that under equilibrium conditions the number of secondary electrons leaving the plate will be equal to the number of primary electrons which are incident. However, if the potential of the piezo- electric plate is varied rapidly compared to the time required to reach equilibrium, then the number of secondary electrons will increase and decrease as the piezoelectrically generated po- tential increases and decreases. If the poten- tial variations are not too large, the relation- ship between the variations in the secondary electron current and the piezoelectric potential is linear. In this way the piezoelectric poten- tial variations of a small region of the piezo- electric plate are transferred to the beam of secondary electrons which are in turn collected by an electrode or an electron multiplier which am- plify the variations. The signal is processed, and, finally, the image is displayed on a tele- vision monitor.

The resolution displayed by this image de- pends on the acoustic frequency, the thickness of the piezoelectric plate and the acoustic velocity in the piezoelectric plate relative to that of water. The plate is usually operated at its fun- damental resonance frequency and so is one half wavelength thick. Although the plate could be any odd multiple of one half wavelength thick, the resolution has been found to decrease as the plate is made thicker. [4] For maximum resolution, the acoustic velocity in the plate should be about the same as that of water so that the maximum angle of incidence for acoustic plane waves will be as close to 90° as possible.

In the past the chief parameter of the piezo-electric plate that was of concern was its mechanical strength. The high mechanical strength required to support an atmosphere of pressure usually meant that the piezoelectric plate had to be made of quartz. The other desirable parameters of an ideal acousto-electric transducer could not be chosen. Even with a quartz plate that is one half wavelength thick at 5 MHz, the maximum allowable diameter is only about 50 mm unless a supporting grid is used. Since the acoustic velocity in quartz is almost four times that in water, a relatively small acceptance angle results, and the resolution is about 3 times the acoustic wavelength measured in water. To put this in perspective if one assumes a 4:3 aspect ratio raster with a diagonal of 50 mm and an image element size of 1 mm (3λ at 5 MHz), the total number of image elements per frame is 1.2×10^3. A typical television frame on the other hand contains a little over 10^5 image elements. Thus, the images presented by a 50 mm diameter Sokolov tube are relatively coarse.

One of the chief reasons for the interest in the Sokolov tube is its potentially high sensitivity along with its real time capability. The limiting source of noise is the shot noise associated with the scanning electron beam. If the operating frequency is 3 MHz and a total of 10^4 image elements are scanned in 1/30 sec by a 10 ua electron beam, one can obtain a sensitivity of 10^{-9} w/cm^2. This assumes that the minimum observable signal to noise ratio is ten. It should be kept in mind that if the size of the raster scanned by the electron beam, the frame time, etc. remain fixed, the sensitivity varies as $1/f^4$ where f is the acoustic frequency. The $1/f^4$ dependence

comes about because the intensity necessary to
generate a given piezoelectric voltage increases
as f^2 and the total number of resolution elements,
and hence the bandwidth, also increases as f^2.

METAL FIBER FACEPLATES

The metal fiber faceplates that are inter-
posed between the vacuum chamber and the electroacoustic transducer are made of wires with a
cladding of glass that has been fused together to
form a vacuum tight faceplate. There are several
combinations of glass and metal that we have successfully used to make the faceplates. Currently,
we believe the preferable combination is a low lead
glass such as Kimball glass No. G-12 and Sylvania
#4 alloy wire. The finished faceplate has 0.05 mm
wires on approximately 0.15 mm centers. A microscope picture of a small area of one of the plates
is shown in Fig. 1. The resulting mosaic is not

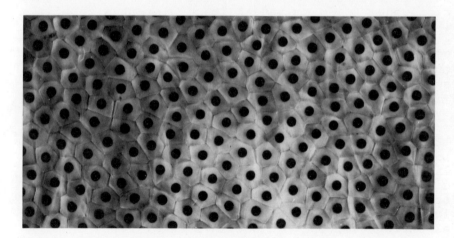

Figure 1.   Microscope picture of a small region of
            a metal fiber faceplate. The wires are
            0.05 mm in diameter and the average
            center to center spacing is about 0.15
            mm.

completely regular but since the spacing of the
wires is much smaller than the anticipated resolution, it does not affect the acoustic image
quality. In principle, these plates can be made
relatively large, we estimate that a diameter of
300 mm is feasible. In practice, we are limited
in the size we can achieve in a single faceplate
by the equipment that is used to fuse the mosaic
of individual glass clad wires. The maximum size
single plate that can currently be made with this
equipment has a cross section of about 75 mm x
150 mm. Therefore, in order to make a 140 mm
diameter faceplate, two 75 mm x 150 mm plates are
butt sealed together using a solder glass. The
resulting seam is about 1/2 mm wide and does not
seem to cause any significant artifacts in the
acoustic image. Three different sizes of faceplates are shown in Fig. 2. Their diameters are
50 mm, 75 mm, and 140 mm.

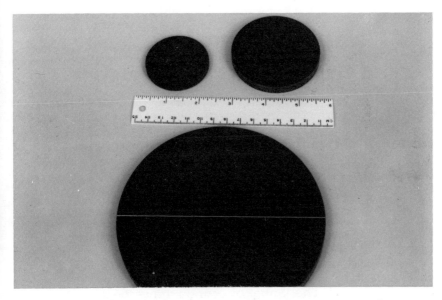

Figure 2.   Assorted metal fiber faceplates with
            diameters of 50 mm, 75 mm and 140 mm.
            Note the solder glass seam in the
            140 mm diameter plate.

Figure 3. Sokolov tube with a 140 mm diameter metal fiber faceplate as it
appears immediately after removal
from vacuum system.

In order to seal these faceplates onto a
glass envelope and have the resulting joint be
vacuum tight, it was necessary to overcome a
number of technological problems. We believe
that the tubes can now be assembled consistently
and reliably. A tube with a 140 mm faceplate
prior to its having plugs for the electron gun and
electron multiplier and a mounting ring for the
piezoelectric plate attached is shown in Fig. 3.

PERFORMANCE OF IMAGE TUBE

Before constructing the 140 mm diameter tube,
several smaller tubes were made. The imaging
properties of these smaller tubes have been investigated more extensively than those of the large
tubes and most of the measurements concerning the
performance of the Sokolov tubes have been made on
these smaller tubes. In order to use the tube as
an acoustic image detector, a piezoelectric plate
is placed over the metal fiber faceplate and held
in place with an O-ring around the periphery. We
have used both PZT-4 and quartz plates as transducers with comparable results. All of the work

reported here was done at an acoustic frequency
of 5 MHz and the images were formed by shadow
casting, with the object no more than a few centimeters from the faceplate, as opposed to using a
lens system and refocusing an image of the object
onto the faceplate.

We were especially interested in comparing
the resolution that was obtained using the tubes
containing a metal fiber faceplate to that obtained
with a tube having only a piezoelectric faceplate.
Two different comparisons were made. First a test
object was selected which contained detail that
was close to the resolution limit of the tube
when only a piezoelectric plate was used. When
the acoustic images of the test object obtained
with each of the tubes were compared, they were
found to be of similar quality as far as the resolution was concerned. The second test was to
compare the measured angular response of a tube
using a metal fiber faceplate and a PZT-4 plate to
the theoretically predicted angular response for
a PZT-4 plate alone.[5] The results of this
measurement are shown in Fig. 4.

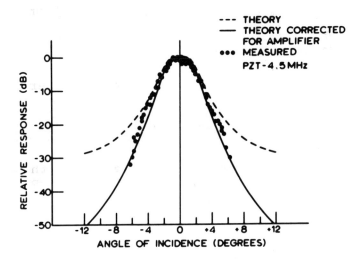

Figure 4.   Comparison of theoretical and
            measured acoustic angular response
            of a PZT-4 faceplate.

It was necessary to correct the theoretical curve
to allow for the limited bandwidth of the amplifier used in the measurement. Note that the
measured points are in reasonably good agreement
with the predicted curve. There is a slight
broadening in the measured points which is attributed to the fact that the incident acoustic wave
was slightly spherical rather than being a perfect
plane wave. These results demonstrate that the
metal fiber faceplate has not degraded the resolution of the tube beyond that which it would possess
if a piezoelectric plate were used alone.

Among those problems that occur in the operation of a Sokolov tube is the phenomenon of charge
spreading. The potentials associated with the
piezoelectrically induced charge pattern on the
back of the faceplate tend to spread out and
cause the edges of objects to appear smeared.
Since the extent of the spreading is dependent on
the strength of the signal, the phenomenon can
lead to distortion if the incident insonification
is not uniform in intensity. In electronic image
tubes this problem is solved by placing a fine
mesh screen close to the target so that it estab-
lishes an equipotential surface close to the
target upon which the electric field lines can
terminate. (6) This mesh also tends to eliminate
the redistribution of secondaries that are not
collected by the electron multiplier. A mesh of
this type has been utilized in one of the tubes
with a metal fiber faceplate and has functioned
as expected.

When a Sokolov tube is operated, it is immediately noticed that when an object is introduced
into the sound field, the background pattern
appearing at the edges of the object is altered.
Furthermore, when the object is moved from one
region of the sound field to another, the background pattern changes in an unpredictible and
annoying fashion. If the insonifying transducer
is moved off axis, high contrast fringes are
observed. It was concluded that this is the
result of a signal which is the average of the
signals generated throughout the entire faceplate

being capacitively coupled into the output. The
fringes and other artifacts result from a mixing
of this average "leakage" signal with the image
containing signal on the secondary electron beam.
This seems to be the same effect that was reported
by Freitag (7) in tubes that did not use an
electron multiplier. This unwanted signal has
been suppressed by using suitable shielding techniques to separate the capacitively coupled signal
from the signal that is impressed on the electron
beam. In Fig. 5, the background pattern for an
off axis transducer is shown before and after the
shielding was used.

Using the Sokolov tube with the modifications
described above, acoustic images of a number of
different test objects have been obtained. Since
the tube is basically meant to be a real time
imaging device, these images do not demonstrate
its true potential but do serve to given an indication of its resolution. In Figs. 6 and 7, are the
acoustic and optical images of thin perforated
aluminum test objects. The small holes in the
object shown in Fig. 7 are about 1 1/2 mm in
diameter.

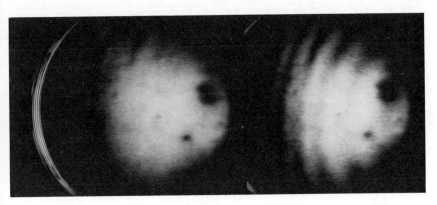

Figure 5.   Comparison of the insonification
            pattern of an off axis transducer
            with (left picture) and without
            (right picture) shielding to elim-
            inate the unwanted reference signal.

Figure 6.   Acoustic image of a perforated
            aluminum plate using 5 MHz
            acoustic waves.

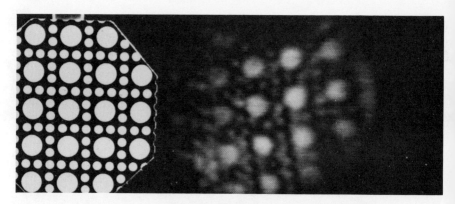

Figure 7.   Acoustic image of a perforated
            aluminum plate using 5 MHz
            acoustic waves. The small holes
            are about 1.5 mm in diameter.

CONCLUSIONS

The conclusion to be drawn from the foregoing discussion is that the metal fiber faceplate does seem to offer a workable solution to the small field of view problem of the Sokolov tube. It does not degrade the resolution and makes it possible to have a mechanically strong vacuum boundary while still being free to select an acoustic to electric transducer that will provide optimum image quality.

ACKNOWLEDGEMENTS

This work would not have been possible without the cooperation of W. Siegmund and M. Smith, of the Fiber Optics Division, who were responsible for manufacturing the metal fiber faceplate material. Gustav Abel was responsible for the glass blowing associated with the tubes and for the solder glass seals. The finishing and optical polishing of the faceplates was skillfully done by Colin Yates. Ken Lawton helped out with some of the faceplate testing procedures. The technical assistance for the project was provided by Richard King. The advice of Tom Polanyi and Melvin Bliven concerning many of the details of electron tube fabrication is gratefully acknowledged. Finally, the author would like to thank Elias Snitzer for many helpful discussions and for his critical reading of the manuscript.

REFERENCES

1. C. N. Smyth, Ultrasonic Imaging Proceedings of a Symposium, April 1965, AD-621372.
   C. N. Smyth, Ultrasonics 4, 15 (1966).

2. D. S. Grasyuk, D. K. Oshchepkov, L.D. Rozenberg, and Yu B. Semennikov, Soviet Physics Acoustics 11, 376 (1966).

REFERENCES

3. Kazan and Knoll, <u>Electronic Image Storage,</u>
    Academic Press (1968), p. 16.

4. C. N. Smyth, F. Y. Poynton and J. F. Sayers,
    Proc. IEE, <u>110</u>, p. 16 (1963).

5. M. Ahmed, R. L. Whitmen and A. Korpel,
    presented at the 85th Meeting of the Acoustical
    Society of America, Boston, Mass., April 1973.

6. Kazan and Knoll, <u>Electronic Image Storage,</u>
    Academic Press (1968), p. 29.

7. W. Freitag, Jenaer Jahrbuch <u>1</u>, p. 228 (1958).

REAL TIME ACOUSTICAL IMAGING WITH A 256 x 256 MATRIX
OF ELECTROSTATIC TRANSDUCERS

P. ALAIS

Institut de Mécanique Théorique et Appliquée

Université de PARIS VI   FRANCE

INTRODUCTION

Electrostatic transducers for ultrasonic applications
are a natural extension of the classical condenser microphone.
For this latter device, numerous studies have shown the possibility of obtaining an internal polarization due to the
electret effect in a dielectric sheet. In the ultrasonic version also, electrostatic transducers may be polarized alternatively by an external circuit or by using the electret effect. Their interest in the acoustical imaging field is that
it is very simple to build complex arrays with printed circuits. The difficulty lies rather in the correct interrogation of the array. A first solution is to build a square array of N^2 independant transducers and to interrogate them
with parallel circuits or with a suitable electronic commutation [1] . A much simpler solution has been proposed
which requires 2 N circuits only. In this case, although it
is still possible to use electret transducers according to
NIGAM, the operation with external polarization is preferable. This technique was presented in Vol.4 of this series
[2] . New refinements and progress in experimental results
are shown here.

THE ELEMENTARY TRANSDUCER

A unique electrostatic transducer is built simply by
setting a thin sheet of dielectric material between two elec-

trodes (fig.1). The front one is moved by the incident acoustic radiation and may be just a metalization of a plastic
sheet which constitutes a protecting transparent window. The
back plate is preferably of high acoustical impedance and
may be a massive piece of metal. As this sandwich is obtained only by mechanical contact without any bonding material,
there remains at both the dielectric-metal interfaces a mean
air gap due to surface irregularities. For this reason, the
compressibility of the sandwich is related not only to the
compression of the dielectric material but also and essentially to the variations of the air gap. Associated resto-
ring forces derive from the compression of the air and from
the flexion energy of the dielectric sheet and eventually
of the front metalized window. Obviously, the acoustical impedance offered by the sandwich to the incident acoustic radiation must be as low as possible to obtain the greatest displacement of the front electrode which attains at the maximum
twice the displacement amplitude of the incident wave.

As for ordinary condenser microphones, the polarization
of this device may be obtained from an external circuit or by
using an electret film. The first case offers the possibility of electrically commuting the electromechanical conversion
of the transducer, which is fundamental for the realization
of the 2 N array, and we shall consider only this situation
in the following.

It must be checked that this electrostatic transducer is
essentially a displacement transducer, the maximum sensitivity of which is given as the ratio of voltage to displacement,
$V/\xi = E_o$, where E_o is the electric field induced in the
air gap. If V_o is the polarization voltage,

(1) $E_o = V_o/e'$ $e' = \dfrac{1}{\varepsilon_r} e_d + 2 e_g$

where e' is the equivalent air thickness of the condenser,
e_d and ε_r the thickness and the relative permittivity of
the dielectric material and e_g the mean air gap at each interface. So, we may write the maximum acoustical sensitivity
of the electrostatic transducer as the ratio of voltage to
the pressure of the incident wave :

(2) $\dfrac{V}{P} = \dfrac{2 E_o}{\rho c \omega}$

and the actual response :

(3) $\dfrac{V}{P} = \dfrac{2 E_o}{\rho c \omega} \; g \, (\omega) \, h \, (\omega)$

where $g \, (\omega)$ is a mechanical corrective factor related to the mechanical impedance of the sandwich and $h \, (\omega)$ an electrical corrective factor related to the electrical impedance Z of the external circuit which may be simply written,

(4) $h \, (\omega) = \dfrac{j \, Z \, C \, \omega}{1 + j \, Z \, C \omega}$

where C is the capacity of the transducer.

We have studied experimentally in water an elementary transducer built as one of the N^2 transducers of our array i.e with two printed circuits. The back one has a rigid glass epoxy support and the front one may be made with a thin mylar support. Both of them consist only of a single strip of 0.7mm width. They are disposed orthogonally so that the area of the transducer obtained at the overlapping region is approximately 0.5 mm². It appeared that, for our special problem of building a 2 N array, the best dielectric material was not a polyimide film but rather a condenser paper. The first reason for our choice is that the paper has a lower resistivity than plastic dielectrics and prevents the formation of homo-charges due to electrical discharges in the air gap. It has been shown previously that these charges which constitute one

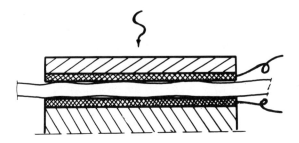

Fig.1 The electrostatic transducer

type of electret charges are responsible for a residual transduction in the absence of external polarization which may be
a serious problem in the 2 N array operation. To show evidence of this phenomenon, an acoustic pulse is emitted towards
the transducer at a repetition rate of 16 Hz and a synchronized square polarization pulse is applied. Fig.2 shows the
response of the transducer with the polarization pulse applied
at the right time (a), just before (b) and just after (c). It
may be checked that the residual response is nearly the same
when the polarization pulse has been applied one hundred microseconds before (a) or sixty milliseconds before (c), i.e.
the period of the recurrent process. In fact, the commutation
factor defined as the quotient of the normal transduction to
the residual transduction may be much higher than in this
example and may attain typically values of a few hundred with
paper, if the working field E_o is far lower than the field
for which electrical discharges appear in the air gap. Another
quality of the paper is its high relative permittivity which
allows use of higher thicknesses and permits to obtain a better uniformity between the different transducers of the array.
This advantage appears specially when the mean air gap is reduced. Instead of the 10 μ air gap obtained by mechanical contact at the ambient pressure, we have reduced this gap to
1.5 μ by lowering the air pressure in the transducer to a
few percent of the atmospheric pressure. Several experiments
done with papers of 5, 10 and 15 μ thickness have shown that
in these conditions the air gap is almost independent of the
paper thickness and the relative dielectric permittivity of
the paper is of the order of 5. We have retained as optimal
values for the 2 N array a paper thickness of 15 μ which gua-
rantees a good uniformity of the N^2 transducers and a working
field E_o of 60 V/μ , which leads to a polarization voltage
of 360 V, the equivalent air thickness being in this case approximately 6 μ . In these experiments, as in the 2 N array,
the factor h (ω) was low, of the order of 10^{-2}, due essentially to the capacitance of the connections in front of the
capacitance of the transducer which is about 1.5 to 2.5 pF
for a surface of 0.5 mm^2. Several measurements of the actual
sensitivity V/p in water and of h (ω) have shown that, with
adequate front printed circuits, it is possible to attain
good values of the mechanical factor g (ω), i.e typically
0.8 at 1 MHz and 0.5 at 2 MHz. In our case the compressibility of the residual air is of no importance and the elastici-
ty of the sandwich must probably be due to the flexion of the
dielectric sheet essentially. Under these conditions, the

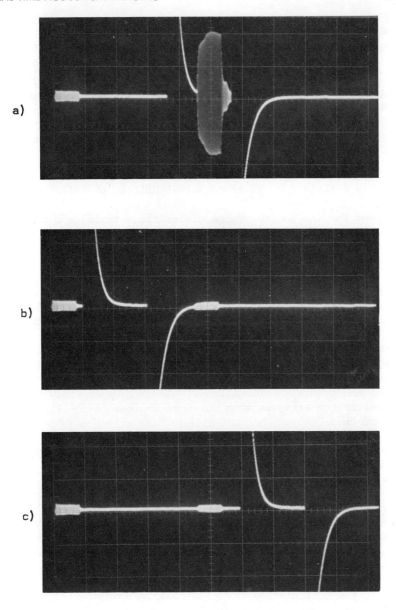

Fig. 2 Response of an elementary transducer to an acoustical pulse with a synchronized polarization pulse applied at the right time (a) just before (b) and just after (c).

measured sensitivity without electrical reduction

(5) $V/p = \dfrac{2\,E_o}{\rho c \omega}\; g\,(\omega)$,

attains − 120 dB re. 1 V/ μB at 1 MHz and − 130 dB re. 1 V/μB
at 2 MHz. These values correspond to the measurements reported by NIGAM [1] , taking into account the fact that in our
experiment the mean air gap is thinner and the electric field
higher.

THE ELECTROSTATIC ARRAY : PRINCIPLE OF OPERATION

Except for the choice of the dielectric material and of
the front printed circuit, the technology of the 256 x 256
array is the same as in [2], i.e. it is simply obtained
by setting a dielectric sheet between two printed circuits
which have 256 parallel metalized strips of 0.7 mm width with
a 1 mm center-to-center spacing and are disposed orthogonally (fig.3). The 256 columns of the back circuit are connected
to 256 parallel receiving circuits. When one line of the front
circuit is polarized, the others being grounded, only the
transducers corresponding to the polarized line are working
and may furnish local information to the receiving circuits.
The 256 lines may be polarized sequentially at frequencies up
to 5 kHz and complete information of the ultrasonic field at
the level of the array is thus obtained in less than fifty
milliseconds.

In our first version, we succeeded in getting good holograms at low frequencies of the order of C.5 MHz but were
unable to use higher frequencies because of electronic problems originating from the receiving circuits. The sensitivity was also limited by significant acoustical noise and by the
multiplexing technique. New refinements have been added to the
electronic circuitry to surmount these difficulties : a modification of the coupling circuitry between amplifiers and synchronous detectors (fig.4) have permitted us to operate at frequencies up to 3.5 MHz without any electronic difficulty. Furthermore, the multiplexing of the 256 values stored by the memory condensers which must be performed at a high frequency
of 2.5 MHz to preserve real time operation is effected by integrated multiplexers which permit a dynamic range of 35 dB.
But, a major improvement was to use a pulsed acoustic illumination synchronized with the recurrent process of reception

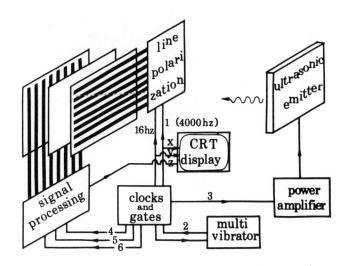

Fig. 3   Simplified block diagram

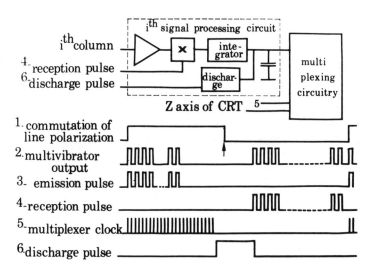

Fig. 4   Signal processing circuitry and
          timing waveforms

Fig.5   Experimental arrangement

line by line. Fig. 4 shows the timing of operation. The same
pulsed multivibrator delivers through appropriate gates the
60 μs pulse that drives the ultrasonic transducer, and the
100 μs reception pulse which is used for the synchronous detection. This timing requires the illuminating transducer to
be set at discrete distances from the array so that the acoustic signal arrives during the reception time. The complete
synchronisation of both emission and reception with the 250 μs
period of the repetition process avoids any jitter and prevents
acoustical modes from developing in the experimentation tank.
In this way, the major part of the acoustical noise which was
previously observed is eliminated. The multiplexing of the
values stored by the condensers used as analog memories is
initiated at the same time as the emission pulse. It is achie-

ved in nearly 100 μs at a frequency of 2.5 MHz. At the end
of this operation which furnishes the video signal for a line,
the condensers are discharged and the commutation of line polarization is effected. Fifty microseconds are allowed for
this process before the reception pulse so that during this
last operation only receiving circuits are operating. The
gain of the receiving circuits is such that the saturation
of multiplexers is obtained for an incident acoustical flux
of 500 mW/cm^2 et 2 MHz. The sensitivity of the array is in
fact still limited by the multiplexers which impose, because
of the high multiplexing frequency, a noise of the order of
200 mV. Under this conditions, the dynamic range of reception
goes from 200 μW/cm^2 to 500 mW/cm^2 at 2 MHz and from 20 μW/cm^2
to 50 mW/cm^2 at 1 MHz. The device operates successfully from
0.5 to 3.5 MHz with a decreasing sensitivity as the frequency increases. Fig.5 shows the new experimental arrangement.

EXPERIMENTAL RESULTS

To obtain a correct acoustic illumination and straight
fringes we have built a large transducer of the dimensions
of the array, i.e 25 cm x 25 cm. It is a mosaic of 25
5 x 5 cm ceramic transducers which are air-backed and coated
on the front side with an epoxy thickness of 0.7 mm. This
mounting allows correct operation at the nominal resonant frequency of the ceramics, 2 MHz, but also a sufficient emission
power for the array down to 0.5 MHz. The mosaic transducer is
operated through adapting transformers by a power amplifier
that is able to deliver near 1 kW of electrical power, so that
it is possible to attain the saturation acoustical intensity
for frequencies between 0.5 and 2 MHz.

Fig. 6 shows the fringes obtained at 2 MHz with the transducer oriented obliquely. Fig. 7 is the hologram obtained under the same conditions of 45 mm high letters made of expanded polystyrene absolutely opaque to ultrasonic radiation
disposed at 10 cms from the array. The optical reconstruction
shown in fig. 8 has been obtained through He Ne laser radiation with a hologram optically reduced to 6 x 6 mms.

We have also developed a very simple imaging technique for
thin objects disposed near the array illuminated under the same conditions. It consists simply in replacing the video holo-
graphic signal A cos φ by $|A \cos \varphi|$ which may be obtained
through a multiplication by a 10 MHz signal and a detection.

Fig. 6   Fringes observed at 2 MHz.

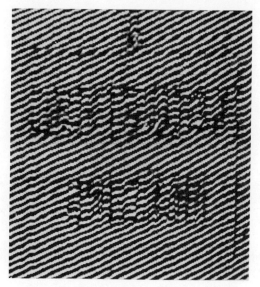

Fig. 7   Acoustical Hologram of 45 mm high letters
            made of expanded polystyrene disposed at
            10 cms from the array (2 MHz).

Fig. 8   Optical reconstruction with 6328 A converging
          laser beam of the hologram of fig.7 after it
          was demagnified to 6 x 6 mms.

Fig. 9   Acoustic image of the same 45 mm high letters
          disposed near the array (2 MHz).

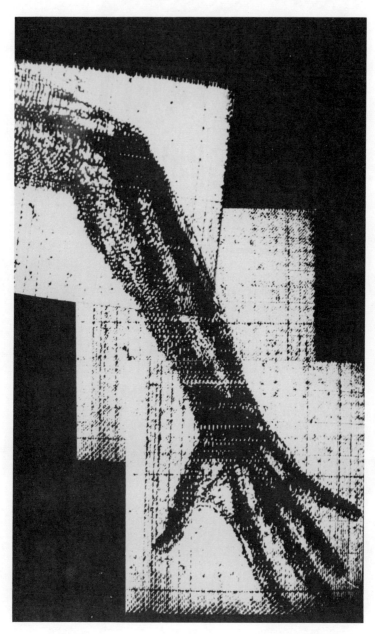

Fig. 10 Acoustic image of a human arm disposed near the array (2 MHz).

Logarithmic amplification permits one to increase the dynamic range of the relative transparency restitution. Through this very crude signal treatment we have obtained better images of relatively thin objects such as the preceding letters, than with optical holographic reconstructions (Fig. 9). Fig. 10 shows a composite of three different pictures obtained in the same way of a human arm. Yet another advantage of this technique is that it preserves real time operation and permits one to observe from a moving object, details difficult to extract form a still photograph, this being a very well known property of human brain signal treatment.

We have begun a study of the possibilities of imaging through lenses with the same technique. Different liquid lenses have been constructed with silicon oil. Fig. 11 shows an image of 20 mm high letters carved in a piece of plastic, obtained with a 20 cm diameter lens. But these first results were rather disappointing because of difficulties encountered in obtaining a good acoustical illumination. Fig. 11 was obtained from a point source with a 30 cm diameter lens used to condense radiation on the other lens.

Fig. 11   Acoustic image of 20 mm high letters carved in a piece of plastic obtained with a 20 cm diameter liquid lens (1.6 MHz).

CONCLUSIONS

The 2 N array technique with electrostatic transducers has proved to be efficient in real time imaging. In the experimental device which has been described here, both the sensitivity and the dynamic range of reception are limited by the multiplexers. We intend to develop a slow technique which could be used in parallel with this fast technique. In this way, it might be possible to avoid the noise introduced by the multiplexers and to increase the sensitivity and the dynamic range by at least 20 dB. Under these conditions the dynamic range of reception would be of the order of 55 dB and would cover acoustic intensities from $2 \mu W/cm^2$ to $500 \ mW/cm^2$ at 2 MHz and $0.2 \ \mu W/cm^2$ to $50 \ mW/cm^2$ at 1 MHz. For this type of operation, a digital restitution of holograms would probably be the most adequate process. On the other hand, real time operation even with lower sensitivity, presents numerous advantages and there is a great hope to obtain from this array good images either from lenses or from an holographic technique with a real time optical restitution.

ACKNOWLEDGEMENTS

The author wishes to thank R. LALIMAN and C. SASSIER for technical assistance given in this work.

REFERENCES

1. A.K. NIGAM, K.J. TAYLOR, and G.M. SESSLER " Foil Electret Transducer arrays for real time acoustical holography "
   Acoustical Holography Vol IV pp 173-194.

2. P. ALAIS " Acoustical imaging by electrostatic transducers "
   Acoustical Holography Vol IV pp 237-249.

MODIFIED SOKOLOV CAMERA UTILIZING CONDENSER-MICROPHONE
ARRAYS OF THE FOIL-ELECTRET TYPE.

Anant K. Nigam* and J. C. French

Bell Laboratories

Murray Hill, New Jersey 07974

ABSTRACT

This paper demonstrates the feasibility of an Electret
Sokolov Camera (ESC) which employs electron-beam scanning of
a subdivided-backplate foil-electret array. The mechanism of
electron-beam reading is described and the different stable
operating modes are identified. In one operating mode, the
threshold ultrasound sensitivity of the ESC is estimated as
better than 2×10^{-6} watts/cm^2 for operation in water at 1MHz.
This compares favorably with other real-time imaging systems
such as the subdivided-foil-subdivided-backplate array and
the conventional Sokolov camera (which employs a piezoelectric-
plate array). However, unlike the conventional Sokolov camera,
the ESC does not suffer from limitations of physical and
angular aperture. Furthermore, the transducing sensitivity
of the ESC can be optimized independently of the material of
the backplate. This added flexibility offers promise for
further enhancement of the sensitivity.

* Present address: CBS Laboratories, High Ridge Road
                    Stamford, Conn. 06905.

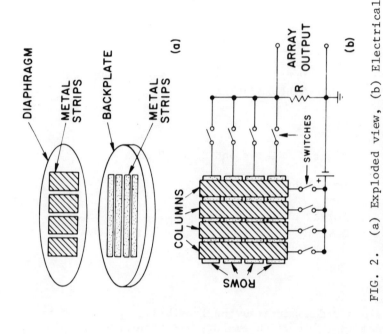

FIG. 1. (a) Cross-sectional view of a 4×4 element, subdivided-backplate (N^2) foil-electret array, (b) showing backplate subdivisions.

FIG. 2. (a) Exploded view, (b) Electrical connections for a 4×4 element subdivided-foil-subdivided-backplate (2N) array. Columns not being sampled are grounded to eliminate electrical cross-coupling.

TWO-DIMENSIONAL ARRAYS

In an earlier paper[1] two-dimensional condenser-transducer arrays, operating in either the externally-biased or
self-biased (electret) mode, have been described. These
arrays possess large angular apertures, have high sensitivity over frequencies extending into the MHz range for both
air and water operation and are relatively simple and inexpensive to produce.

Basically two design concepts have been evaluated. The
first design -- the subdivided-backplate array requires N^2
subdivisions of the backplate and an equal number of electrical connections to sample a N×N field. The second design,[1,2]
the subdivided-foil-subdivided backplate array, requires
only 2N elements and 2N electrical connections to sample a
N×N field. These arrays are shown schematically in Figs. 1
and 2.

The N^2 array, operated sequentially with a matrix of
switches[1], possesses a threshold sensitivity of approximately
10^{-8} watts/cm^2 in air and 10^{-11} watts/cm^2 in water at 1MHz.
The interelement crosstalk is also very small (better than
-35dB)[1], however, large size arrays require custom-designed
IC switch-matricies consisting of N^2 switches (as is the case
for piezoelectric arrays of the N^2 design)[3].

The 2N array requires a significantly reduced number of
elements and switches, however, the net sensitivity is reduced
due to electrical loading of elements -- a drawback which is
inherent in this design[1,4]. For a 200×200 element array of
the 2N design the loading leads to higher threshold sensitivity of about 10^{-8} watts/cm^2 at 1MHz in water.

Another two-dimensional array of interest here is the
Sokolov Camera which employs electron-beam scanning of a
piezoelectric-plate array. The Sokolov Camera has the advantage that it does not require any switches for sampling the
array elements, however, its sensitivity is slightly lower
than that for the 2N array. In addition, the conventional
Sokolov camera suffers from the following two additional
limitations:

(1) Angular aperture limitation -- inherent due to the
     large impedance mismatch between a piezoelectric-
     plate array and water.

(2) Physical aperture limitation -- inherent if the
sensitivity and lateral resolution of the piezoelectric-plate array are optimized.

Some of the earlier work of notable mention as regards
to overcoming these limitations is by Semennikov et. al.[5],
Smyth[6], Jacobs[7] and DuBois[8]. The basic approach towards increasing the sensitivity has been to investigate different
piezoelectric materials[9] (which are also mechanically strong).
Studies in increasing the angular aperture of the piezoelectric array have primarily been directed towards use of a
lens in front of the array or by depositing a quarter-wave
layer on the piezoelectric plate. A possible implementation
of the latter technique was recently demonstrated by Jacobs.[10]

Studies in increasing the physical aperture have basically
considered a reinforced support for the piezoelectric plate.
A ribbed supporting structure has been extensively employed
with its attendant image degradation. Use of a mechanically
strong Charlotte plate[11] has been suggested and recently implemented.[12,13]

It is, however, recognized that the limitations of the
Sokolov Camera mentioned above arise from the use of a piezoelectric-plate detector. Accordingly, the present paper considers another approach in which the piezoelectric-plate array
is replaced by a condenser-transducer array of the N^2 design
(see Fig. 1). In the following text the basic considerations
for assessing the feasibility of this proposed modification
are presented followed by a discussion of the potential improvements afforded by the modification.

ELECTRON-BEAM SCANNING OF AN ISOLATED TARGET ELEMENT

In the proposed Electret Sokolov Camera (ESC), the isolated metal backplate elements of the N^2 array (see Fig. 1)
are interrogated by a monoenergetic electron beam of current
I and energy $E=e(V-V_k)$ where e is the electron charge and
V and V_k are the potentials of the target and the cathode
respectively. The target emits secondary electrons, measured
as a current I' at a collector whose potential $V_c > V$ such that
all secondaries are collected. The parameter of interest here
is the coefficient σ of total secondary emission,

$$\sigma \equiv I'/I \qquad (1)$$

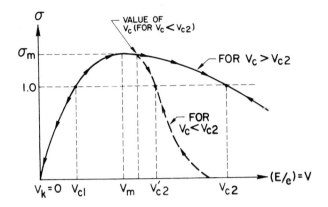

FIG. 3 Dependence of secondary emission coefficient σ on incident beam energy $E=eV$ ($V_k=0$). V_c denotes collector potential and V_{c1}, V_{c2} the crossover potentials. Arrows denote which of the two stable states, $V=V_{c2}$ or 0, is attained starting from any chosen initial value of E.

This coefficient depends on several factors[14] such as the material of the target, its surface condition and the incident beam energy E. Considering targets with polished surfaces of high purity, the typical σ vs E curve appears as shown by the solid-line in Fig. 3. An important feature of this curve is that σ shows a maximum value σ_m at some pre-scribed energy $E=E_m=eV_m$ of primary electrons. Both σ_m and E_m are characteristics of the material of the target. For most semiconductor and metallic targets σ_m is in the range 0.5-2. Dielectric materials such as quartz, glass, polymers, etc., show a value of σ_m in the range 1-3. For several other dielectrics, such as the alkali-halide compounds and alkali-earth materials, σ_m is in the order of 10-30. The corresponding values of E_m for metals, semiconductors and dielectrics lie in the range from several tenths of a KeV to a few KeV. Extensive data on metals, semi-conductors and dielectrics may be found in the literature.[14,15] In contrast, relatively fewer number of metallic alloys have been investigated. Some data on magnesium and beryllium alloys is found in reference 14.

For an isolated target element for which $\sigma_m > 1$ there are two stable values for the target potential V. These correspond to $V = V_k (=0)$ and $V = V_{c2}$ (see Fig. 3) providing that $V_c \not> V_{c2}$. At these two points the equality $I = I'$ is satisfied.[16] Note that the equilibrium potential $V = V_k$ corresponds to the case when the target has been charged to cathode potential by the primary beam and $I = I' = 0$.

When the collector potential $V_c < V_{c2}$ the equilibrium point $V = V_{c2}$ is altered because a portion of the emitted secondaries fall back on the target and the net current I' is less than I. However, because σ increases with decreasing V in the vicinity of $V = V_{c2}$ (see Fig. 3) a new equilibrium potential $V = V'_{c2} < V_{c2}$ is attained for $V_{c1} < V_c < V_{c2}$ where the net current I' equals I. The exact shape of the curve σ vs E remains unaltered for the range $E < eV_c$ and, for the range $E \not> eV_c$ is drawn from a study of the energy distribution of the secondary electrons. A typical modified curve is shown by the broken-line in Fig. 3.

For simplicity in notation, the equilibrium point $V = V_k = 0$ of the target is called the cathode potential stabilized or CPS state, the point $V = V_{c2}$ is called the free potential stabilized or FPS state and the point $V = V'_{c2}$ as the anode (collector) potential stabilized or APS state. Electron-beam sampling of the ultrasonically induced ac voltages of the backplate elements (targets) of a foil-electret array may be performed at any one of these stable potential states. The ac voltage of the backplate element(s) modulates the secondary current as illustrated in Fig. 4.

In the FPS or APS mode of operation the perturbation potential ΔV continuously modulates the equilibrium secondary current as shown in Fig. 4. In the CPS mode of operation the primary beam is incident on the target element only during a part of the positive half-cycles of the perturbation voltage. In addition, since $\sigma < 1$, during each positive half-cycle, the target element is charged further to a higher negative voltage. Accordingly, a rectified output is obtained which gradually decreases with sampling duration (see Fig. 4). Thus, while the device can be operated continuously in the FPS or APS mode, for prolonged operation in the CPS mode the backplate elements have to be periodically discharged and restored to the equilibrium cathode potential. This can be done in real time as demonstrated by Smyth[6] for a conventional Sokolov camera.

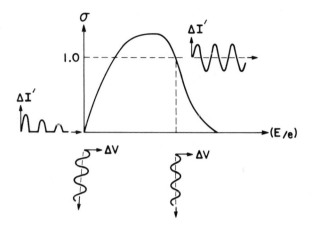

FIG. 4 Modulation of secondary emission current (σI) by target potential variations for the APS/FPS mode of operation (right) and the CPS mode of operation (left).

It was mentioned above that in the APS mode of operation $V > V_c$ and some of the secondary electrons fall back on the target. For a multielement target some electrons fall on adjacent elements causing a "redistribution effect" which is undesirable. Experience with orthicons and vidicons, however, has shown that the redistribution effect is substantially reduced if a fine-mesh barrier grid is placed very close to the target surface and the collector is maintained at a positive potential with respect to the grid. In this case, secondary electrons escaping through the holes of the mesh are unable to fall back on the target surface because of the accelerating field existing between the mesh and collector. In addition, those secondary electrons with insufficient energy to escape through the mesh return to the target very close to the point from which they were emitted. (It is important to note that in this case the barrier mesh acts as the effective collector in determining the equilibrium potential V'_{c2} of the target.)

THRESHOLD SENSITIVITY AND NOISE

To estimate the threshold ultrasound sensitivity of the

ESC, it shall be assumed that operation is in the APS (or
FPS) mode and that signal detection is at a collector grid.[17]
A simple analysis, which closely follows that developed by
Semennikov et. al.,[5] is presented below to illustrate the
basic parameters which limit the signal-to-noise ratio in
the ESC.

The predominant source of electrical noise is the shot-
noise in the secondary current. Considering an output bandwidth B and ignoring space-charge smoothening effects, this
is given by

$$i_n = (2eIB)^{\frac{1}{2}} \tag{2}$$

The signal current i, from a scanned backplate element
is given by[5]

$$i = SI\Delta V \tag{3}$$

where S is the transconductance of the secondary beam in
amperes per volt per unit incident primary current I_o,

$$S \equiv \frac{\partial I'}{\partial V} \bigg|_{I = I_o} \tag{4}$$

and is a function of the secondary emission characteristics
of the backplate material. The analytical dependence of S
on material parameters is not fully understood at present and
it is necessary to determine S experimentally for each target
material.

It is seen from Eqs. (2) and (3) that while the noise
signal i_n is proportional to $I^{\frac{1}{2}}$, the signal current i is
proportional to the first power of I. Thus the desirability
of keeping I large. There is, however, an upper limit to
the value of I because (a) the electron beam tends to defocus
at high beam currents, and (b) due to the beam current a
charging current $I_c = \Delta V$ IσS flows through the capacitance C
of the array element which diminishes the open-circuit voltage ΔV_o to a value ΔV:

$$\Delta V = \frac{1}{1 - j(IS\sigma/\omega C)} \Delta V_o \tag{5}$$

From Eqs. (2), (3) and (5) the threshold open-circuit
voltage $|\Delta V_o| = V_T$, which is necessary to produce a unit

signal-to-noise ratio at the collector, is obtained as

$$V_T = S^{-1} \{2eB[1+(S\sigma I/\omega C)^2]/I\}^{\frac{1}{2}} \quad (6)$$

This expression shows a minimum value of V_T for an optimum value $I=I_{opt}=\omega C/S\sigma$ of the primary beam current.[18] Substituting $I=I_{opt}$ leads to

$$V_T = 2S^{-\frac{1}{2}} [eB\sigma/\omega C]^{\frac{1}{2}}, \quad I=I_{opt} \quad (7)$$

The minimum threshold V_T given by Eq. (7) can be expressed in terms of the threshold ultrasound intensity I_T, which produces a unity signal-to-noise ratio at the collector, as follows,

$$I_T = (V_T/\rho)^2/Z_m \quad (8)$$

where ρ is the sensitivity of array elements and Z_m is the specific acoustic impedance of the medium in contact with the array.

For a typical array,[1,20] designed to operate at 1MHz in water, the element size is 0.6 mm, the field of view is about $60°$, $C \approx 2pF$ and $\rho = -120dB$ re. 1V/µbar. Assuming a backplate material for which $\sigma = 1.2$ and $S = 0.1$ µA/V per 1µA of beam current, a value of $V_T \approx 1.9$ mV is obtained from Eq. (7) and from Eq. (8), at 1MHz in water,

$$I_T \approx 2\times10^{-6} watts/cm^2 \quad (9)$$

for a bandwidth of 6MHz. A smaller bandwidth results in a lower value for I_T at the expense of scanning speed. (For lack of experimental data on the value of S a conservative estimate has been heuristically employed in the calculation above.)

PRELIMINARY EXPERIMENTAL RESULTS

For evaluating the feasibility of the proposed ESC, a 7-inch demountable Sokolov tube was obtained.[21] This tube employs electrostatic deflection of the electron beam with a

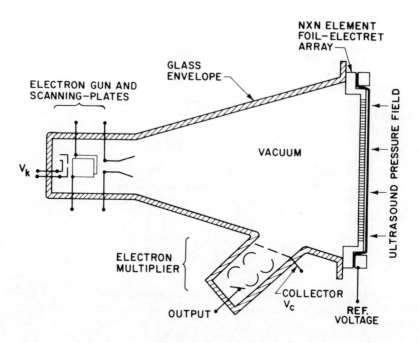

FIG. 5   Schematic cross-sectional view of proposed Electret
         Sokolov Camera (ESC).

collector-electron multiplier setup for detection. An
electret array of the N^2 design is placed in front of the
tube by means of O-rings and held under constant evacuation
of the tube (see Fig. 5).

     For experimental studies a 20×20 element array of the
N^2 design was constructed as outlined earlier.[1] The experimental setup is shown in Fig. 6.

     Tests were conducted in the APS mode of operation with
the cathode at approximately -600 volts and the target
(backplate) voltage stabilized at ground potential by means
of a remotely placed barrier mesh (as supplied with the tube).
Scanning beam current was maintained in the range of 20-100µa.
The collector mesh was maintained at approximately 100 volts
and approximately 60dB of gain was provided by the electron
multiplier. Subsequent elements in the circuit were an

FIG. 6 Experimental setup showing a 20×20 element backplate.
Electret foil and clamping ring have been removed and are
seen in the foreground.

ac-coupled preamplifier, a broadband amplifier and a bandpass filter which provide additional gain capability of
about 40dB. The output bandwidth was about 6MHz and, for
these preliminary tests, the unrectified video signal was
used directly to intensity modulate a display scope. A
60 frame per second scanning rate with 262 lines/frame was
employed for scanning and display.

Prior to ultrasonic excitation, a 1MHz electrical signal was applied directly to each backplate element, one at
a time, from a variable impedance oscillator. This test
revealed that the stray capacitances (to ground) of the
backplate elements were in the range 10-30pF. (It is felt
that the stray capacitances can be reduced and made uniform
across the array by repositioning of the barrier mesh.) The
tests also showed that target signal levels of 1mV were
clearly displayed on the video monitor. This agrees with
the anticipated value for V_T calculated above in the computation for Eq. (9).

A 12.5 μm Teflon FEP foil charged to about 10^{-8} C/cm^2
was placed over the backplate and an ultrasonic source operating at 100 KHz in air was placed in front of the array.
Preliminary results, showing a bright glow on the video display, have demonstrated the feasibility of the ESC.

The 20×20 element backplates employed for the preliminary tests show rapid deterioration with age. The glue layer
between the elements was found to become porous, thus destroy-
ing the vacuum in the tube. New backplates employing a different epoxy glue having low vapor pressure are currently
being designed. Concurrently several other backplates are
being tested. Results from a backplate containing only four
elements (about 2mm in size) is shown in Fig. 7.

Further work is currently in progress to (1) optimize
backplate design, (2) investigate secondary emission characteristics of a few materials,(3) optimally reposition the
barrier grid and (4) to adapt the equipment for underwater
imaging.

SUMMARY AND OUTLOOK

The Electret Sokolov Camera (ESC) can be designed for
large angular apertures because each individual electret

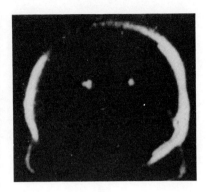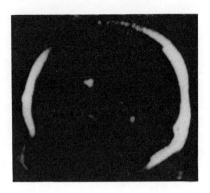

FIG. 7 Results of a feasibility study at 1MHz on a test
backplate containing four elements. Only two elements
                illuminated at one time.

element can be made accordingly smaller in size. With decreasing element size, due to the finite stray capacitance
loading, the sensitivity of the elements also decreases,
however, it appears that angular apertures up to 60° are
feasible without appreciable loss in sensitivity.[20]

     The ESC can also be designed for large physical aper-
tures because, in the present design, strength considerations for the backplate are essentially separated from parameters (such as[1,20] spacing between foil and backplate, re-
siliency of the medium entrained in this space, thickness
and charge level of the foil, etc.,) that determine the
sensitivity of the electret elements. In addition, the uniformity in sensitivity across the face of the array[1] is
essentially unchanged regardless of array size.[1]

     Another important flexibility in the ESC is provided
by the fact that the transducing sensitivity of the electret is independent of the backplate material. This permits use of backplate materials which are most favorable
only as regards their secondary emission characteristics.
Either the backplate elements can be directly made of this
optimum material or the material deposited at the rear of
the backplate elements. This added flexibility appears
to hold great promise for further enhancing the sensitivity
of the ESC.

ACKNOWLEDGEMENTS

The senior author would like to thank Dr. G. M.
Sessler for several stimulating discussions and to his
and Dr. J. L. Flanagan's assistance in obtaining release
of the equipment for continued work at CBS Laboratories.
Thanks are also due to J. E. Jacobs, Professor Northwestern University, for supplying a metal-fiber impregnated
glass-backplate and to A. J. Chapman, Professor Georgia
Institute of Technology, for supplying samples of a unique
melt-grown oxide-metal composite backplate.

REFERENCES

1. A. K. Nigam, K. J. Taylor and G. M. Sessler, in "Acous-
    tical Holography, Vol. IV," Plenum Press, New York (1972)
    p. 173; A. K. Nigam and G. M. Sessler, Appl. Phys. Letters
    21, 229 (1972).

2. P. Alais, in "Acoustical Holography, Vol. IV," Plenum
    Press, New York (1972) p. 237.

3. An assessment of the complexity of interconnections for
    the N^2 array is provided in References 1 and 20. See
    also M. G. Maginness, J. D. Plummer and J. D. Meindl,
    this volume.

4. Another drawback of the 2N array appears to be higher
    interelement crosstalk due to residual sensitivity of
    elements. The results of references 1 and 2 above,
    indicate that this drawback is eliminated by suitable
    choice of foil-material having the appropriate bulk-
    resistivity and/or by applying opposite polarity bias-
    voltages during alternate scanning frames.

5. Iu. B. Semennikov, Sov. Phys. - Acoust. 4, 72 (1958);
    P. K. Oschepkov, L. D. Rozenberg, Iu. B. Semennikov,
    Sov. - Phys. Acoust. 1, 362 (1955).

6. C. N. Smyth et al, Proc. IEE 110, 16 (1963).

7. J. E. Jacobs, H. Berger and W. J. Collis, IEEE Trans.
    Son. and Ultrason. SU-10, p. 83 (1963); J. E. Jacobs,
    IEEE Trans. Son. and Ultrason. SU-15, 146 (1968).

8. J. L. DuBois, IEEE Trans. Son. and Ultrason. SU-16, 94 (1969); J. L. DuBois, in "Acoustical Holography, Vol. II," Plenum Press, New York (1970) Chapter 6.

9. See for example Yu. B. Semennikov, Sov. Phys.-Acoust. 7, 56 (1961).

10. J. E. Jacobs, this volume.

11. H. F. Charlotte, Electronic Engg. 29, 373 (1957).

12. Iu. B. Semennikov et al., Sov. Phys.-Acoust. 12, p. 376 (1966).

13. R. E. Addison, this volume.

14. L. N. Dobretsov and M. V. Gomoyunova, Emission Electronics (Translated from the Russian by I. Shecht-man), Keter Press, Jerusalem, Israel, Chap. VII (1971); H. Bruining. Physics and Applications of Secondary Electron Emission, Pergamon Press, New York, Chap. VII (1954).

15. R. Kollath, Handbuch der Physics 21, 282 (1956); O. Hachenber and W. Brauer, Adv. in Electronics and Electron Physics 11, 413 (1959).

16. At the point $V=V_{c1}$ the equality $I=I'$ is also satisfied, however, this point does not correspond to stable equilibrium for small perturbations in the target potential V.

17. The signal may also be detected at the front electrode of the foil (Fig. 1), however, it appears that this scheme may suffer from crosstalk if large stray capacitance loading is evident on the backplate elements.

18. The results of a more detailed analysis (see Ref. 19 below), in which several other parameters as well as space charge smoothening-effects have been considered, indicate that for certain target (backplate) materials V_T has another minimum (for a second value of I) which may correspond to its lowest value.

19. H. W. Jones, in "Acoustical Holography, Vol. IV," Plenum Press, New York (1972) p. 401.

20. A. K. Nigam, J. Acoust. Soc. Amer. (to be published).

21. From James Electronics Inc., Chicago, Ill.

A SOLID PLATE ACOUSTICAL VIEWER FOR UNDERWATER DIVING

Joseph A. Clark

Catholic University of America

Washington, D.C.

Karyl-Lynn K. Stone

Visual Acoustics

Chevy Chase, Md. 20015

INTRODUCTION

Underwater viewing is fraught with problems, as anyone who has ever attempted to do so with the unaided eye or optical instruments well knows. The problems are caused by the behavior of light in the water. Light is effected by three phenomena: refraction, absorption, and scattering. Figure I illustrates causes, effects, remedies, limitations, and solutions for these phenomena. One can see the importance of going to a method that is not light dependent such as acoustics. At the present time, with the exception of a head mounted sonar system under development in Panama City, [1] the diver, in optically opaque water, has nothing but the "feel and pray" method of "seeing"; this is neither appealing nor secure. Several types of acoustical imaging systems have been considered or used in underwater viewing. However, to the authors' knowledge, no feasible designs have been developed that can withstand the rough conditions of open water, be compact enough for a diver to carry, and give a picture of high enough resolution to be recognizable to

	Refraction	Absorption	Scattering
Causes	Light travels faster in air than water. When it hits the water's surface, it is bent.	Suspended and dissolved matter in water. Blue-green color of water acts as filter.	Suspended matter increases with density and size of particles.
Effects	Magnification by 1/3. Distortion of objects.	Loss of reds, oranges, yellows at shallow to moderate depths. Color cast of everything with with blue-green tinge.	Viewing looks hazy. Loss of contrast. Image degradation.
Remedies	Corrective lenses.	Strobe lighting to restore color. Filters to eliminate excessive blues, greens, permitting more reds, oranges and yellows.	Close to subject as possible with wide angle lens. Little debris as possible between subject and camera.
Limitations	Some distortion still present.	Below certain depth no filter helps. Limited by strength of lights.	Loss of depth of field. So close, can't get much in picture. Very trubid water, no visibility.
Solutions	Less abberation with high index of refraction lenses available acoustically.	Less absorption with sound waves.	Less scattering with longer wave length sound waves.

Figure 1   Phenomena Effecting Light in Water

the eye. Our system, theoretically, can accomplish all three tasks. This system can open new worlds to both the diver, who is forced to work in a world of deep, dark, murky water, and the vicarious observer on land.

A novel feature of the acoustical imaging system we are developing is the use of a type of plate wave to focus the acoustically sensed image. We call this new system the solid plate viewer.

Many systems have been developed, beginning with SONAR, that convert acoustic waves into images directly visible to the eye. The illustrated diagrams below distinguish the solid plate viewer from prior developed systems. Each system contains an imaging device and conversion device to transform the original acoustic waves to optical images which are displayed at its output, as is shown in Figure 2. In our system, the solid plate lens system, the acoustic waves are first converted into a type of plate wave, which we

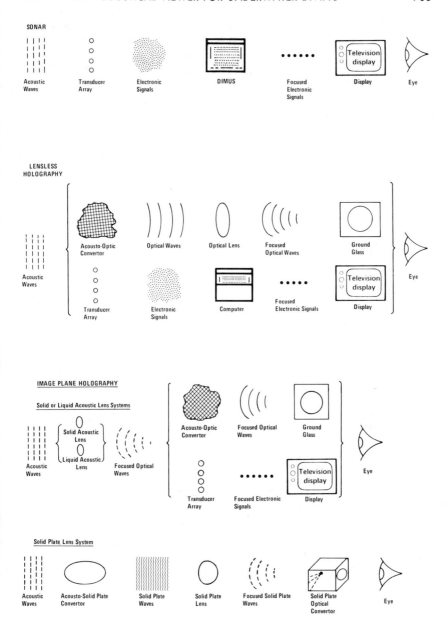

Figure 2   Comparison of Acoustical Imaging Systems

identify as the solid plate wave. This is done by an acousto-solid plate (a-sp) converter. The solid plate waves are focused by the solid plate (sp) lens, thereby forming a solid plate image. This image, in turn, is converted to an optical image by the solid-plate optical (sp-o) convertor. This optical image can be photographed or viewed directly.

In this paper, we will present the theory of operation of the components of the solid plate viewer, some of our initial experimental results, a design of the solid plate viewer for underwater diving, and an evaluation of the viewer's predicted performance.

THEORY OF SYSTEM'S OPERATION

In order to develop a theoretical model of the wave be-havior in the solid plate viewer, we have defined a category of waves as solid plate waves. These are waves that propa-gate across a plate surface in only one mode under a broad range of loading conditions. Several well known types of waves, which we classify as solid plate waves, are shown in Figure 3. Each of these waves differ in deformation or in-ertial forces but are similar in their transverse wave motion and in our ability to control their wave speed by other para-meters besides bulk material properties[2,3]. Some parameters are plate thickness, and applied tension. This characteristic is important because greater changes in the wave speed are possible by varying plate thickness and applied tension than by selecting bulk material properties.

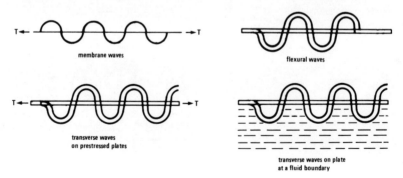

Figure 3   Examples of Solid Plate Waves

The next set of figures illustrate how solid plate waves perform the operations of an acoustical imaging system. Figure 4 illustrates the conversion of acoustic waves into solid plate waves. As the acoustic waves, with velocity c, incident at grazing angle θ, strike the a-sp convertor plate, the plate is deformed as the acoustic waves are converted into solid plate waves. These waves travel on the plate with a velocity c_p. $c_p = c/\cos\theta$ expresses the condition of matched velocities. For acoustic waves to be converted to solid plate waves, they must strike at approximately the angle θ where $\theta = \arccos(c/c_p)$. However, if a matched impedance condition is satisfied, the range of angles is quite broad.

Figure 5 illustrates the focusing of the solid plate waves into a solid plate image[4]. The lens for focusing the sp waves is formed by attaching a second plate to the a-sp convertor plate with a joining strip shaped in an arc. This second plate, the sp lens, has a wave velocity c_{pl} differing from that of the a-sp convertor plate. Its focal length (f) can be calculated by the lens formula: $(n/f) = (n-1)/r$, where $n = c_p/c_{pl}$ and r is the radius of curvature of the joining strip arc.

Therefore, a wave, propagating on the a-sp convertor plate, is brought to focus by the sp lens and a solid plate image is formed.

Figure 6 illustrates the conversion of the solid plate image into an optical image by the sp-o convertor. This convertor is an optical spatial filtering system (5,6,7,8). A point light source illuminates the sp image on the sp lens through a collimating lens. A mirror surface of the sp lens causes the transverse vibrations of the sp image to phase modulate the incident light beam. The reflected light beam comes to focus at a point just in front of the optical spatial filter. This optical spatial filter, placed at the focal point, blocks the unmodulated component of the light beam, and passes the modulated component. A lens placed after the filter brings the modulated beam to a focus, forming an optical image of the solid plate wave image on the screen. If, instead, we view the sp lens directly with our eyes, then the second lens is that in our eyes, and the optical image will appear on the sp lens.

Figure 7 is a schematic diagram of the complete solid

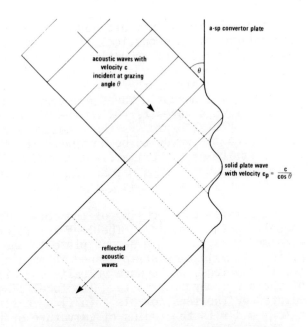

Figure 4   Acousto-Solid Plate Convertor

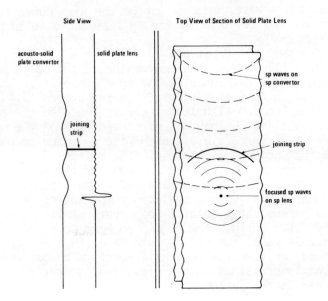

Figure 5   Focusing of Solid Plate Waves into a Solid
           Plate Image

plate underwater acoustical viewer. The conversion of the acoustical field into a visible image is illustrated. A sound source insonifies a region in the water. Waves are scattered from a representative object in the insonified region. These scattered acoustic waves are converted to solid plate waves on striking the a-sp convertor plate. These waves are focused to an sp image by the sp lens. The sp image on the sp lens is converted to an optical image by the sp-o convertor. This optical image, projected on the screen, is visible to the eye.

Figure 8 is a photograph of a prototype of a diver's viewer under construction. The components are described in detail. The sound source is a 3/4" diameter 600 kHz transducer. The unit will be either hand-held by another diver or mounted on the viewing system. The a-sp convertor is a plate on which sp waves of high velocity and high impedance propagate. The plate is made of .025" thick aluminum and is tuned to detect sound waves incident on the plate at 18° from normal. It is connected to the sp lens by a plastic joining strip. The sp lens is a prestressed aluminum-coated mylar film. Low velocity, low impedance sp waves propagate in this plate.

The sp-o convertor contains the following. The point light source consists of a 1 milliwatt laser and a 60x microscope objective. Power for the laser is obtained via a cable from a surface support vessel. The collimating lens is a 2" diameter, 8" focal length plano-convex lens, located 1/2" away from the sp lens. The spatial filter is a 1/2" diameter, 1/8" wide transparent ring in an otherwise opaque piece of film. The screen on which the optical image is projected by the imaging lens is a 2" diameter, fiber optics faceplate. Since the system is for underwater use, it is encased in a watertight, pressure proof housing. Our housing is a 6" inside diameter, plexiglas cylinder, 1/4" thick and 2' long. There are two 2" diameter viewing areas, one for a Nikonos II camera. The diver's viewer is extended a few inches to facilitate the diver positioning him (her) self, since with his (her) mask, his (her) eyes will be 2'-3' from the viewer. There is also a mount for the sound system and handles for holding the system in the water. The photograph illustrates the compactness, ease of control, and ruggedness available with solid plate acoustical imaging.

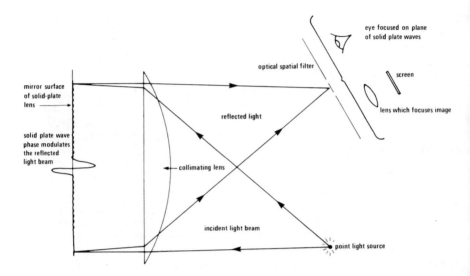

Figure 6   Solid Plate-Optical Convertor

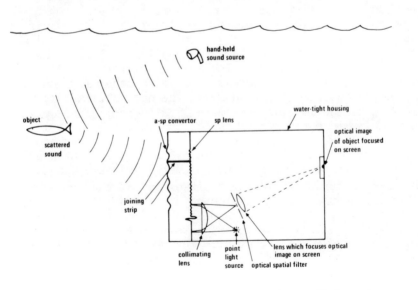

Figure 7   Operation of Solid Plate Viewer Underwater

Figure 9 illustrates our laboratory model of the diver's viewer. In this model, the a-sp convertor and sp lens are the same. However, they are mounted on the wall of a small water tank. The sp-o convertor differs in that a 50 milliwatt laser is used and the beam is expanded to 5" in diameter. Now both the focused traveling waves on the sp lens and the sp image are observed.

These waves produce several orders of diffracted light in the plane where the spatial filter is located. These diffraction patterns of the solid plate waves are shown in Fig. 10. The velocity and wavelengths of these waves, as experimentally determined from the spacing of orders in the diffraction pattern, is $\lambda = 10^{-4}$ m and c = 20 m/sec. This corresponds to a lens with an index of refraction of 100. The spatial filter is used to stop or pass any part of the diffraction pattern. When the filter is adjusted to pass only light from the +1 diffraction order, the focused traveling waves are observed as illustrated in Fig. 11. When only the -1 diffraction order is passed, diverging waves traveling in the opposite direction are observed.

TECHNICAL ADVANTAGES OF THE SYSTEM

Better resolution is achieved because the diffraction limit on the minimum angle of resolution (α min) is: α min (solid plate viewer) = (λ/D) cos θ rather than α min (lateral acoustical viewers) = (λ/D), where θ is the grazing angle (Fig. 4). Since it is theoretically possible to tune the a-sp convertor plate to grazing angles larger than 45°, the improvement in angular resolution is more than 1.4 times.

There is an increase in instrumental sensitivity, demonstrated by a comparison of the lens amplification gain of the acoustical lens with the solid plate lens. The maximum gain achievable by demagnification of an image is: G (solid plate lens) = 20 log (D/λsp) rather than G (acoustical lens) = 40 log (D/λ_a). The acoustical lens can achieve this gain only if it can be designed with an f_1 aperture without spherical aberration. Solid plate wavelengths can be reduced

Figure 8   Prototype of Diver's Viewer

Figure 9   Laboratory Model of Solid Plate Viewer

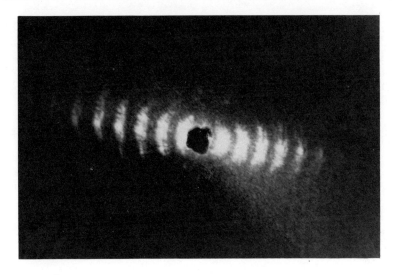

Figure 10   Diffraction Patterns of Solid Plate Waves

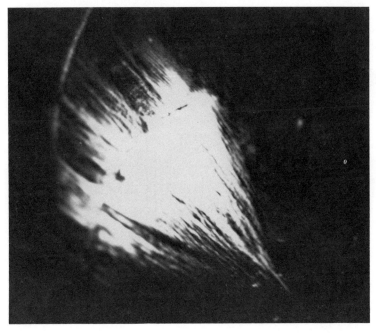

Figure 11   Image of Focused Solid Plate Waves

to less than .01 times the acoustic wavelength. Therefore,
the solid plate lens has improved gain over an f₁ acoustical
lens with an aperture less than 100 acoustic wavelengths.
Spherical aberration errors are less for a high index of
refraction solid plate lens than for an acoustical lens of
equivalent curvature.

We have also eliminated image degradation caused by internal multipath effects. Waves through either solid or
liquid lenses can travel along different paths. This results
in several blurred images superimposed on the focused image.
Solid plate waves can only travel along one path through the
solid plate lens.

The solid plate image corresponds to a longitudinal
field of view, similar to range and bearing presentations of

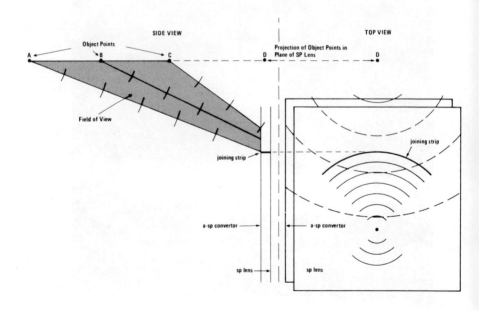

Figure 12   Large Depth of Field with Solid Plate Viewer

sonar images, rather than the lateral field of view of bulk
type acoustical and optical lenses. The large depth of field
of the solid plate image is demonstrated geometrically in
Figure 12. Here the a-sp convertor plate intercepts a
spherical wave scattered by an object point along an arc
whose center of curvature is located in the plane of the
plate above the object point. Therefore, solid plate waves
have the same center of curvature (D) for acoustic waves
scattered by all objects along the line between points (A)
and (C). And they all focus to the same point (E) in the solid
plate image, producing the large depth of field noted above.
This eliminates the need to focus.

The advantages of eliminating focusing are numerous,
since a diver is often plagued by rough seas, currents, and
subject movement. Usually, high resolution acoustical imaging
systems have a large aperture and narrow depth of field,
making focusing adjustment critical. Also, the entire object
is not usually in focus. The ability of the solid plate
viewer to overcome the focusing limitations present in other
current acoustical imaging systems should be a significant
contribution to the acoustical imaging problem.

ACKNOWLEDGEMENT

This research was supported by the Advanced Research
Projects Agency of the Department of Defense and was mon-
itored by ONR under Contract No. N00014-67-A-0377-0014.
The authors would like to express their appreciation to
F. A. Andrews, C. Ravitsky, M. Lasky and H. Fitzpatrick for
their continuous encouragement, and to A. Afkar for his assistance in the experimental development.

REFERENCES

1.   Warner, H. L., "Ultrasonic Imaging for Underwater
      Viewing," J.A.S.A., 53, #1, 307A, (1973).

2.   Clark, J. A., and Durelli, A. J., "Optical Stress
      Analysis of Flexural Waves in a Bar," J. Appl. Mech.
      37, Series E, 331-338 (1970).

3.   Clark, J. A., Durelli, A. J., and Laura, P. A.,

"On the Effect of Initial Stress on the Propagation of Flexural Waves in Elastic Rectangular Bars," J.A.S.A. 52, #4, 1077-1086 (1972).

4. Clark, J. A., "Circular Lenses for Longitudinal Acoustical Viewing Systems," J.A.S.A., paper JJ4, Spring Meeting (1973).

5. Clark, J. A., Durelli, A.J., and Parks, V. J., "Shear and Rotation Moiré Patterns Obtained by Spatial Filtering of Diffraction Patterns," J. Strain Analysis 6, #2, 134-142 (1971).

6. Durelli, A. J., Chichenev, N. A., and Clark, J. A., "Developments in the Optical Spatial Filtering of Superposed Crossed Gratings, Exp. Mech. 12, #11, 496-501 (1972).

7. Clark, J. A., "Focused Optical Imaging Systems for Vibration Measurements," J.A.S.A. 53, #1, 338A (1973).

8. Clark, J. A., and Mehta, A. V., "Experimental Analysis of the Motion of a Vibrating Membrane by Optical Spatial Filtering," J.A.S.A., paper O1, Spring Meeting (1973).

9. Booth, N., and Saltzer, B., "An Experimental Holographic Acoustic Imaging System," Acoustical Holography, Vol. 4, Plenum Press, 371-379 (1972).

LIST OF PARTICIPANTS

Iwao P. Adachi
Minolta Company
3105 Lomita Blvd.
Torrance, Ca.   90505

Robert Addison
American Optical
P.O. Box 2201
Framingham, Mass.   01701

Mahfuz Ahmed
Zenith Radio Corporation
6001 West Dickens Avenue
Chicago, Illinois   60639

Pierre Alais
Universite Paris
Laboratoire de Mecanique
2 Pl. de la Gare de Ceinture
78 St. Cyr L'Ecole France
Paris VI, France

R.E. Anderson, M.D.
University of Utah
Dept. of Radiology
Salt Lake City, Utah

Weston Anderson
Varian Associates
611 Hansen Way
Palo Alto, Ca.   94303

Yoshinao Aoki
Hokkaido University
N12, W8 Dept. of Electronics
Sapporo (060), Japan

Norm Austin
Mgr. Tech. Operations
Gould Inc., ISD
384 Santa Trinita Avenue
Sunnyvale, Ca.   94086

M.C. Bartz
Beckman Instruments, Inc.
2500 North Harbor Blvd.
Fullerton, Ca.   92634

William Beaver
Varian Associates
611 Hansen Way
Palo Alto, Ca.   94303

Klaus Biedermann
Institute of Optical Research
Royal Institute of Technology
S-10044 Stockholm 70, Sweden

Gerald V. Blessing
Naval Weapons Lab., GBF
GBF, Dahlgren, Va.   22448

Helge Bodholt
Simrad
Horten, Norway

715

Newell O. Booth
Naval Undersea Center
Code 6513
San Diego, Ca.

W.P. Chase
Beckman Instruments, Inc.
2500 North Harbor Blvd.
Fullerton, Ca.   92634

Roger Bradford
University of Maryland
214 Park Terrace Ct., S.E.
Vienna, Va.   22180

Fu-Pen Chiang
SUNY at Stony Brook
Dept. of Mechanics
Stony Brook, New York   11777

Richard G. Brandt
Office of Naval Research
1030 E. Green Street
Pasadena, Ca.   91106

Chris L. Christman
U.S. Public Health Service
BRH 12709 Twinbrook Parkway
Rockville, Maryland   20852

Herman P. Briar
Aerojet Solid Propulsion Co.
P.O. Box 13400
Sacramento, Ca.   95847

Tien S. Chou
Honeywell, Inc.
2 Forbes Road
Lexington, Mass.

P.S. Bringham
Lawrence Berkeley Lab
Hearst Avenue - Bldg. 4
Berkeley, Ca.   94530

Carol Clark
McDonnell Douglas Corp.
700 Royal Oaks Drive
Monrovia, Ca.

Minguzzi Bruno
Selenia
via Tiburtina Rm. 12.4
00131 Rome, Italy

J.A. Clark
Catholic University
Washington, D.C.   20017

L. Doug Clark
Varian Associates
611 Hansen Way
Palo Alto, Ca.   94303

Olof Bryngdahl
Xerox
3180 Porter Drive
Palo Alto, Ca.   94304

Michel J-M. Clément
Dept. of Electronics and
   Electrical Engineering
University of Tech.
Loughborough (Leics.), England

C.B. Burckhardt
F. Hoffmann-La Roche & Co.
Grenzacherstr.
Basel, Switzerland

James H. Cole
TRW Systems
One Space Park
Redondo Beach, Ca.   90277

Billy M. Coleman
Hercules Inc.   (NDT Methods)
Bacchus Works, Box 98
Magna, Utah   84044

Dale Collins
Holosonics
2950 George Washington
Richland, Wash.   99352

Bonnar Cox
Stanford Research Institute
333 Ravenswood Avenue, L1110
Menlo Park, Ca.   94025

Michael J. Cudahy
Marquette Electronics, Inc.
3712 West Elm Street
Milwaukee, Wis.   53209

James A. Cunningham
TRW Systems
R1/1213
One Space Park
Redondo Beach, Ca.   90278

Fleming Dias
Hewlett Packard Company
1501 Page Mill Road
Palo Alto, Ca.   94304

John F. Dreyer
Vari Light Corp.
9854 Zig Zag Road
Cincinnati, Ohio   45242

Darryl E. Edgars
Diversified Services Co.
Bank of America Tower
    Suite 426
One City Blvd. West
Orange, Ca.   92668

John A. Edward
General Electric Company
HMES - Court Street
    Plant 4 - 39A
Syracus, New York   13201

Reginald Eggleton
Indianapolis Center for
    Advanced Research
ICFAR, Interscience Res. Div.
410 Beauty Avenue
Indianapolis, Ind.   46202

H.M.A. El-Sum
El-Sum Consultants
74 Middlefield Road
Atherton, Ca.   94025

Ken Erikson
Actron
McDonnell Douglas
700 Royal Oaks Drive
Monrovia, Ca.

James O. Ewing
Bendix Electrodynamics Div.
11600 Sherman Way
North Hollywood, Ca.   91605

N.H. Farhat
University of Pennsylvania
EE Dept.
200 South 33rd Street
Philadelphia, Pa.   19104

Stephen Farnow
Stanford University
114 Tennyson
Palo Alto, Ca.

John B. Farr
Amoco Production Co. Research
Box 591
Tulsa, Oklahoma   74102

Wayne R. Fenner
The Aerospace Corp.
Electronics Res. Lab.
P.O. Box 92957
Los Angeles, Ca.   90009

Jeffrey A. Fenton
Naval Intel. Support Center
4301 Suitland Road
Washington, D.C.   20390

James Fienup
Stanford University
P.O. Box 8789
Stanford, Ca.   94305

John Flynn
Orange County Medical Center
Dept. of Radiology
101 City Drive South
Orange, Ca.   92668

Donald L. Folds
Naval Coastal Systems Lab.
Panama City, Florida   32401

M.D. Fox
University of Connecticut
Box U-157
Storrs, Conn.   06268

Frank Fry
Indianapolis Center for
   Advanced Research
Interscience Research Division
410 Beauty Avenue
Indianapolis, Ind.   46202

Ronald J. Fredricks
Lear Siegler Inc.
4141 Eastern S.E.
Grand Rapids, Mich.   49508

Hugh Frohbach
Stanford Research Institute
333 Ravenswood Avenue, K1080
Menlo Park, Ca.   94025

Iwao Fujimasa
Institute of Medical Elect.
University of Tokyo
7-3-1 Hongo, Bunkya-ku
Tokyo, Japan

Herman Gabor
SUNY at Buffalo
Dept. Computer Science
4226 Ridge Lea Road
Amherst, New York   14226

Robert W. Gill
Stanford University
McCullough Bldg. Room 26
Stanford, Ca.   94305

Paul Goldberg
Smith Kline Instruments, Inc.
440 Page Mill Road
Palo Alto, Ca.   94306

Samuel C. Goldman
Picker Corporation
333 State Street
North Haven, Conn.   06473

Albert Goldstein
Univ. of Kansas
Medical Center
Rainbow at 39th Street
Kansas City, Kansas   66103

Joseph W. Goodman
Stanford University
Durand 127
Stanford, Ca.   94305

Lou Ann Granger
Bendix
15625 Roxford
Sylmar, Ca.

Ed Greaves
Holosonics, Inc.
2830 George Washington Way
Richland, Wa.   99352

Philip S. Green
Stanford Research Institute
333 Ravenswood Avenue K1088
Menlo Park, Ca.   94025

James F. Greenleaf
Mayo Clinic
512 5th Avenue S.W.
Rochester, Minn.   55901

Dick A. Hall
G.D. Searle & Co.
Inst. Group
2000 Nuclear Drive
Des Plaines, Ill.   60066

Michael E. Haran
Bur. of Radiological Health
5600 Fishers Lane
Rockville, Md.   20852

Stephen D. Hart
Naval Research Laboratory
(Code 8435)
4555 Overlook Avenue, S.W.
Washington, D.C.   20375

J.J. Hartog
Lockheed Res. Laboratory
3251 Hanover Street
Palo Alto, Ca.   94304

James F. Havlice
Stanford Microwave Lab.
Stanford University
Stanford, Ca.   94305

Robert Heimburger
ICFAR, Interscience Res. Div.
410 Beauty Avenue
Indianapolis, Ind.   46202

Kenneth W. Henry
Div. 8344
Sandia Laboratories
Livermore, Ca.   94550

Bruce Herman
Bur. of Radiological Health
12320 Twinbrook Parkway
Rockville, Md.   20852

B.P. Hildebrand
Battelle Northwest
Battelle Blvd.
Richland, Wash.   99352

G.S. Kino
Microwave Laboratory
Stanford University
Stanford, Ca.   94305

P.B. Kivitz
16350 Matilija Drive
Los Gatos, Ca.   95030

Richard Knapp
Stanford Integrated Circuits
  Laboratory
AEL 121
Stanford, Ca.

G.C. Knollman
Lockheed Research Laboratory
3251 Hanover Street
Palo Alto, Ca.   94304

Justin L. Kreuzer
Perkin Elmer Corp.
Main Avenue
Norwalk, Conn.   06856

John Landry
University of California
  Santa Barbara
Dept. of Elec. Eng.
Santa Barbara, Ca.

J. Larson
Hewlett Packard
1501 Page Mill Road. IL
Palo Alto, Ca.   94304

A.L. Lavery
US DOT/TSC
Kendall Square
Cambridge, Mass.   02142

Chin-Hwa Lee
EE Dept., UCSB
Santa Barbara, Ca.   93106

P.T. Lee
15901 W. Nine Mile Road
Southfield, Mich.   48075

Wendell Lehr
Varian Associates
611 Hansen Way
Palo Alto, Ca.   94303

Ross A. Lemons
Hansen Microwave Lab.
Stanford University
Stanford, Ca.   94305

Sheldon Leonard
Litton Medical Products
515 East Touhy Avenue
Des Plaines, Ill.   60018

H.G. Loos
Cleveland State University
Cleveland, Ohio

Margitta Lütkemeyer-Hohmann
Krupp Atlas-Elektronik
28 Bremen 44, Postf. 44 85 45
Sebaldsbrücker Heerstr, 235
28 Bremen, Germany

Peter J. McCartin
E.I. du Pont de Nemours & Co.
Photo Products Dept.
Bldg. 352, Exp. Station
Wilmington, Del.   19898

Edward M. McCurry
Children's Hospital of
  San Francisco
3700 California Street
San Francisco, Ca.

B.J. McKinley
University of California
Lawrence Livermore Lab.
P.O. Box 808, L-415
Livermore, Ca.   94550

Max G. Maginness
Stanford Electronics Labs
Stanford University
Stanford, Ca.   94305

James Meindl
Elec. Eng. Dept.
Stanford University
Stanford, Ca.   94305

A.F. Metherell
McDonnell Douglas
700 Royal Oaks Drive
Monrovia, Ca.

Reuben S. Mezrich
RCA Labs
Princeton, New Jersey

Eric B. Miller
Duke University
4216 Garret Road A-1
Durham, N.C.   27707

Charles Mosher
Varian Associates
611 Hansen Way
Palo Alto, Ca.   94303

R. Mueller
Bendix Res. Labs.
Bendix Center
Southfield, Mich.

Anant K. Nigam
CBS Labs
High Ridge Road
Stamford, Conn.

Toyota Noguchi
Matsushita Research Institute
Ikuta, Kawasaki, Japan

Stephen Norton
Dept. of Applied Physics
Box 42
Stanford University
Stanford, Ca.   94305

Craig Nunan
Varian Associates
611 Hansen Way
Palo Alto, Ca.   94303

Charles P. Olinger
University of Cincinnati
C-3 Stroke Clinic
Cincinnati General Hospital
Cincinnati, Ohio   45229

Peter R. Palermo
Zenith Radio Corporation
6001 West Dickens Street
Chicago, Illinois   60639

John Pavkovich
Varian Associates
611 Hansen Way
Palo Alto, Ca.   94303

Harvey F. Peck
Lockheed Missiles and Space
  Corporation
Organization 84-34, B/154
Sunnyvale, Ca.

Soo-Chang Pei
University of California
Department of Elect. Eng.
Santa Barbara, Ca.   93106

Dietlind Pekau
Siemens AG,
München -Fl Opt 17-
8 München, Postfach 700076
Germany

Ronald C. Peterson
Holosonics
4340 Redwood Highway
Suite 128
San Rafael, Ca.   94903

Vermesse Philippe
28 Arena de l'Ewroje
Marcqen Baroeul, France

David Phillips
Duke University
201 West Markham Avenue
Durham, No. Carolina   27701

E.J. Pisa
Actron/Mc Donnell Douglas
700 Royal Oaks Drive
Monrovia, Ca.

James D. Plummer
Stanford University
McCullough 22, Elect. Labs
Stanford, Ca.

M.D. Pocha
Stanford University
#N 858 Coleman Avenue
Menlo Park, Ca.   94025

Robert A. Poirier
Div. of Technological Appl.
Dept. of Health, Education,
  and Welfare
National Institutes of Health
Bethesda, Md.   20014

Richard L. Popp
Cardiology Division
Stanford Medical Center
Stanford, Ca.   94305

Roger S. Powell
National Heart and Lung
  Institute
Westwood Building
Room 6A15B
Bethesda, Md.   20014

John P. Powers
Electrical Engineering Dept.
Naval Postgraduate School
Monterey, Ca.

Norman J. Pressman
University of Pennsylvania
5241 Norma Way
Building 20, #208
Livermore, Ca.   94550

Con D. Rader
Beckman Instruments, Inc.
1630 State College Blvd.
Anaheim, Ca.   92806

S. David Ramsey, Jr.
Stanford Research Institute
333 Ravenswood Avenue, K1084
Menlo Park, Ca.   94025

W. F. Ranson
Auburn University
Auburn, Ala.   36830

Barry Reed
U.S. Public Health Service
546 - 7th Street, S.E.
Washington, D.C.   20003

Don W. Reid
Varian Associates
611 Hansen Way, C-052
Palo Alto, Ca.   94303

Stewart M. Reiner
825 Nicholas Blvd.
Elk Groove Village, Ill.

N.E. Ritchey
Syntex Medical Instruments
3401 Hillview Avenue
Palo Alto, Ca.   94303

Charles G. Roberts
Stanford University
96 A Escondido Village
Stanford, Ca.   94305

Lewis H. Rosenblum
Picker Corporation
333 State Street
North Haven, Conn.   06473

Robert Rousseau, M.D.
133 Iowa Drive
Santa Cruz, Ca.   95060

Leonard Russo
Naval Research Laboratory
(Code 5492)
Washington, D.C.   20375

Louis F. Schaefer
Stanford Research Institute
333 Ravenswood Avenue, K1074
Menlo Park, Ca.   94025

Bernard Shacter
Nat. Inst. of Gen. Medical
  Sciences
National Institutes of Health
Bethesda, Md.   20014

N.K. Sheridon
Xerox Corporation
3180 Porter Drive
Palo Alto, Ca.   94304

John D. Shott
Stanford Integrated Circuits
  Lab., AEL 121
Stanford, Ca.

Evan H. Shu
Collins Medical Center
1605 South Washington Street
Seattle, Washington   98144

Joseph Siedlecki
U.S. Navy
Pentagon
Washington, D.C.

Buddy L. Smith
Smith Clinic
1402 Linda Drive
Daingerfield, Texas   75638

Donald R. Smith
Smith Clinic
1402 Linda Drive
Daingerfield, Texas   75638

Roy A. Smith
TRW Systems
One Space Park
Redondo Beach, Ca.   90277

Stephen W. Smith
Duke University
Bio-Medical Eng.
Durham, No. Carolina   27701

Elias Snitzer
American Optical
Framingham, Mass.   01701

Gordon E. Stewart
The Aerospace Corporation
P.O. Box 92957
Bldg. 120, Room 2005
Los Angeles, Ca.   90009

Karyl-Lynn Stone
Visual Acoustics
4242 E.W. Hwy.
Chevy Chase, Md.

Joe Suarez
Stanford Research Institute
333 Ravenswood Avenue, K1072
Menlo Park, Ca.   94025

Jesus T. Suero
Veterans Adm. Hospital
Muskogee, Okla.   74401

Alan L. Susal
Stanford University
2370 Greer Road
Palo Alto, Ca.   94303

Jerry L. Sutton
Naval Undersea Center
Code 6513
San Diego, Ca.   93132

Don Sweeney
Purdue University
School of Mech. Eng.
Lafayette, Ind.

T.S. Tan
Stanford University
Radio Astronomy Institute
Stanford, Ca.   94305

Ken A. Thompson
McDonnell Douglas Corp.
700 Royal Oaks Drive
Monrovia, Ca.

R. Bruce Thompson
Science Center
Rockwell Int.
P.O. Box 1085
Thousand Oaks, Ca.   91360

Jim Thorn
Naval Undersea Center
San Diego, Ca.   92132

Fredrick L. Thurstone
Duke University
Biomedical Engr. Dept.
Durham, No. Carolina   27706

Neal Tobochnik
UCLA
Medical Physics Division
Dept. of Radiological Sci.
Los Angeles, Ca.   90024

Hans Toffer
United Nuclear Industries
P.O. Box 490
Richland, Washington   99352

Nie But Tse
UCSB
6711 El Colegio Apt. 54
Goleta, Ca.   93017

Victor Vaguine
Varian Associates
Radiation Division
611 Hansen Way
Palo Alto, Ca.   94303

John R.M. Viertl
General Electric Co. R&D
Bldg. 37, Room 680
Schenectacy, New York   12301

David Vilkomerson
RCA Labs
Princeton, New Jersey   08540

Olaf T. von Ramm
Duke University
Dept. of Bio-Medical Eng.
Durham, No. Carolina   27705

Glen Wade
University of California
Dept. of EE
Santa Barbara, Ca.   93106

James T. Walker
Stanford University
64 F Escondido Village
Stanford, Ca.   94305

Keith Y. Wang
University of Houston
Dept. of Electrical Eng.
Cullen College of Engineering
Houston, Texas   77004

Peter C. Wang
IBM - San Jose
F97-015
San Jose, Ca.   95193

J.L. Weaver
Lockheed Research Laboratory
3251 Hanover Street
Palo Alto, Ca.   94304

Denis C. Webb
Naval Research Labs
Code 5257
Washington, D.C.   20375

Steven Wetterling
Stanford Electronics
Box 7782
Stanford, Ca.   94305

Harper J. Whitehouse
Naval Undersea Center
Code 6003
San Diego, Ca.   92132

Robert L. Whitman
Zenith Radio Corporation
6001 West Dickens Avenue
Chicago, Illinois   60639

Kenneth A. Wickersheim
Spectrotherm Corporation
3040 Olcott Street
Santa Clara, Ca.   95051

H.K. Wickramasinghe
University College London
Dept. of Elec. Engineering
Torrington Pl, W.C.I.
London, England

Herbert L. Williams
501 N. Central Avenue
Chicago, Illinois   60644

D. Wilson
Hewlett-Packard
1501 Page Mill Road
Palo Alto, Ca.   94304

D.K. Winslow
Stanford University
Hansen Labs
Stanford, Ca.   94305

Thomas G. Winter
Department of Physics
University of Tulsa
Tulsa, Oklahoma   74104

Michael Wollman
UCSB
Dept. of Elect. Eng.
Santa Barbara, Ca.

Hideo Yamamoto
Canon USA Inc.
c/o Antex
1059 East Meadow Circle
Palo Alto, Ca.   94303

Joel W. Young
Naval Undersea Center
San Diego, Ca.   92132

Leslie M. Zatz
959 Mears Court
Stanford, Ca.   94305

Gene Ziliuskas
Bendix Electrodynamics
15825 Roxford Street
Sylmar, Ca.

INDEX